Appropriate Technology in Vector Control

Editor

C. F. Curtis, Ph.D.
Department of Medical Parasitology
London School of Hygiene
and Tropical Medicine
London, England

CRC Press
Taylor & Francis Group
Boca Raton London New York

CRC Press is an imprint of the
Taylor & Francis Group, an informa business

First published 1990 by CRC Press
Taylor & Francis Group
6000 Broken Sound Parkway NW, Suite 300
Boca Raton, FL 33487-2742

Reissued 2018 by CRC Press

Library of Congress Cataloging-in-Publication Data

Appropriate technology in vector control/editor, C. F. Curtis
p. cm.
Includes bibliographical references.
ISBN 0-8493-4755-6
1. Vector control. I. Curtis, C. F.
RA639.3.A66 1989
614.4'3—dc20 89-23911

A Library of Congress record exists under LC control number: 89023911

Publisher's Note
The publisher has gone to great lengths to ensure the quality of this reprint but points out that some imperfections in the original copies may be apparent.

Disclaimer
The publisher has made every effort to trace copyright holders and welcomes correspondence from those they have been unable to contact.

ISBN 13: 978-1-315-89072-2 (hbk)
ISBN 13: 978-1-351-06982-3 (ebk)

Visit the Taylor & Francis Web site at http://www.taylorandfrancis.com and the
CRC Press Web site at http://www.crcpress.com

ACKNOWLEDGMENTS

The Editor is particularly grateful to Hazel Smith, Ludvik Gomulski, Ron Page, Philip Corbet, and Lisa Oke for help in preparing the manuscript.

THE EDITOR

Dr. Chris Curtis is a member of the British Medical Research Council's External Scientific Staff and is based at the London School of Hygiene and Tropical Medicine where he is Reader in Medical Entomology.

He graduated in Zoology at Oxford in 1961 and obtained a Ph.D. in Genetics at Edinburgh, Scotland, in 1966. He worked on genetic methods for tsetse control at Langford, Bristol U.K., and was geneticist at the W.H.O./I.C.M.R. Research Unit on the Genetic Control of Mosquitoes, New Delhi, India, from 1972 to 1975. Subsequently, he carried out theoretical work on the management of insecticide resistance, with related practical work in Sri Lanka, Sudan, and Zanzibar. He now works mainly on appropriate (intermediate) technology for mosquito control with two on-going field projects in Tanzania and with some field experience in Latin America also. He has published over 130 papers and reviews.

CONTRIBUTORS

Christian Boudin
7 Cours Saint André
38800 Pont de Claix, France

Roberto Briceño-León
Laboratorio de Investigaciones Sociales
Universidad Central de Venezuela
Apartado 47795
Caracas 1040A, Venezuela

Pierre J. Carnevale
Antenne O.R.S.T.O.M./O.C.E.A.C.
B.P. 288, Yaoundé, Cameroun

Kai-Lok Chan
Department of Zoology
National University of Singapore
Kent Ridge, Singapore 0511

Moh-Seng Chang
Medical Department
Jalan Tun Abg. Hj. Openg
93590 Kuching, Sarawak, Malaysia

J. Derek Charlwood
Swiss Tropical Institute Field Laboratory
Box 53, Ifakara, Tanzania

Mario Coluzzi
Istituto di Parassitologia
Università di Roma
Rome, Italy

Philip S. Corbet
Department of Biological Sciences
University of Dundee
Dundee, DD1 4HN, U.K.

Christopher F. Curtis
Department of Medical Parasitology
London School of Hygiene and Tropical Medicine
London WC1E 7HT, U.K.

P. K. Das
Vector Control Research Centre
Indira Nagar, Pondicherry 605006, India

Pierre Gazin
Antenne O.R.S.T.O.M./Institut Pasteur
B.P. 1857, Yaoundé, Cameroun

J. David Gillett
Department of Medical Parasitology
London School of Hygiene and Tropical Medicine
London WC1E 7HT, U.K.

Ludvik M. Gomulski
Liverpool School of Tropical Medicine
Pembroke Place
Liverpool L3 5QA, U.K.

Jean-Paul Gouteux
O.R.S.T.O.M.
213, Rue La Fayette
75010 Paris, France

Patricia M. Graves
Queensland Institute of Medical Research
Herston, Brisbane, Queensland, Australia

Jean-Michel Halna,
O.R.S.T.O.M.
213 Rue La Fayette
75010 Paris, France

M. Ismail Hossain
Department of Zoology
Dhaka University
Dhaka, Bangladesh

Kazuyo Ichimori
Department of Medical Zoology
Teikyo University School of Medicine
Kaga, Itabashi-Ku, Tokyo, Japan

Toshiaki Ikeshoji
Department of Agricultural Biology
University of Tokyo
Bunkyo-Ku, Tokyo 113, Japan

Nils Kolstrup
N-9130
Hansnes, Norway

Takeshi Kurihara
Department of Parasitology
Teikyo University School of Medicine
Kaga, Itabashi-Ku, Tokyo, Japan

Marshall Laird
Unit on Vector Pathology
Memorial University of Newfoundland
St. John's, Newfoundland, Canada

Peter A. Langley
Tsetse Research Laboratory
University of Bristol
Department of Veterinary Medicine
Langford, Bristol BS18 7DU, U.K.

Claude Laveissière
O.R.S.T.O.M.
B.P. 1434
Bouaké, Côte d'Ivoire

Li Zuzi
Guangdong Province Anti-Parasitic Diseases Institute
176 Xin Gang Xi Road
Guangzhou, China

Liao Guohou
Guangxi Institute of Parasitic Disease Control
80, Tao Yuan Road
Nanning, Guangxi, China

Steven W. Lindsay
Medical Research Council Laboratories
Fajara, Banjul, The Gambia

Jonathan D. Lines
Department of Medical Parasitology
London School of Hygiene and Tropical Medicine
London WC1E 7HT, U.K.

Lu Baolin
Department of Vector Biology and Control
Institute of Microbiology and Epidemiology
P.O. Box 130, Beijing 100800, China

Carol P. MacCormack
Bryn Mawr College
Bryn Mawr, PA 19010, U.S.A.

Giancarlo Majori
Department of Parasitology
Istituto Superiore di Sanità
V. Regina Elena 299
Rome, Italy

Thomas N. Mather
Department of Tropical Public Health
Harvard School of Public Health
Boston, Massachusetts, 02115, U.S.A.

Caroline Maxwell
Ministry of Health
P. O. Box 236
Zanzibar, Tanzania

Japhat N. Minjas
Department of Parasitology and Medical Entomology
Faculty of Medicine
Muhimbili Medical Centre
P.O. Box 65011
Dar es Salaam, Tanzania

Peter Morgan
Blair Research Laboratory
P.O. Box 8105, Causeway
Harare, Zimbabwe

Jean Mouchet
O.R.S.T.O.M.
213 Rue La Fayette
75010 Paris, France

Enala T. Mwase
Samora Machel School of Veterinary Medicine
University of Zambia
P.O. Box 32379, Lusaka, Zambia

Kato J. Njunwa
Amani Medical Research Centre
National Institute for Medical Research
Box 4, Amani, Tanga, Tanzania

K.N. Panicker
Vector Control Research Centre
Indira Nagar, Pondicherry 605006, India

Lionel-Henri Pazart
B.P. 2682
21058 Dijon Cedex, France

Rupert G. Pegram
FAO Project TCP/Z1M/8853
Box 8101, Causeway, Harare, Zimbabwe

Boonluam Phanthumachinda
Department of Medical Sciences
Yodse, Bangkok 10100, Thailand

P. K. Rajagopalan
Vector Control Research Centre
Indira Nagar, Pondicherry 605006, India

D. Raghunatha Rao
Centre for Research in Medical Entomology
P.O. Box 5, Sree Sathya Sai Nagar
Madurai 625003, Tamil Nadu, India

Alfons H. J. Renz
Tropenmedizinisches Institut
Universität Tübingen
Tübingen 7400, West Germany

Rachel Reuben
Centre for Research in Medical Entomology
P.O. Box 5, Sree Sathya Sai Nagar
Madurai 625003, Tamil Nadu, India

Alain Richard
O.R.S.T.O.M.
213, Rue La Fayette
75010 Paris, France

Vincent Robert
Antenne O.R.S.T.O.M./O.C.E.A.C.
B.P. 288
Yaoundé, Cameroun

Guido Sabatinelli
Dipartimento Cooperazione Sviluppo
Ministero Affari Esteri, Rome, Italy

C. J. Schofield
Tropical Diseases Research
World Health Organization
1211 Geneva 27, Switzerland

Anthony Sebastian
Department of Medical Research
Ministry of Health
5, Zafar Shah Road
Dagon P.O., Rangoon, Burma

Lee S. Self
World Health Organization
Regional Office for the Western Pacific
P.O. Box 2932, Manila, Philippines

Robert Snow
CRC-KEMRI Coastal Unit
Box 428, Kilifi, Kenya

Glynowr Vale
Tsetse and Trypanosomiasis Control
P.O. Box 8283, Causeway
Harare, Zimbabwe

David J. T. Webb
Overseas Division, Building Research Establishment
Garston, Watford, WD2 7JR, U.K.

Graham B. White
Department of Medical Parasitology
London School of Hygiene and Tropical Medicine
London WC1E 7HT, U.K.

Tony Wilkes
Ubwari Field Station
Box 81, Muheza, Tanga, Tanzania

Wu Neng
Guangxi Institute of Parasitic Disease Control
80, Tao Yuan Road
Nanning, Guangxi, China

TABLE OF CONTENTS

Chapter 1

APPROPRIATE TECHNOLOGY AND WHY IT IS NEEDED

C. F. Curtis

The phrases "intermediate technology" and "appropriate technology" were coined by Schumacher[1] in his seminal book *Small Is Beautiful: A Study of Economics As If People Mattered.* By these phrases he meant small-scale technology of the kind that communities, especially in tropical developing countries, can manage for themselves with minimal inputs from outside. The present book explores this kind of approach to the control of arthropod (insect and tick) vectors of several human diseases. Most of the examples are drawn from mosquito control, with a few from the control of tsetse, ticks, etc; the snail intermediate hosts of schistosomiasis are not here considered to be "vectors" and are not discussed in this book, though appropriate technology may have an important role to play in schistosomiasis control. The chapters in this book are generally introduced by a brief statement about the present importance of the diseases which the vectors transmit, but it is assumed that the reader knows the basic biological facts about the vector-pathogen relationships or will consult one of the standard textbooks.

Up to now most vector control has been attempted by centrally organized insecticidal spray teams, and these have had their successes, for example, in the 1950s and 1960s in greatly reducing malaria prevalance in almost all parts of Asia by DDT spraying in houses, and in the 1970s and 1980s in almost eliminating onchocerciasis transmission in large parts of West Africa by the aerial spraying of temephos in the riverine breeding sites of the blackfly vectors.[2] This book is in no sense an attack on such programs, but it does start from the position that the methods that have been used for vector control over the last 40 years have now encountered a number of very serious problems, including the following:

1. Many vector populations have evolved physiological resistance to the cheap organochlorine insecticides and in a few areas to organophosphates and carbamates also.[3]
2. There is a tendency of insects not to rest for long enough on sprayed surfaces to pick up a lethal dose because of either (a) the excito-repellent effect of insecticides such as DDT, (b) the evolution of a change in behavior as a result of the exposure of many generations to sprayed houses,[4] or (c) a preexisting tendency of some species to exit from houses soon after blood feeding.
3. There is difficulty of access by spray teams to houses[5] and increasing objection of householders to entry of unknown spraymen into their homes and to the unsightliness and smell of spray deposits, especially now that organophosphates such as malathion have to be used, and the bonus effect of killing nuisance insects such as bedbugs is often absent owing to the evolution of resistance by these insects.
4. There is difficulty of organizing and paying for trained spray teams which have to be supplied and transported to cover a whole region in time for the disease transmission season. For example, organizational defects have been given much of the blame for the resurgence of malaria in India in the 1970s,[6] and 45% of the central government health budget and a similar amount from the state governments is spent on malaria control.[7] The Onchocerciasis Control Programme in West Africa is dependent on a generous grant from the World Bank, and although the cost of about $16 million annually[8] is about that of a couple of fighter aircraft, funding of the program would be beyond the resources that the governments of the countries concerned would feel able to spend on control of a single disease.

5. There are risks of insecticides to nontarget organisms, for example:

   a. Accidental poisoning of people by pesticides (mostly, however, with chemicals intended for control of agricultural pests rather than vectors) (see Chapter 12)
   b. Killing of cats, chickens, bees, silkworms, or fish which sometimes occurs as accidental effects of antivector spraying[7,9]
   c. Damage to natural enemies of vectors sometimes leading to surges in pest populations a few weeks after spraying (see Chapter 7)
   d. Damage to wildlife such as raptorial birds (almost entirely from agricultural and not antivector spraying)

This book reviews ways in which appropriate technology may overcome some of these problems. It takes a critical standpoint and tries not to be carried away by attractively simple ideas which, in fact, do not work. For example, cutting vegetation is often advocated, or even enforced, as a simple method in which the community can supposedly participate in mosquito control, though studies dating from the 1940s provide strong evidence that this procedure is useless.[10]

Many research groups have contributed to this book. Rather than producing a mere bundle of papers, an attempt has been made to weld together related contributions to make a series of coherent chapters, summarizing the "state of the art" in various topics. The chapters are under multiple authorship, but, where appropriate, the contributors responsible for particular subsections are indicated by footnotes.

This book does not support the lobby which condemns all pesticides and perhaps the most promising ideas contained in the book concern the rational placement of the new, safe, and quick-acting synthetic pyrethroid insecticides (products of high technology) where the target species are bound to contact them — on bednets (Chapter 2), tsetse baits (Chapter 3), or even mouse nesting material (Chapter 5). Other themes in the book are ways in which communities can be given the incentive to manage their environment to make it unsuitable for mosquito breeding (Chapters 6, 7, and 8) and to prevent houses and latrines from becoming havens for bugs, flies, cockroaches, or mosquitoes (Chapters 10 and 11). Chapters 4, 5, and 9 suggest ways in which repellents and genetic control might be made relevant in the villages of the Third World. Finally, Chapters 12 and 13 discuss ways in which methods of proven effectiveness can be scaled up from field trials to routine use by the billions of people who will live in the tropics in the next century.

REFERENCES

1. **Schumacher, E. F.,** *Small Is Beautiful: A Study of Economics As If People Mattered*, Abacus, London, 1974.
2. Ten Years of Onchocerciasis Control in West Africa, OCP/GVA/85.1B, World Health Organization, Geneva, 1985.
3. Resistance of Vectors and Reservoirs of Disease to Pesticides, WHO Tech. Rep. Ser. No. 737, World Health Organization, Geneva, 1986.
4. **Li, M., Liang, T., Zhang, H., and Chen, T.,** [The bionomics of *Anopheles minimus* in the Zhougsha area of Hainan Island and its relation to DDT spraying], *Annu. Bull. Parasitol. Soc. Guangdong Province*, 180, 1983 (in Chinese); *Trop. Dis. Bull.*, 85(Abstr.), 662, 1988.
5. **Chadee, D. P.,** Effect of 'closed' houses on the *Aedes aegypti* eradication programme in Trinidad, *Med. Vet. Entomol.*, 2, 193, 1988.
6. **Sharma, V. P. and Mehrotra, K. N.,** Malaria resurgence in India: a critical study, *Soc. Sci. Med.*, 22, 835, 1986.
7. **Sharma, V. P.,** Community-based malaria control in India, *Parasitol. Today*, 3, 222, 1987.

8. **Prost, A. and Prescott, N.,** Cost effectiveness of blindness prevention by the Onchocerciasis Control Programme in Upper Volta, *Bull. W.H.O.*, 62, 795, 1984.
9. **Curtis, C. F. and Lines, J. D.,** Insecticides in the management of insect vectors of tropical diseases, *Insect. Sci. Appl.*, 8, 709, 1988.
10. **Ribbands, C. R.,** Effects of bush clearance on flighting of West African anophelines, *Bull. Entomol. Res.*, 37, 33, 1946.

Chapter 2

IMPREGNATED BED NETS AND CURTAINS AGAINST MALARIA MOSQUITOES

C. F. Curtis, J. D. Lines, P. Carnevale, V. Robert, C. Boudin, J.-M. Halna, L. Pazart, P. Gazin, A. Richard, J. Mouchet, J. D. Charlwood, P. M. Graves, M. I. Hossain, T. Kurihara, K. Ichimori, Li Zuzi, Lu Baolin, G. Majori, G. Sabatinelli, M. Coluzzi, K. J. Njunwa, T. J. Wilkes, R. W. Snow, and S. W. Lindsay

TABLE OF CONTENTS

I. UNTREATED AND TREATED NETTING AS PROTECTION AGAINST MOSQUITOES

Malaria is by far the most important insect-borne disease. For example, in Ghana it is top of the list of causes of loss of days of healthy life, and no other insect-borne disease ranks in the top 25 on this list.[1] The *Anopheles* vectors of malaria may be attacked in their aquatic larval stages (see Chapters 7 and 8) or as adults in houses where most of them come to bite people. The latter approach has the advantages that:

1. Houses are generally easier to find and treat than are the seasonally shifting puddles etc. where *Anopheles* typically breeds.
2. Insecticides persist for 3 to 6 months on solid surfaces, but much more frequent insecticidal treatment of water bodies is generally necessary.
3. The maturation of malaria parasites to the sporozoite stage in female mosquitoes requires 10 d or more during which time they return to houses on several occasions for blood feeding: the cumulative effect of a moderate risk of being killed on each occasion may amount to a very small chance of living long enough to develop mature sporozoites,[2] whereas any puddle which is missed in larvicidal operations will yield adult mosquitoes with a good chance of surviving to a dangerous age.
4. *Anopheles* mosquitoes and other biting insects are a nuisance, and most people need no persuasion to use available precautions against immediate insect nuisance, whereas a sustained campaign by a community against *Anopheles* larvae is likely to require considerable education and motivational effort because its effects are delayed and indirect.

For rural areas the spraying of residual insecticides on the inside of the walls and roofs of houses has been the recommended measure against malaria mosquitoes for several decades,[3] but this method has now encountered several problems as explained in Chapter 1.

There is now much interest in whether these problems could be avoided by improved methods of personal protection which could be organized at the community level by a primary health care system with minimal input from outside.

Bed nets are widely used against nuisance mosquitoes in, for example, China, Thailand, and Latin America, and in certain ethnic groups in Papua New Guinea and Africa where they also may be favored for privacy and protection against scorpions, rat droppings, snakes, etc.[4] Since most *Anopheles* species bite at night, it has long been assumed that nets must be useful against malaria.[5,6] However, until recently this assumption had not been tested. In parts of China,[7] and at a boarding school in Kenya,[8] evidence has been obtained of nets being effective against malaria, but it has long been realized that in areas holoendemic for malaria (especially in much of tropical Africa) it will be necessary to lower the number of biting *Anopheles* by a large factor before one would *begin* to reduce the percentage of the human population with malarial parasites.[9]

In many parts of the world it is far from obvious that bed nets will give sufficient protection to control malaria because of the following reasons:

1. Nets may be torn or not properly tucked in.
2. Mosquitoes may feed through nets on limbs which touch the net during the night (a particular problem with hammock nets in Latin America).[10]
3. Mosquitoes may enter nets from underneath where beds with woven string or maize stalks for "mattresses" are in use.
4. People may be bitten before they go to bed.
5. If people get up before dawn, mosquitoes may be "waiting for their chance" to obtain a blood meal.

TABLE 1
Numbers of Fed Anophelines Found inside Untreated and Permethrin-Treated Nets

	Number of fed females in nets		
Anopheles species	Untreated	Treated	Ref.
balabacensis	180 (15)	17[a] (15)	12
gambiae s.1.	265 (79)	7 (51)	13
gambiae s.1.	81 (256)	0 (255)	14

Note: No. of net nights of observation shown in parentheses.

[a] Only 4 were alive.

TABLE 2
Numbers over 5 Weeks of *An. arabiensis* Entering, Feeding, and Surviving in Trials[17] in Veranda Trap Huts with or without Nets, with or without Holes (20 × 10 cm), and with or without Impregnation with 0.2 g Permethrin/m²

Net		Total entered	No. fed	No. fed and survived
None (Control)		403(100%)	347[a](100%)	342[a](100%)
Holes	Permethrin			
No	No	223(55%)	62[b](13%)	61[b](19%)
Yes	No	408(105%)	324[a](93%)	311[a](93%)
Yes	Yes	320(79%)	124[b](36%)	54[b](13%)
No	Yes	263(66%)	75[b](22%)	36[b](11%)

Note: Percentages of the control without a net are shown in parentheses. No. in the same column followed by different letters differ significantly at the 5% level by analysis of variance.

Torn nets are known to be a good place for entomologists to collect blood-fed anophelines (e.g., Bryan[11] and see data on untreated nets in Table 1). Port and Boreham[15] and Charlwood[16] demonstrated that the use of bed nets reduced the number of fed anophelines even if the nets were torn. However, when eight holes 20 × 10 cm were cut in a net the number of *Anopheles arabiensis* feeding was no less than without nets (see Lines et al.[17] and the data on untreated nets with holes in Table 2).

The first mention of nets with a mosquito-repellent effect appears to be by Herodotus[18] in the Fifth Century B.C. who described the behavior of fishermen living in marshy areas in Egypt as follows:

Every man of them has a net, with which he catches fish by day, and for the night he sets it round the bed where he rests, then creeps under it and so sleeps. If he sleep wrapped in a garment or cloth, the gnats bite through it; but through the net they do not even try at all to bite.

Perhaps fish oil on the nets had a repellent effect and prevented the "gnats" (i.e., mosquitoes) trying to bite through the nets.

The idea of deliberately treating netting with insecticidal or repellent compounds appears to have originated independently in the U.S.S.R.[19-21] using lysol and *d*-alpha-pinene from juniper oil and with the American[22] and German[23] armies in World War II who used DDT on bed nets against mosquitoes and the vectors of sandfly fever. The use of synthetic volatile repellents such as dimethyl phthalate and diethyl toluamide on netting is described in Chapter 4. Organophosphate insecticides were tested on netting with only moderate success.[24]

The development of synthetic pyrethroids, such as permethrin, which are photostable analogues of natural pyrethrum, with low mammalian toxicity, is described by Elliot et al.[25] They are highly insecticidal as well as being excito-repellent. Permethrin has been used for the impregnation of clothing as a protection against mosquitoes[26,27] and ticks[28] (see also Chapter 5). At about the same time the impregnation with pyrethroids of the materials from which mosquito nets are made started to be studied in the laboratory.[29-32] Based mainly on these two lines of work a World Health Organization Panel[33] recommended field trials of pyrethroid impregnation of nets. Following this recommendation and/or independent invention of the same idea, numerous projects have been initiated around the world.

The objectives of the impregnation of bed nets with insecticide are twofold.[34,35]

1. The individual effect — to improve the personal protection conferred by nets by overcoming the problems of damaged or badly used nets.
2. The mass effect — to use insecticide more efficiently, than is the case with house spraying, to shorten the mean lifespan of the vector by placing the insecticide on a surface which mosquitoes are bound to encounter in their attempts to bite sleeping people. The treated net could be considered as a trap baited with human odor. This effect would benefit nearby nonusers of nets or people when outside their nets.

The objection has been raised that bed nets are expensive and ineffective against mosquitoes biting before people go to bed, and that they restrict ventilation in a hot climate. In response to these criticisms a number of studies have started on impregnated broad mesh netting or other coarse material which could be used as "curtains" over windows, doors, and eaves gaps to mosquito-proof a whole house.

II. LABORATORY EXPERIMENTS ON PYRETHROID-IMPREGNATED NETTING

A. CHEMICAL AND PHYSICAL STUDIES*

In most of the research studies impregnation of nets has been done by dipping them into an emulsion made by mixing an emulsifiable concentrate of the pyrethroid with water. To obtain a predetermined dose it is necessary to determine how much liquid is absorbed per unit area of the fabric. It has been recommended[36] that one should measure the surface area of each net and pour out the amount of diluted emulsion to wet the net just completely. This is still the method preferred in China (Figure 1) to avoid transferring dirt from one person's net to another. However, it is easier to dip clean nets in a large volume of emulsion and wring them out (Figure 2). It was found by weighing that the amount of liquid left after dipping and wringing by hand was almost the same as if liquid was added until the point of saturation. In some dipping processes (e.g., the dipping of sheep in acaricide) there is selective uptake ("stripping") of the active ingredient, thus necessitating the addition of compensating amounts of concentrate as the operation proceeds. To test for such an effect,

* C. F. Curtis, M. I. Hossain, Li Zuzi, R. W. Snow, and S. W. Lindsay

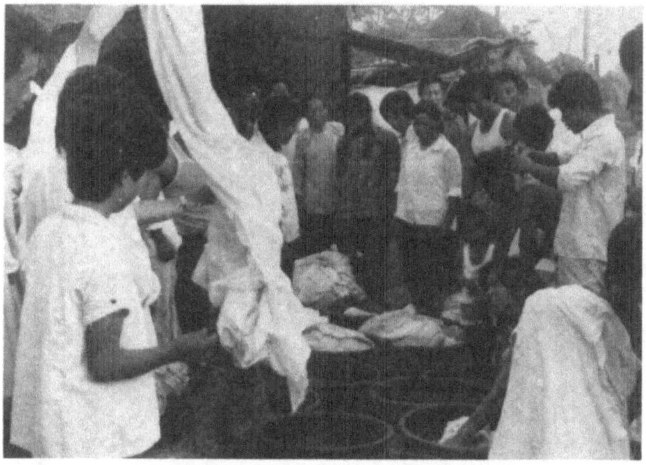

FIGURE 1. The impregnation of nets with deltamethrin by the people of Yiyu county, Jiangsu Province, China. Note that a separate volume of emulsion is measured out for each net, the volumes being one of three which correspond to the major size categories of nets in use in the area. (Photograph from Lu Baolin.)

FIGURE 2. Impregnation of nets with permethrin by Tanzanian villagers by dipping into a large volume of emulsion and wringing out by hand wearing rubber gloves. (Photograph by C. F. Curtis.)

TABLE 3
Effect of Sequential Dipping of Pieces of Fabric in a Single Volume of Emulsion[37]

Fabric	Wt. of liquid absorbed (g/m²)	Permethrin content (g/m²) Expected[a]	Permethrin content (g/m²) Observed[b]
Nylon netting	45.0	0.19	0.27
Polyester netting	40.9	0.17	0.17
Cotton sheeting	124.4	0.52	0.53
•			
•			
Nylon netting[c]	49.8	0.21	0.28

[a] Assuming passive absorption.
[b] By gas liquid chromatography.
[c] Note that this sample of nylon netting which was dipped at the end of a series of nine samples of fabric gave very similar results to the nylon netting dipped at the beginning.

TABLE 4
Tensile Strength (Vertical/Horizontal) of Cotton Netting after Impregnation with Insecticides

Insecticide	Dosage (g/m²)	Tensile strength (kg/5 cm × 20 cm) at indicated no. of months after impregnation: 7 mo	14 mo	20 mo
Delta-methrin	0.025	24.5/25.6	23.8/24.3	24.0/25.6
DDT	2	23.0/22.6	24.0/24.2	20.6/23.6
Control	Water	23.2/24.2	25.2/24.8	26.1/27.3

Note: Data are from the Guangzhou Municipal Industrial Products Assessment Center.[39]

a weighed series of different types of fabric were dipped in a single volume of emulsion, squeezed until they stopped dripping, and then reweighed.[37] The amount of liquid taken up by cotton was much greater than by synthetic fabrics. The amount of permethrin expected to be taken up on the null hypothesis of no "stripping" was calculated from the liquid absorbed and the concentration of the emulsion. The amount of permethrin found by chemical analysis agreed quite well with these expectations (Table 3). This indicates that a series of nets of the same fabric can justifiably be dipped in a bulk mix, and this has been done simply and quickly in villages in Tanzania (Figure 2) and The Gambia.[38] The villagers dried the nets on their beds which had the advantages of keeping the permethrin out of the strong tropical sun which might have denatured some of it, and it also deposited some permethrin on the mattresses which probably helped in the suppression of bedbug infestations (see Section V.B).

In order to check whether impregnation with deltamethrin or DDT could cause structural deterioration in cotton netting, samples were tested for tensile strength at intervals after impregnation in Guangzhou, China.[39] The results compare well with the standard 23/23 (vertical/horizontal) and showed no signs of deterioration of the netting due to the insecticides (Table 4).

The frequency with which bed nets are washed varies from weekly among the Ndjuka

TABLE 5
Chemical Content of Polyester Netting Impregnated with Various Pyrethroids; Half of Each Sample Was Hand-Washed in Cold Soapy Water[41]

Pyrethroid	Washed/unwashed	Expected dose[a] (g/m^2)	Observed dose (g/m^2)
Cyfluthrin	Unwashed	0.13	0.11
	Washed		0.03
Cypermethrin	Unwashed	0.10	0.09
	Washed		0.03
Deltamethrin	Unwashed	0.026	0.02
	Washed		0.007
d-Phenothrin	Unwashed	0.19	0.04
	Washed		0.01
Permethrin	Unwashed	0.21	0.19
	Washed		0.09
Lambda cyhalothrin	Unwashed	0.033	0.022
	Washed		0.012
Permethrin	Unwashed	0.24	0.66
(97°C treatment at pH 3.2)	Washed		0.39

[a] Based on weight of liquid absorbed and concentration of emulsion.

(''Bushnegro'') people of Suriname,[10] to fortnightly among the Mandinkas of The Gambia,[4] to annually at the time of the Chinese New Year in China.[40] Chemical analysis shows that following hand washing in cold, soapy water about half the initial content of a variety of pyrethroids is lost (Table 5).[13,41] Under the conditions used in these studies there was a rapid loss of *d*-phenothrin. Impregnation with permethrin in the high temperature and low pH conditions used for moth-proofing carpets[42] caused selective absorption of the insecticide, in contrast to the results of Table 3 with cold-water dipping. The consequences of the chemical losses on washing to the insecticidal effect of the deposit are described in the next section.

B. BIO-ASSAYS ON PYRETHROID-IMPREGNATED NETTING*

The toxic effect of samples of impregnated netting can be tested by confining mosquitoes over the samples for a standard exposure period in such a way that they spend as much as possible of this period walking or resting on the net. With pyrethroids the toxic effect generally takes the form of early knockdown followed by death of some individuals and recovery of others. It seems useful to record the short-term knockdown and eventual mortality, though it may be that in a tropical house even temporary knockdown of a mosquito would lead to its being consumed by a scavenging ant.

Plotting the probits of the percentage knockdown or mortality against the log of the dosages (at constant exposure time) or against the log of the exposure times (at constant dosage) generally produces a straight regression line. From this the ED_{50} or ET_{50}, i.e., the dosage or time for 50% (or any other percentage) toxic effect, can be read off.

Tables 6 and 7 show data on the knockdown effect of two different pyrethroids on two types of netting for four important vector species in China.[43] It is concluded that permethrin is considerably more effective on polyester than on cotton (i.e., the exposure time required for 50% knockdown is *less* on polyester). There is little difference between the fabrics when deltamethrin is applied, and, on cotton, deltamethrin is much more effective than permethrin.

* C. F. Curtis, M. I. Hossain, Li Zuzi, K. J. Njunwa, T. J. Wilkes, P. Carnevale, G. Majori, and S. W. Lindsay

TABLE 6
Time of Exposure in Minutes for 50% Knockdown (KT_{50}) of Deltamethrin-Impregnated Nets to Four Important Species of Mosquitoes in China[43]

Species	Nets	Dosage (mg/m²) 0.05	0.5	5.0	10.0
Anopheles anthropophagus	Cotton	23.5	13.7	8.9	5.9
(= *An. lesteri anthropophagus*)	Polyester	33.4	15.9	14.5	10.4
An. sinensis	Cotton	40.1	20.0	8.4	6.9
	Polyester	34.3	17.5	10.4	7.9
An. minimus	Cotton	16.6	13.5		6.6
	Polyester	17.6	9.6		7.0
Culex quinquefasciatus	Cotton	44.5	25.9	9.7	8.8
	Polyester	41.7	22.0	8.0	8.8

TABLE 7
KT_{50} of Permethrin-Impregnated Nets to Four Important Vector Species in China[43]

Species	Nets	Dosage (mg/m²) 0.005	0.05	0.5	5.0	10.0
Anopheles anthropophagus	Cotton		52.9	25.7	17.8	14.2
	Polyester		57.7	21.5	11.7	8.9
An. sinensis	Cotton		60.5	44.3	32.2	23.6
	Polyester		32.5	22.7	8.7	7.7
An. minimus	Cotton	29.2	28.4	17.3	3.9	
	Polyester	18.3	14.1	11.4		
		Dosage (mg/m²) 10.0	**20.0**	**50.0**	**100.0**	
Culex quinquefasciatus	Cotton	75.2	32.2	16.6	13.8	
	Polyester	22.9	15.1	11.9	7.7	

Comparing the species, *Anopheles minimus* is highly susceptible and *Culex quinquefasciatus* is much more tolerant. Durability of both the pyrethroids was reported to be better on cotton than polyester.[43]

Using bioassays with *Aedes aegypti* and 1-h exposure Hervy and Sales[29] found that permethrin was equally effective on cotton and synthetic netting, but deltamethrin was more effective on cotton than synthetic.

In testing for the lethal effect of pyrethroids it is desirable to use a short exposure period so that knockdown does not begin until exposure is completed. This avoids some mosquitoes spending part of the intended exposure period lying on their backs on or off the impregnated netting, thus absorbing either more or less insecticide than if they spent the whole period absorbing it through their tarsi. To ensure accurate timing of the start and finish of exposure, it is convenient to use World Health Organization (WHO) insecticide resistance test kits in which the mosquitoes can be blown into and out of the exposure tube lined with the impregnated fabric.[44] Figure 3 gives the results of such tests[37] which show the following:

1. Polyester is a better substrate for permethrin than is cotton (in agreement with Tables 6 and 7), and nylon is similar to polyester.

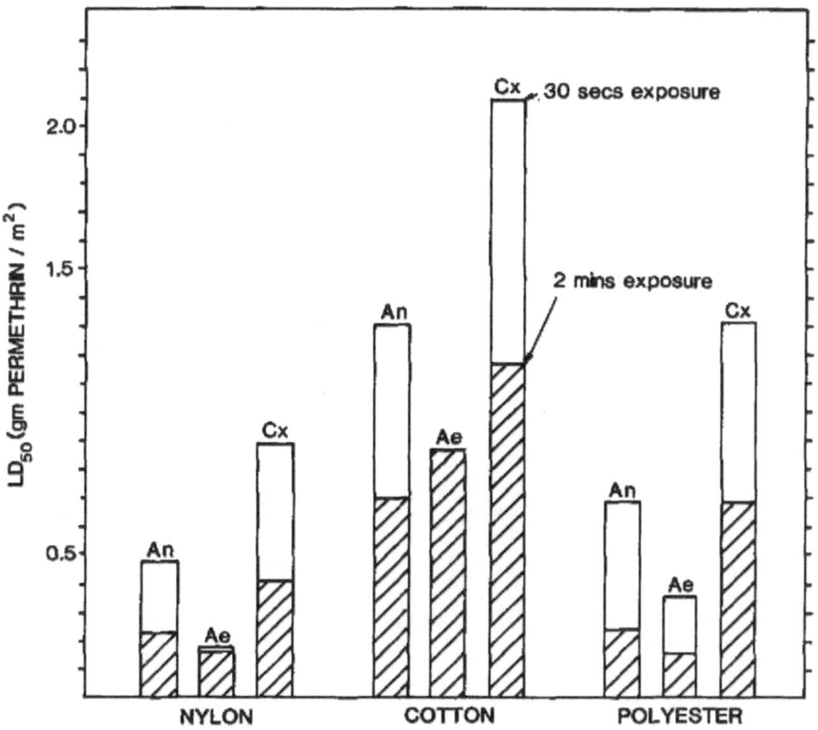

FIGURE 3. LD_{50} values from tests of permethrin on cotton, nylon, and polyester nets with *Anopheles gambiae*, *Aedes aegypti* and *Culex quinquefasciatus*.[37] Exposure times were 30 s or 2 min.

2. *Anopheles gambiae* is more susceptible than *Cx. quinquefasciatus* (in agreement with the contrast of anophelines with *Cx. quinquefasciatus* shown in Tables 6 and 7), and *Aedes aegypti* is generally even more susceptible than *Anopheles gambiae*.
3. The LD_{50} values following 30-s exposure are less than four times as great as those following 2-min exposure, i.e., there is not a simple interchangeability of dose and time, short exposures being disproportionately more effective.

Figure 4 shows the results of studies with the important vector species *Anopheles dirus*.[39] Exposure was for 30 min, and survival 24 h later was observed. A dosage of 25 mg deltamethrin per square meter continued to give 100% mortality for over 300 d and was still very effective after 540 d. By contrast 2 g DDT per square meter lost most of its toxicity after only 70 d. More recent data have shown longer persistence of deltamethrin on netting than on building materials.[40]

It might be expected that abrasion, accumulation of dirt, etc. on a bed net in use in a house would reduce the effectiveness of a pyrethroid deposit. To test for such changes without cutting up a bed net, one can use WHO bioassay plastic cones attached by rubber bands to the net (Figure 5). When mosquitoes are introduced they are generally more willing to rest on the net than the smooth plastic surface of the cone. Table 8 shows no decline in toxicity of permethrin on nylon net to *An. gambiae* over a 5-month period of use of bed nets in Tanzanian villages (including nets in rooms with cooking fires which deposited soot on the nets) and curtains made of sisal fiber.[45] After 6 months there were indications of a loss of toxicity. In tests on nets in Burkina Faso mortality had declined after 5 months, especially on the lower parts of nets which are much handled by people as they get into and out of bed.[46]

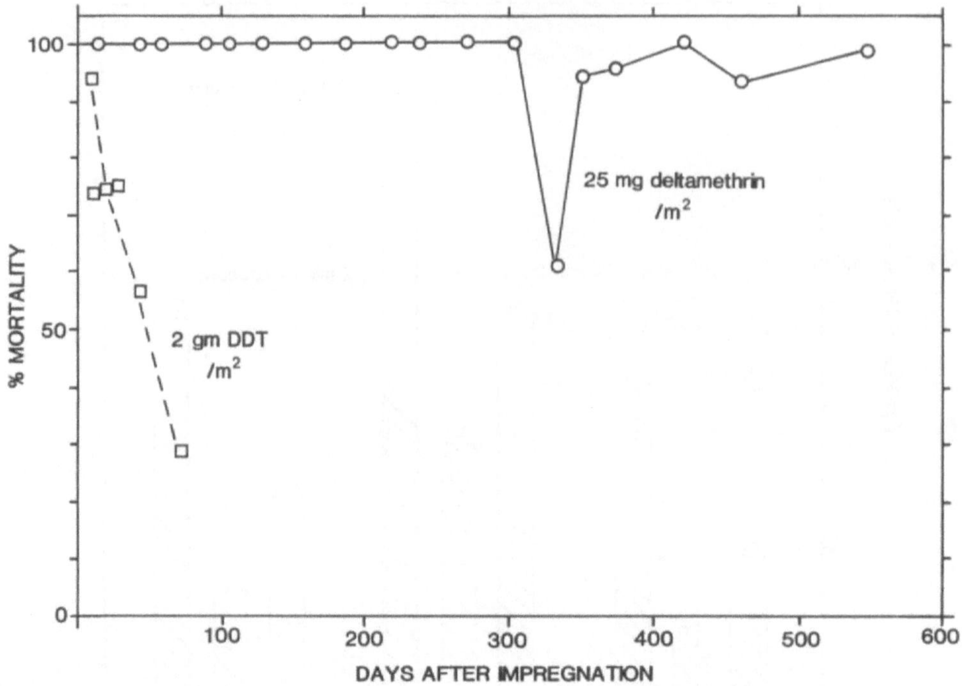

FIGURE 4. Mortality of *Anopheles dirus* in laboratory tests following impregnation of cotton netting with deltamethrin or DDT. Exposure was for 30 min and the sample size was about 70 for each test.[39]

FIGURE 5. Bioassaying on a net in a rural Tanzanian house. (Photograph by C. F. Curtis.)

Permethrin-impregnated cotton netting curtains in use in Burkina Faso showed a decline from initially giving 100% knockdown within 8 min of exposure, to 100% knockdown within 14 min of exposure after 3 months of use, to 64% knockdown after 30 min exposure after 11 months of use.[47]

Figure 6 shows the mortalities in bioassays with *An. gambiae* of the washed and unwashed

TABLE 8
Mortality of *Anopheles gambiae* after 3 min Exposure and 24 h Holding

Months since impregnation	Untreated bed nets	Bed nets (0.2 g permethrin/m^2) Clean	Sooty	Sisal curtains (approx. 1.0 g/m^2)
1	4%(25)	90%(30)	96%(24)	100%(31)
2	12%(85)	93%(86)	85%(80)	93%(29)
3	9%(21)	86%(21)	100%(21)	—
4	9%(44)	95%(57)	67%(21)	86%(28)
5	8%(40)	100%(56)	100%(40)	93%(15)
6	9%(32)	71%(138)	—	—

Note: Exposure was to nylon bed nets or sisal curtains impregnated with permethrin and in use in three Tanzanian villages[45] (numbers tested shown in parentheses).

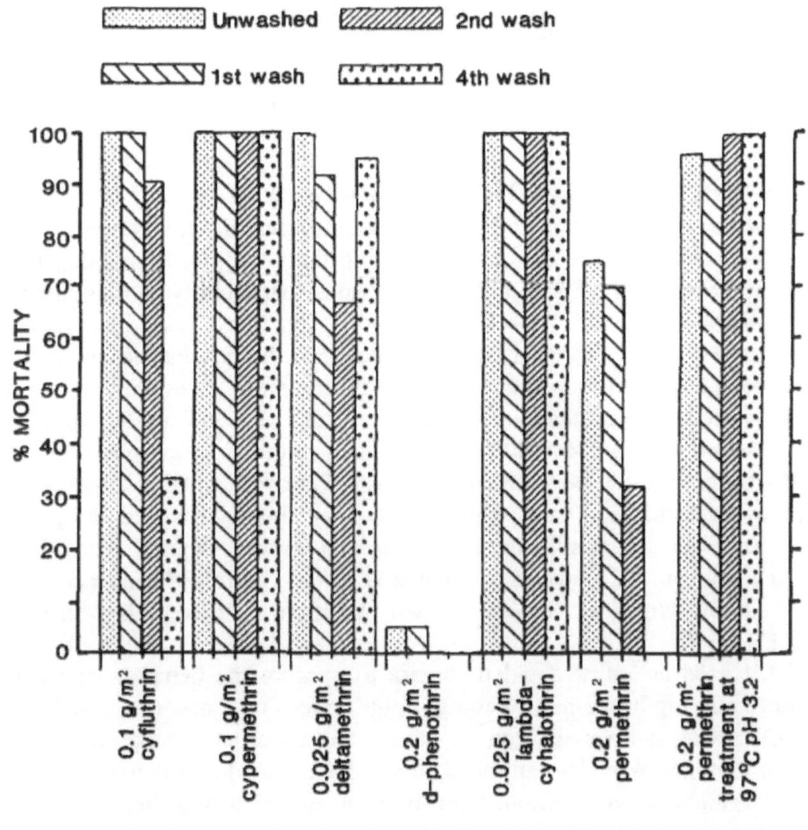

FIGURE 6. Mortality of *Anopheles gambiae* after 3-min exposure to various pyrethroids on polyester netting which was either unwashed or washed up to four times.[41]

nets whose pyrethroid content has been shown in Table 5.[41] Though the loss of each compound tested after one wash was around 50%, there was a reduction of the kill due to permethrin to virtually zero after four washes, but with cypermethrin and lambda cyhalothrin, mortality remained at 100%. This indicates that with these very powerful pyrethroids even the low initial doses applied represented considerable "overkill" and one could afford to

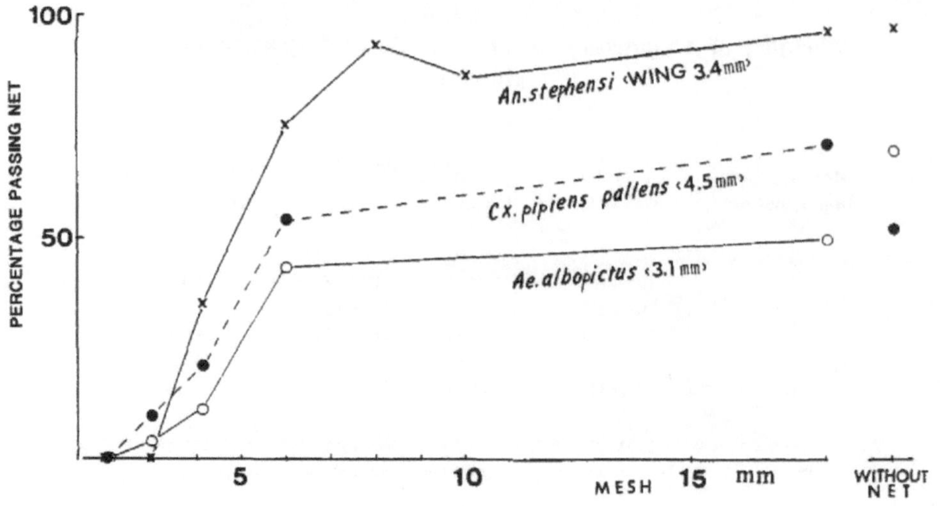

FIGURE 7. Proportions of three species of mosquito passing unimpregnated netting of various mesh sizes. The length of each wing of each species is indicated.[49]

wash out a large fraction of the active ingredient without losing the ability to kill 100% of mosquitoes with 3-min exposure.

C. BEHAVIORAL STUDIES IN THE LABORATORY*

The bioassay type of experiment does not give information on what will happen when free flying mosquitoes have the options of avoiding the netting or attempting to penetrate it and to feed after doing so. To study these questions, particularly as they apply to broad mesh netting, Kurihara et al.[48,49] and Hossain and Curtis[50] have used interconnecting laboratory cages with a piece of netting interposed between them, a bait in one and provision for releasing mosquitoes into the other. After release the mosquitoes are left overnight, and the numbers in each compartment, alive and dead, and fed and unfed, are recorded. In reporting the results, mosquitoes which fed and then returned to the release compartment are counted with those in the bait compartment as having succeeded in penetrating the net. Figure 7 shows that unimpregnated 6-mm mesh only slightly reduced the percentage of mosquitoes succeeding in entering the bait cage, and above that mesh size, there was no appreciable reduction in the percentage compared to the control with no net.[49] The data[50] for untreated netting with *An. gambiae* shown in Figure 8 agree well with that for *An. stephensi* in Figure 7.

Itoh et al.[51] used an infrared video camera to observe the behavior of *Culex pipiens pallens* on encountering broad-mesh netting. With a mesh of 8 mm (i.e., slightly less than the wingspan), all mosquitoes alighted on the net and spent an average of 23 s resting on it and walking through. With 16-mm mesh only 6.3% alighted on the net, and with 32-mm mesh all flew straight through. Thus, with mesh of around 8 mm there would be a good chance of mosquitoes picking up an insecticidal deposit from a netting barrier.

Figures 8 and 9 show the effects of impregnation of broad-mesh nylon netting with the pyrethroids permethrin or phenothrin or the organophosphate fenitrothion. Five different species of mosquito were used in these experiments.

Figures 8 and 9 show that against anophelines and *Cx. quinquefasciatus* impregnated mesh of 6 to 10 mm is effective in preventing feeding on the bait when any of the insecticides were used (except at the lowest doses of phenothrin or fenitrothion). Most of the anophelines

* T. Kurihara, K. Ichimori, C. F. Curtis, and M. I. Hossain

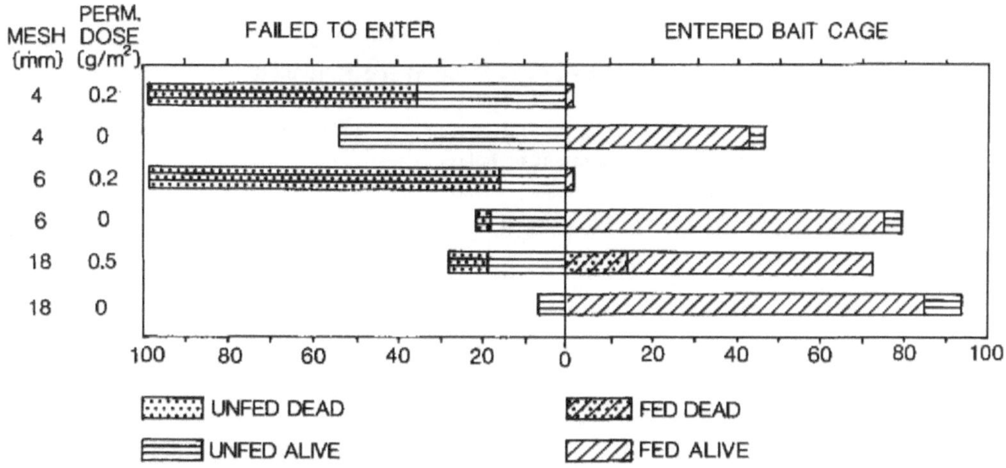

FIGURE 8. Effect of various netting barriers on the ability of *Anopheles gambiae* to reach a bait animal, feed, and survive (about 70 mosquitoes tested with each treatment).[50]

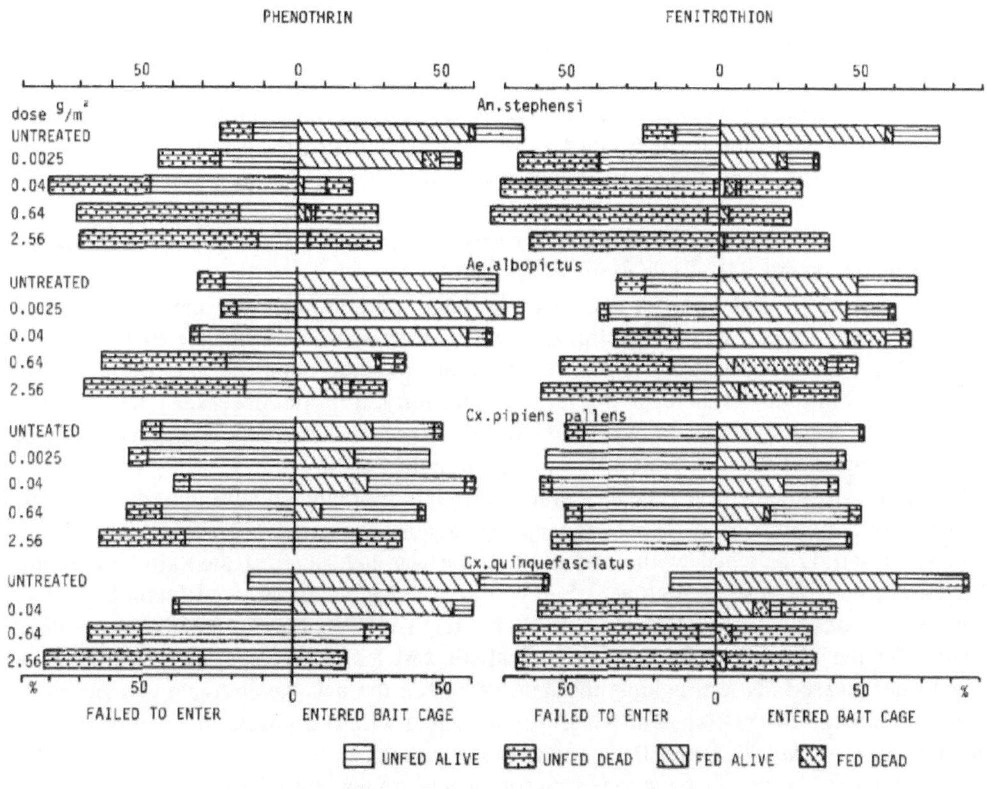

FIGURE 9. Effect of impregnated or unimpregnated 10-mm mesh netting on the ability of four species of mosquito to reach a bait animal, feed, and survive.[49]

were killed without having succeeded in passing the netting, but in the case of the pyrethroids, a considerable number remained alive in the release cage, indicating repellency before a lethal dose had been picked up. With the higher doses of fenitrothion, however, there was almost 100% mortality of *An. stephensi* and *Cx. quinquefasciatus*. *Ae. albopictus* and *Cx. pipiens pallens* were much less responsive to the treated net, the feeding rate remaining

TABLE 9
Tests with an Observer Sitting under a 1-m Cube Net in a Mosquito-Proof Room into Which *Anopheles gambiae* Were Released[52]

	Series A: 4-mm mesh; no holes		Series B: 1.5-mm mesh; 2 holes (5 × 10 cm)	
Permethrin (g/m²)	0	0.2	0	0.2
Mosq. released	68	48	31	27
Time spent on net (geom. mean and range)	44 s (10 s—8 min)	15 s (5 s—2.5 min)	2.23 min (30 s—10 min)	40 s (5 s—2 min)
% Entering net	56	16	19.4	18.5
% Entering and feeding	51	0	19.4	3.7
% Fed on arm through net	—	—	64.5	0
% Knockdown at 30 min	0	100	0	100
% Mortality at 24 h	0	100	0	100

Note: In the tests with the 1.5-mm mesh net the observer pressed an arm against the net.

almost unchanged from control levels except at the very high dose of 2.56 g/m^2 of phenothrin or fenitrothion, which caused high mortality of *Ae. albopictus* and inhibition of feeding in *Cx. pipiens pallens*.

The causes of the diversity in response between the species are unknown. Fortunately, the night biting species (anophelines and *Cx. quinquefasciatus*) respond well. *Ae. albopictus* being a day biting species would probably not in any case be a suitable target for impregnated netting.

As a more realistic test of treated nets an observer sat under a net in a mosquito-proof room into which *Anopheles gambiae* were released.[52] Mosquitoes were observed to remain standing and probing on the net for considerably longer periods when the net was untreated than when it was treated with permethrin, which is a striking consequence of the excito-repellent effect of this compound (Table 9). Mosquitoes could enter the nets through their broad mesh or holes cut in them, and a large proportion fed on the observer when the nets were untreated. When the nets were treated, fewer entered the nets, only one individual fed, and all were knocked down within 30 min and eventually died.

To investigate whether mosquitoes can feed through impregnated netting a series of tests were first done by pressing a tube of mosquitoes against the arm. A piece of permethrin-impregnated netting attached to the end of the tube was held against the skin.[52] This test indicated that even a dose as high as 2.5 g/m^2 did not completely prevent feeding, though the fed mosquitoes died later. However, in such a test situation the mosquitoes do not need to search for the blood source. As a more realistic test the observer under the treated or untreated net pressed his arm against the netting. When the net was untreated, many mosquitoes fed through the netting, but when it was treated with 0.2 gm permethrin per square meter none did so (Table 9, Series B).[52] The contrast with the above experiment is presumably because when the mosquitoes have to search for the arm on which to feed they almost always pick up a harmful dose before they have found the skin, but when no searching is needed no time is wasted on the net before starting to feed.

Experiments with laboratory-reared mosquitoes released into a room exclude a number of factors operating in the field. Logically the next stage in experimentation would be in experimental huts with wild mosquitoes allowed to enter from outside.

III. EXPERIMENTAL HUT STUDIES*

It has long been realized that in studies with species of mosquito which exit from houses soon after feeding and/or with excito-repellent insecticides (such as DDT or pyrethroids), one has to make provision to catch mosquitoes exiting during the night and not simply rely on counts of mosquitoes found resting in the house in the morning, which may only include a fraction of the mosquitoes which came to bite during the night. Exit traps are therefore placed to cover windows, eaves gaps, etc., but one cannot leave all apertures covered, otherwise there would be no opportunity for wild mosquitoes to enter. Solutions[53] to this problem include:

1. Covering the entry ports with louvers or baffles which allow mosquito entry but make exit less likely
2. Covering two out of the four eaves gaps with veranda traps so that the mosquitoes caught can be assumed to be half of those which exited
3. Closing entry ports during the course of the night
4. Use of light traps to catch and retain any mosquitoes which pass barriers such as impregnated curtains
5. Catching mosquitoes as they bite humans

In the tropics one has to build the huts on ant traps so that killed mosquitoes can be found and recorded before scavenging ants remove them.

In a trial in Burkina Faso with cotton nets, with or without holes cut in them, and with or without impregnation with 0.08 g permethrin per square meter, Darriet et al.[46] came to the following conclusions based on the capture of 4137 *Anopheles funestus* and *An. gambiae* in experimental huts over a 5-month period:

1. In the huts with impregnated nets the rate of entry of mosquitoes was reduced by about 70% compared with control huts with untreated nets.
2. Of the female mosquitoes 95% succeeded in blood feeding in the presence of the intact unimpregnated nets, indicating the ineffectiveness of these untreated nets, but this rate was reduced by 20% when the nets were impregnated, whether or not they had holes.
3. The mortality was increased by 20% with impregnated nets (with or without holes) compared with controls; one third of the mortality occurred during the night, and the remainder was delayed until the following day.
4. The rate of exit from the huts was 25 to 30% in the controls, but 97% with impregnated nets.

Table 2 shows results from veranda trap huts in Tanzania[17] with nylon nets with or without holes cut in them and with or without impregnation with 0.2 g/m^2 permethrin. As in the work in Burkina Faso permethrin impregnation restored the effectiveness of damaged nets in terms of reduction in the number of live fed mosquitoes. However, whereas in Burkina Faso the effect was mainly due to a reduction in the number of mosquitoes entering the hut, in Tanzania the effect was mainly due to a reduction in blood feeding in mosquitoes which had entered, and in killing before or after feeding. Possibly the differences in the mode of action are due to differences in permethrin dosage, the fabric, and the size of the nets used.

In Tanzania the nets were retested at intervals. Seventeen months after impregnation and when the nets with holes had been in domestic use for 5 months, there was a reduction

* P. Carnevale, J. D. Charlwood, C. F. Curtis, Li Zuzi, G. Majori, G. Sabatinelli, M. Coluzzi, T. Kurihara, and K. Ichimori

TABLE 10
Test for Diversion of *Anopheles arabiensis* in Tanzania from One Host to Another by Untreated or Permethrin(0.2 g/m^2)-Treated Intact Nets[17]

Net	No. children in net	Observed total no. fed mosquitoes	Calculated no. bites per protected child	Calculated no. bites per unprotected child
None	Neither	752	—	376[a]
Untreated	Both	158	79	—
Untreated	One	527	79	448[b]
Treated	Both	57	28.5	—
Treated	One	202	28.5	173.5[c]

Note: The table shows the total number of fed mosquitoes found in the mornings in the veranda trap huts with different sleeping arrangements of two children over a 5-week period. a is less than b but more than c, i.e., there was diversion of mosquitoes to the unprotected child from an untreated net, but the treated net gave some degree of protection to a child sleeping outside but close to it.

(compared to the control without nets) in the number of live, fed *An. arabiensis* by 83%.[54] After another 6 months domestic use, by which time the net was black with soot and full of holes, this reduction factor was 59%.[55]

In both Burkina Faso and Tanzania the treated nets caused a much larger percentage of both the fed and unfed mosquitoes to exit from the huts and to be caught in the exit traps. This was presumably mainly due to the exicto-repellent effect of permethrin, but in the case of the unfed individuals it could also be due to their failing to obtain a feed because of the nets and going in search of a feed elsewhere. If there was diversion of mosquitoes from people with nets to less privileged people (especially children) without nets, it could be counterproductive to provide nets for some but not all members of a community, because this might lead to an increase in biting on the most vulnerable groups. A test for such diversion within a house was carried out in the veranda trap huts in Tanzania.[17] The results (Table 10) indicated that in close proximity to untreated nets the biting rate was increased above what it would have been in the absence of nets. However, near to permethrin nets the biting rate was *reduced*, presumably because more of the mosquitoes rested on the treated net and were harmed by it before they had a chance to bite. This is reassuring as far as it goes, but it remains to be investigated whether there is diversion of feeding from a house fully equipped with nets to other houses without nets.

To reduce the risk of a diversion effect and to increase the "mass effect" of impregnated nets against malaria, it is very desirable that the treated nets kill a high percentage of the mosquitoes which enter a treated house. This is what was found using deltamethrin-impregnated nets in a trial with louver trap huts in China.[39] Penned water buffaloes or oxen, not humans, were used as the hosts (Table 11). The experiment compared untreated, deltamethrin- and DDT-treated nets. The nets were either lowered to reach the floor or partially lifted to simulate an ill-used net. The experiments were timed to coincide with the peak seasons of 1983, 1984, and 1985 for *Anopheles sinensis*, a malaria vector species, which is nevertheless predominantly animal biting. Table 11 shows that in the 2 years after impregnation with deltamethrin, the overall mortality with the fully lowered net was 95.2% (range 89.1 to 99.7%). Few fed and the majority had exited into the trap before dying there. With the partially lifted deltamethrin-impregnated net, a large proportion of the mosquitoes fed, but the average mortality was still as high as 82.8% (range 75.2 to 98.6%). Fewer reached the exit trap, and many died in the hut (including under the net) and were collected from the floor.

With the fully lowered net, 18 to 41 d after impregnation with DDT, the total mortality was 54.8%, and with the partially lifted net it was only 26.9%. There was little mortality with the untreated net even when it was fully lowered, when deaths from starvation might have been expected.

TABLE 11
Tests in Louver Trap Huts of Deltamethrin- and DDT-Impregnated Nets against *Anopheles sinensis* near Guangzhou, China[39]

Treatment of net	Days after impregnation	No. mosquitoes collected	Overall mortality (%)	Percent fed	Percent in exit trap
25 mg Deltamethrin/m^2					
Fully lowered	18—41	955	99.7	0.42	79.4
	161—173	953	98.1	1.26	85.1
	385—392	1382	89.1	22.60	69.0
	511—514	71	91.1	9.86	87.3
	723—725	1556	96.4	1.48	91.8
	Total	4917	95.2	7.30	81.6
Partially lifted	18—41	169	94.1	62.1	61.5
	161—173	430	98.6	27.4	45.3
	385—392	516	75.2	88.9	33.5
	511—514	39	84.6	92.3	71.8
	724—726	1535	81.0	99.0	25.9
	Total	2689	82.8	83.2	33.4
2 g DDT/m^2					
Fully lowered	18—41	281	54.8	3.9	53.0
Partially lifted	18—41	93	26.9	79.6	71.0
	164	100	14.0	82.0	92.0
Untreated					
Fully lowered	18—726	1593	19.0	41.2	59.2
Partially lowered	18—726	2627	1.4	97.6	16.2

Similar louver trap huts in rice fields have been used in Yangzhou, China to study the behavior of the Japanese encephalitis vector *Culex tritaeniorhynchus* in relation to pyrethroid-impregnated nets.[56] Collections from the exit traps every 2 h throughout the night showed that most *Cx. tritaeniorhynchus* left the house during the night even with the untreated net, but the percentage leaving before midnight was much higher when the nets were treated. There were virtually no living mosquitoes left inside the huts when the nets were treated.

Another trial against *Cx. tritaeniorhynchus* was carried out in Toyama Prefecture, Japan in 1984.[57] "Curtains" of wide-mesh nylon net were erected in a pigsty, and the number of mosquitoes passing the curtains was evaluated by collection in a light trap hanging within the curtains. These were 1.8 m high; there was no netting "roof". The roof of the building was 4 to 6 m high, and the light trap was hung 1.7 m from the floor, i.e., just below the level of the top of the curtains. The curtains surrounded 3 × 4-m sections of the pigsty, each containing five to six pigs which attracted the mosquitoes. Netting, with 10 mm and 40 mm mesh, untreated or impregnated with 0.5 g phenothrin per square meter, were compared. Table 12 shows that only the treated smaller mesh reduced the number of mosquitoes which passed the netting, but both sizes of treated net reduced the percentage which fed and the mosquito survival. It seems likely that a vapor effect of the phenothrin was at least partly responsible since, according to the above-mentioned video study,[51] mosquitoes would have been expected to fly straight through the 40-mm mesh netting.

Further studies have been carried out with broad-mesh cotton netting curtains against mosquitoes of the *An. gambiae* complex in experimental huts in Burkina Faso[47] and Tanzania.[58] In both cases impregnation with 1 g permethrin per square meter reduced the number of mosquitoes biting, but among those that did bite and survive, there was a strong tendency for them to be driven into the exit traps either due to a delayed effect of permethrin picked up at the time of entry, or when mosquitoes use the curtains as a resting site after feeding. Details of the results from Burkina Faso are shown in Table 13(A). In these experiments a

TABLE 12
Comparison of Collections of *Culex tritaeniorhynchus* in Japan in Light Traps Surrounded by Phenothrin-Treated Broad Mesh Nets, Untreated Nets, or No Net[57]

Net	No. collected	Percent fed	Percent survived
40-mm mesh, treated	5382	30.6	0.4
10-mm mesh, treated	1661	38.6	5.5
40-mm mesh, untreated	3987	80.5	35.5
10-mm mesh, untreated	5361	75.7	18.7
No net	6308	70.9	21.5

TABLE 13
Tests with Broad Mesh Cotton Netting Curtains over Doors, Windows, and Eaves Impregnated with 1 g Permethrin/m² near Ougadougou, Burkina Faso[47]

(A)	No. *Anopheles gambiae* s.l.[a] caught	Percent fed	Percent mortality	Percent exiting Fed	Percent exiting Unfed
Impregnated curtains	193	66.3	76.2	65.6	67.3
No curtains	389	89.2	3.8	13.8	14.3

(B)		Months since installation of curtains 1	3	11
Impregnated curtains	Inside	0.3	0.2	9.0
	Outside	7.3	4.3	3.7
No curtains	Inside	20.6	10.7	19.7
	Outside	5.7	3.3	8.4

Note: (A) shows the results with a veranda trap; (B) shows the results of biting catches on humans in or outside huts.

[a] The population consisted of about 30% *An. arabiensis* and 70% *An. gambiae* s.s. (of which 85% were of the Mopti chromosomal form and 10% of the Savanna form).

15-cm gap in the curtains was left for the first part of the night to allow mosquito entry. At 02.00 h a veranda trap was fitted to catch mosquitoes which exited before dawn.

Table 13(B) shows that the catch on human baits was greatly reduced as a result of the curtains 1 and 3 months after impregnation, but the effect had largely worn off after 11 months.

Nearby dwelling houses fitted with these impregnated curtains (as shown in Figure 10) were subjected to repeated pyrethrum spray catches of resting mosquitoes. For 11 months these catches were reduced almost to zero, and only a year after installation of the curtains did catches rise to 5% of their initial level. These reductions were presumably due to a combination of inhibition of mosquito entry, and the mortality and stimulation of exit shown in Table 13(A). Untreated curtains of the same type reduced the resting catch by about 55%.

Curtains were less successful than bednets in the tests in Tanzania,[17] perhaps because the curtains were not as well fixed over the eaves as those shown in Figure 10. This series of tests also included the use of 0.2 and 1.0 g permethrin per square meter on both curtains and bednets. The results, in terms of reduction in blood feeding and mortality, did not differ significantly between the dosages when they were freshly applied. It seems likely that higher dosages would remain effective for longer, but this remains to be tested.

FIGURE 10. Impregnated curtains over the eaves gap and doorway of a circular hut near Ougadougou, Burkina Faso. (Photograph from G. Majori.)[47]

Tests in experimental huts in Tanzania with "bed curtains" made from locally grown sisal fiber or fiber from polypropylene sacks, impregnated with permethrin, indicate that they give a highly significant reduction in number of live fed *An. gambiae* s.s.[59] However sisal has been found to be dangerously inflammable[45] and data with polypropylene do not yet allow one to conclude whether this cheap and readily available material is a fully adequate substitute for a nylon bednet.

IV. FIELD TRIALS OF IMPREGNATED NETS FOR MALARIA CONTROL

A. TRIALS IN THE GAMBIA*

In The Gambia there is a sharp peak of *Plasmodium falciparum* malaria transmission from August to November, during the rains. The main vector is *An. gambiae* s.s. which breeds in freshwater although there is some contribution from *An. melas* which breeds in brackish water in mangrove swamps[11] beside the Gambia River. Bed nets, mostly of thick cotton sheeting, are already widely used.[4]

Two studies,[60,61] based on retrospective data, suggested that fewer malaria attacks occurred in children who slept under untreated nets than in children in the same village who did not.

In the malaria seasons of 1985, 1986, and 1987 three intervention studies were conducted in villages on the north bank of the Gambia River near Farafenni to determine whether a reduction could be shown in malaria morbidity in children aged 1 to 9 years provided with the following:

1. Untreated nets to all the inhabitants of villages where nets were previously uncommon (1986 study)[62,63]

* R. W. Snow and S. W. Lindsay

TABLE 14
Effects of Unimpregnated and Permethrin-Impregnated Nets in The Gambia on the Numbers of New Febrile Episodes with Malarial Parasitemia in Children

	New febrile episodes with parasitemia					
	Intervention			Control		
	Ch.[a]	Any[b]	High[c]	Ch.[a]	Any[b]	High[c]
Untreated nets[62]	233	44	26	163[d]	40	29
Permethrin on individuals' nets[64]	205	23	10	184[e]	34	17
Permethrin on all nets[65]	219	16	12	255[e]	49	33

[a] No. of children at the start of the trial.
[b] Any level of parasitemia.
[c] Parasitemia above 5000/μl
[d] Controls had no nets.
[e] Controls had placebo-treated nets.

2. Permethrin impregnation of the nets which they already had, but leaving the majority of the nets in the same village unimpregnated (1985 study)[13,64]
3. Permethrin impregnation of all the nets in their villages (1987 study)[14,65]

The same method of monitoring malaria morbidity was used in all three studies, i.e., weekly interviews with the mothers of virtually all the children throughout the malaria season, and taking of the child's axillary temperature to determine if he or she then had a fever (above 37.4°C). A blood slide was taken where there was evidence of current or recent fever, and the slides were stained and read for malaria parasites. The intention was to record all cases of fever associated with malarial infection which were new occurrences since the previous week. In addition all children were medically examined before and after the malaria season with records being made of splenomegaly and packed cell volume (as an index of anemia).

In the studies with permethrin impregnation the control nets were treated with a placebo so that neither the villagers nor the field staff knew which nets had received the insecticidal treatment.

Table 14 shows the numbers of febrile episodes with parasitemia in the three trials, together with the initial number of children in the intervention and control groups. Conversion of the data to number of child-weeks exposure to risk and statistical testing confirm the clear indication from Table 14 that the untreated nets had no appreciable effect on malaria risk, permethrin treatment of an individual's net gave a small but significant protective effect, and impregnation of all the nets in a village (where the individual protective effect would be expected to be enhanced by the "mass effect" as defined in Section I) gave a considerably greater protective effect, amounting to about 63%.

To explain the apparent conflict between the ineffectiveness of providing untreated nets and the retrospective studies on the effects of having untreated nets,[60,61] it was pointed out[62] that the controls in the latter studies were in the same villages as the children with nets. There could, therefore, have been diversion of extra mosquitoes from children with nets to those without (as illustrated within one hut in Table 10), thus enhancing the difference in risk between children with and without nets. In the intervention study, however, the children with and without nets were in different villages (seven villages with nets and nine without). There was no evidence for diversion of mosquitoes from villages with nets to those without,[63] and the difference between malaria risk in villages with and without nets would therefore be expected to be less than between neighbors with and without them.

In an early study in Mali[66] an effect was reported of deltamethrin-treated nets on splenomegaly, but not other symptoms of malaria. In the 1985 study in The Gambia no significant effect of individual use of permethrin-impregnated nets was found on spenomegaly, packed cell volume, or percentage of children with malaria parasites (parasite rate).[64] However, in the 1987 study when all the nets of certain villages were impregnated, this was associated with significantly lower incidence of splenomegaly and less of a decline of packed cell volume over the course of the malaria season than in the controls.[65] A cross-sectional survey of all the children at the end of the season showed no significant effect of the impregnated nets on overall parasite rate, but a significantly smaller number with high parasitemias. All of these results suggest that the widespread impregnation of nets, while reducing the morbidity due to malaria, did not significantly alter the prevalence of malaria infection. In present-day African conditions this may be the best outcome, as, although the harmful effects of malaria are reduced, the normal buildup of immunity to the disease would presumably not be affected. Complete removal of malaria transmission by a time-limited program may leave the population vulnerable to a disastrous malaria epidemic should the vector control program falter.

Entomological studies conducted during these three trials included not only inspection of nets for mosquitoes (Table 1) and collections of mosquitoes resting in the rooms, but also the use of exit traps over windows or doorways, with records of the numbers of mosquitoes which were unfed, fed on humans, or fed on animals. The intention was to get some idea of the degree of protection from biting conferred by the nets in use in the houses. This differs from most other village-scale trials which have only attempted to measure the "mass effect", i.e., the effect of widespread use of insecticidal nets in lowering the biting population of infective mosquitoes *outside* the nets.

In the 1986 study unimpregnated nets were shown to reduce the proportion of bloodfeeds from humans, and it was estimated that children under these nets received 24 to 60% fewer bites than those unprotected by nets.[63] Such a reduction was clearly insufficient under Gambian conditions to have an appreciable impact on malaria incidence, as shown by the results in Table 14.

The catches in exit traps on rooms containing permethrin-impregnated nets in both 1985[13] and 1987[14] showed that these nets tended to drive mosquitoes outside, but mortality was no greater than in traps on rooms with placebo-treated nets.

In the 1987 study the catches in rooms with placebo-treated nets and in exit traps of freshwater *An. gambiae* s.l. and *An. melas* rose to seasonal peaks a few weeks after the rains (Figure 11).[14] However, the mosquito catches remained at a low level where the nets were permethrin treated. It was estimated that the number of bites per child was reduced by 90% as a result of permethrin treatment. In the permethrin- and placebo-treated villages which originally had the highest transmission, multiplying the numbers of bites per child by the observed sporozoite rates, led to the conclusion that the mean number of infective bites per child during the transmission season was reduced from 2 to 0.1 as a result of permethrin impregnation. This reduction was enough to have a marked effect on the number of malaria attacks as shown by Table 14.

B. TRIAL OF DELTAMETHRIN-IMPREGNATED BED NETS IN BURKINA FASO*

The trial[67] was in the village of Karangasso, 45 km west of Bobo-Dioulasso, in the southwest of Burkina Faso. This region is on the northern edge of the wooded sudan-savanna zone. There are two contrasting seasons: a dry season from November to April and a rainy season from May to October when there is an average rainfall of 1100 mm.

* P. Carnevale, V. Robert, C. Boudin, J.-M. Halna, L. Pazart, P. Gazin, A. Richard, and J. Mouchet

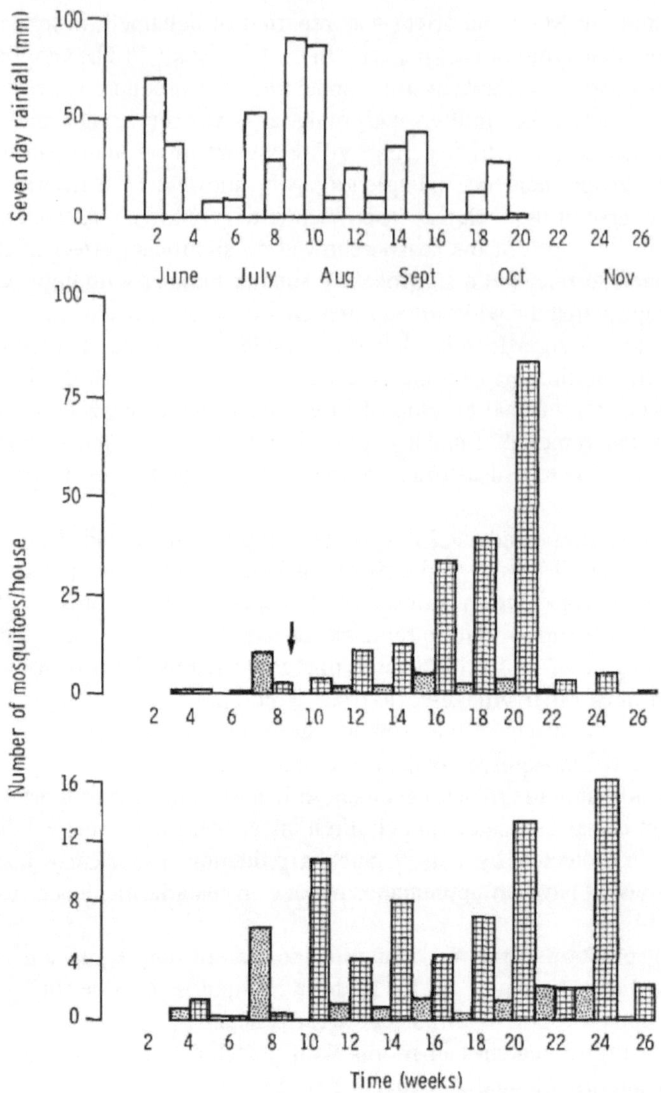

FIGURE 11. Rainfall (top) and numbers of freshwater *Anopheles gambiae* s.l. (middle) and *An. melas* (bottom) in houses in two Gambian villages. In one village there were permethrin-treated nets (data indicated by stippled bars) and in the other placebo-treated nets (data indicated by netted bars). Each bar is based on the sum of the number of mosquitoes caught under bednets, in rooms, and in exit traps fitted over windows or doorways. The time of impregnating the nets is indicated by the arrow. (From Lindsay, S. W. et al., *Med. Vet. Entomol.*, 3(3), 266, 1989. With permission.)

The village of Karangasso has a total population of 2400 and is divided into two hamlets, Koko and Massasso, each of which has a population of about 1200 and is spatially distinct, being separated by 800 m of uninhabited land. In the first year of the trial (May 1985 to April 1986) baseline data were collected in both hamlets. In the second year (May 1986 to April 1987) deltamethrin-impregnated nets were given to all the inhabitants of Massasso, while Koko continued to be kept as a control.

Mothers often sleep with their children, and after a survey of people's sleeping places it was found that 700 impregnated nets would be needed for the people normally resident in Massasso. Single bed nets had an area of 14 m^2, and nets for two or three people were

20 m² in area. The roof and lower border of the nets were made of cotton, but most of the vertical surfaces were of synthetic netting. A dosage of 25 mg deltamethrin per square meter was applied, and the nets were put in position at the beginning of May 1986 before the beginning of the seasonal malaria transmission peak.

1. Entomological Evaluation

The pre- and posttreatment years, 1985—1986 and 1986—1987, were similar in temperature, but in the second year the rainy season was two months longer than in the first although the total rainfall was 34% less in the second year. This caused an increase in *Anopheles funestus* and a decrease in *An. gambiae*, and this was also seen throughout the Bobo-Dioulasso region.

Samples of the *An. gambiae* complex taken in the middle of the rainy season and examined cytogenetically revealed only *An. gambiae* s.s. whose chromosomes were of Savanna and Mopti inversion cytotypes.[68]

The mosquito collections were carried out on humans. In each hamlet there were four catching stations (representative bedrooms). These were sampled once a fortnight from 20.00 to 06.00 h by one man in each room. The catchers were on malarial chemoprophylaxis and were rotated from room to room according to a schedule to minimize the effect of variation between individual catchers. There were 400 man nights of catching over the 2 years, and these yielded 7253 mosquitoes, of which 5700 were anophelines, and of these, 3589 were dissected to record parity from the ovarian tracheoles by the method of Detinova.[69] The salivary glands were checked for sporozoites while still fresh and without staining.

Table 15 shows the results in each year and hamlet, together with the percentage reduction in the second year in Massasso corrected for the change in Koko between the first (K_1) and second year (K_2). There was no appreciable change in the number of *An. gambiae* bites per man per night, but in every other comparison there was a marked reduction in the second year in Massasso, i.e., the presence of the nets reduced mosquito survival and sporozoite rate. In the control hamlet the total sporozoite inoculation rate (number of bites per person per unit time $\times$ the sporozoite rate) for all three vector species rose between the first and second year from 116 to 224 infective bites per year. In Massasso in the first year the annual value was 403, but with the introduction of the nets it declined to 73, which is equivalent to a corrected reduction of 91%. This was due to the reduction in the sporozoite rate in all three species and the reduction in the biting density of *An. funestus*. Figure 12 shows that the reduction in the sporozoite inoculation rate was observable throughout the transmission season.

The reduction in the sporozoite inoculation rate in Massasso is a measure of the "mass effect" of the nets, i.e., the degree of protection which they confer to people not under their nets. People who are under nets must be much better protected, and the 91% reduction in the inoculation rate mentioned above is a considerable underestimate of the actual reduction in transmission among the population of the hamlet.

The hamlets were only 800 m apart, and there is a possibility of exchange of mosquitoes between them, which would have resulted in a further underestimate of the true effect of the impregnated nets. This could only be resolved by working in a much larger area where all the inhabitants used impregnated nets and where immigration of *Anopheles* into the middle part of the treated area would be negligible.

2. Parasitological and Clinical Evaluation

Medical examinations and consultations were carried out every 5 d in the village with anyone who cared to present him or herself for whatever reason. Inhabitants of villages other than Karangasso were examined, but the data are not included with those shown here.

Those presenting themselves were asked in their own language why they came, the

TABLE 15
Trial at Karangasso, Burkina Faso:[67] Number of Bites per Man per Night, % of Females Parous (Parity), Proportion with Sporozoites (Sporo. Rate), and Sporozoite Inoculation Rate (i.e., Biting × Sporozoite Rates)

		Koko (control)		Massasso (nets in 2nd year)		
Parameter	*Anopheles* sp.	1st yr(K_1)	2nd yr(K_2)	1st yr(M_1)	2nd yr(M_2)	Corr.[a] % change
Bites/man/night	*gambiae*	6.05	3.84	9.03	5.84	+2%
	funestus	5.39	8.60	8.25	4.32	−67%
	nili	1.02	1.02	1.93	1.71	−11%
Parity (%)	*gambiae*	68.2	60.9	74.8	38.7	−41%
	funestus	72.7	65.8	77.5	51.2	−27%
	nili	84	82	72	55	−21%
Sporozoite rate	*gambiae*	2.70	6.47	7.36	1.99	−89%
	funestus	2.91	4.22	5.02	1.86	−74%
	nili	0	0	1.82	0	—
	All spp.	19/745[b]	44/966	60/1043	14/835	
Sporozoite inoc. rate	*gambiae*	0.163	0.248	0.665	0.116	−89%
	funest.	0.157	0.362	0.414	0.080	−91%
	nili	0	0	0.035	0	—
	All spp.	0.320	0.610	1.114	0.196	−91%

Note: The % reductions in the second year in Massasso are shown corrected for the changes in Koko between the first and second year.

[a] $[100(M_2 \times K_1)/(M_1 \times K_2)] - 100$.
[b] No. with sporozoites/no. dissected.

symptoms experienced, the occurrence of fever and its periodicity in the preceding days, and drugs, especially antimalarials, taken. The medical examination by a physician included taking the rectal temperature and making a thin blood film from a finger prick. Appropriate treatment was prescribed immediately, usually from the essential drugs available to the physician. The data were entered on standard forms for analysis.

The malaria parasite density in the thin films was determined from examination of 100 fields with a 100× objective, and the results expressed assuming 200 RBC per field and 4 million RBC per microliter.

A total of 2533 consultations were carried out over the 2 years. The proportion of these in the season of high transmission (July to December) differed markedly between the two hamlets in the 2 years (Table 16), and there was also a slight difference between the mean ages of the samples. One would expect these factors to be strongly correlated with parasitological and clinical manifestations of malaria and analyses of variance/covariance were required to isolate the effect of the impregnated nets. To ensure equality of the variance the parasite densities were transformed to logarithms (after adding 1.0 to all values).

The analysis of variance on all subjects (whether positive or negative for malaria) showed that there was no significant difference in parasite density of *P. falciparum* during the first year between the two hamlets (variance ratio, $F = 0.08$, degrees of freedom, d.f. = 1,953, $P = 0.78$). However, there were significant differences between the 2 years both for Koko and Massasso, the parasite densities being lower in the second year in each case. Furthermore, in the second year there was a significant difference between the hamlets ($F = 10.3$, d.f. = 1,1572, P less than 0.0001), the parasite density being lower in Massasso (with im-

TABLE 17
Percentage of Subjects in the Burkina Faso Trial[67] with Attacks of Malaria and Parasites by Age, Hamlet, and Year

		Age			
		Under 1	1—4.99	5—13.99	14 and over
Both	No. subjects	56	187	146	342
hamlets	No. with attacks	3	35	20	5
year 1	% With attacks	5.4	18.7	13.7	1.5
	% With parasites	41.3	57.0	49.9	23.1
Koko	No. subjects	35	83	129	234
	No. with attacks	4	10	11	3
year 2	% With attacks	11.4	12.0	8.5	1.3
	% With parasites	31.1	51.7	31.9	18.6
Massasso	No. subjects	60	203	178	435
	No. with attacks	2	17	4	2
year 2	% With attacks	3.3	8.4	2.2	0.5
	% With parasites	30.4	37.7	46.7	11.8

Note: Data given are for the wet and dry seasons combined.

3. Malaria Attacks

Malaria attacks were defined as follows:

1. Rectal temperature of at least 38°C
2. Parasitemia of at least 10,000 *P. falciparum* per microliter (attacks due to *P. malariae* or *P. ovale* were rarely recorded)
3. Absence of evident nonmalarial causes of fever

This definition is restrictive, but it gives the kind of objective data needed for a study of this kind. Self-reporting of recent fever is not a usable criterion in an area such as this as 79% of children made such reports. If instead of 10,000 parasites per microliter, 5000 or 15,000 had been used, the conclusions would not have been appreciably different.

As might be expected the frequency of malaria attacks was at a maximum in childhood (Table 17) after the loss of maternal antibodies and before the individual's own immunity had built up. The frequency of attacks was lower in each age group in Massasso in the second year than in both hamlets in the first year or in Koko in the second year. To test the statistical significance of this apparent effect of the nets, the data for the second year were analyzed by a step-by-step logistic regression method, taking into account the confounding factors of the subject's age and the season of high or low transmission.

The variable "season" did not play a significant role in this model. The relative risk associated with nonuse/use of a net was found to be 2.42:1 and was significantly different from 1:1 at the 1% level. Thus, the best estimate is that the risk of a malaria attack was reduced by 59% as a result of using an impregnated net. This conclusion is very similar to that from the 1987 trial in The Gambia (see Section IV.A above).

C. TRIAL OF PERMETHRIN-IMPREGNATED NETS IN TANZANIA*

The trial[70] was in villages in the Tanga region of Tanzania in which there is intense perennial transmission of *P. falciparum*, mainly by *An. gambiae* s.s. Two villages were chosen for intervention with nets impregnated with 0.2 g permethrin per square meter, and two were held as controls. Within each of these pairs one was a traditional village with mud

* K. J. Njunwa, T. J. Wilkes, J. D. Lines, and C. F. Curtis

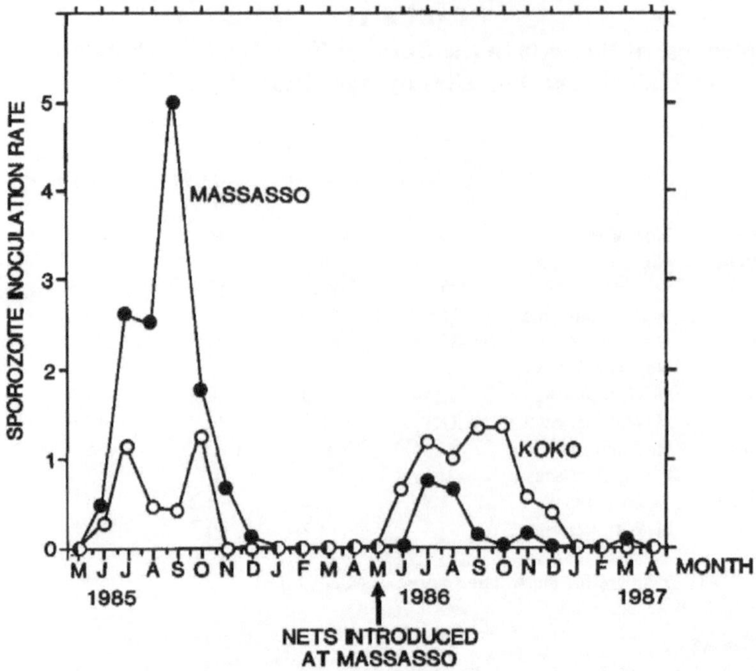

FIGURE 12. Monthly changes in the sporozoite inoculation rate per day due to the three vectors *Anopheles gambiae*, *An. funestus*, and *An. nili* in Koko (control hamlet) and Massasso (impregnated nets from May 1986 Burkina Faso).[67]

TABLE 16
Mean Age and Number of Subjects Sampled in the Burkina Faso Trial[67] in the High Transmission (Wet) Season and the Low Transmission (Dry) Season; Geometric Mean *P. falciparum* Density in All Subjects and in Those Found Positive and Percent of Subjects Positive for *P. falciparum*

		First year		Second year	
		Wet	Dry	Wet	Dry
Koko (control)	Mean age	14.0	21.8	18.9	18.0
	No. subjects	283	119	256	279
	Overall para.[a] density	83.9	6.2	24.0	6.2
	Para. density in +ve	4817.4	1772.2	5825.5	3361.0
	% +ve	52	24	37	23
Massasso (nets in second year)	Mean age	15.6	19.1	14.9	20.1
	No. subjects	150	405	654	387
	Overall para. density	39.6	12.6	12.9	3.9
	Para. density in +ve	6310.7	1978.3	2981.0	1394.1
	% +ve	42	33	32	19

[a] Parasite.

pregnated nets) than in Koko (control hamlet). A similar analysis only of subjects positive for *P. falciparum* led to essentially the same conclusions.

The percentage of subjects positive for malarial parasites (Tables 16 and 17) is not a sensitive parameter in an area of high endemicity such as this and was not markedly affected by introduction of the nets.

TABLE 18
Data on *Anopheles gambiae* s.s. from Four Villages in the Tanga Region of Tanzania

		Control		Intervention	
		U	K	M1	Mn
Light trap catches	Before	33.5(1306)[a]	21.9(787)	13.3(479)	85.3(3326)
	After	12.3(442)	1.6(54)	0.8(26)	15.0(554)
Resting catches	Before	5.7(205)	7.1(236)	1.1(38)	20.6(659)
	After	7.1(329)	2.2(85)	0.1(6)	1.0(62)
Parity (%)	Before	56.1(582)	69.9(555)	45.6(373)	63.7(1190)
	After	67.1(323)	78.3(111)	37.5(48)	45.6(511)
Age grade	Before	1.1(522)	1.7(617)	0.79(334)	1.35(1244)
	After	1.3(475)	1.6(107)	0.44(48)	0.61(503)
Sporozoite rate (%)	Before	5.8(582)	8.2(545)	4.2(306)	5.5(1168)
	After	4.1(481)	5.4(111)	0(48)	0.59(511)
Sporozoite inoculation[b]	Before	1.9	1.8	0.55	4.7
	After	0.5	0.09	0	0.09

Note: The data refer to the 6 months before and the 6 months after introduction of permethrin-impregnated nets into the villages designated Ml and Mn.[70]

[a] *No. mosquitoes counted and dissected shown in parentheses.*

[b] Sporozoite inoculation rate obtained by multiplying light trap catches (which were made in bedrooms) by sporozoite rate.

houses and thatch roofs and the other was of a more modern type with brick houses and tiled roofs. There were very few nets in the villages before the trial began. After collection of baseline data impregnated nets were distributed. After 6 months the nets were washed and reimpregnated by the villagers (Figure 2).

Table 18 summarizes the entomological data for a period of 6 months before intervention with the nets and for 6 months afterward. The main method of catching mosquitoes was the setting of light traps in designated bedrooms provided with unimpregnated nets. The results correlated well with human biting catches made in the same rooms on nights following those on which light traps had been set.[71] The light trap method was preferred because human biting catches tire staff and increase their exposure to the risk of drug-resistant malaria.

The control villages showed a seasonal decline in the numbers caught in the light traps between the first 6 months and the second. However, the survival, as measured by the parity (Detinova) and age grading (Polovodova) techniques[69] and the sporozoite rates showed little change over the year and were particularly high in the traditional village, designated K in Table 18.

After the introduction of permethrin-impregnated nets into two of the villages, the light trap catches declined, but this may be at least partly a seasonal effect. The catches of mosquitoes resting in rooms were greatly reduced by the introduction of nets — due to the excito-repellent effect of the nets already encountered in the experimental hut studies (Section III above). Furthermore, both measures of mosquito survival and the sporozoite rate were greatly reduced after introducing the nets which can be attributed to the killing of large numbers of mosquitoes by contact with the nets. The reductions in sporozoite rate may be compared with those found in Burkina Faso (Table 15). It should be recalled that this effect is over and above the individual protection that people under their nets will have enjoyed.

At the time of writing it is too early to give definitive conclusions about the impact of these very encouraging entomological results on malaria. The indications are that the percentage of children aged 1 to 10 with malarial parasitemia is unchanged from the very high levels which are normal for this area of holoendemic malaria. However, the percentage with high parasitemia (more than 100 parasitized red cells per 200 white cells) appears to have been markedly reduced in the villages with the treated nets.

D. PAPUA NEW GUINEA*

Several studies of permethrin-impregnated nets have been carried out in the Madang area of Papua New Guinea, where there is intense perennial transmission of both *P. falciparum* and *P. vivax*. The vectors are *Anopheles punctulatus, An. koliensis,* and *An. farauti,* the last being mainly coastal and tending to rest out of doors in contrast to the more inland distribution and indoor resting behavior of the other two species.

Millen[72] compared malaria prevalence in the inland villages, one with untreated nets and one with nets impregnated with 0.2 g permethrin per square meter. At a survey 7 months after net impregnation, prevalence of *P. falciparum* in 1 to 4 year olds was 50% in the village with treated nets, compared with 77.7% in the control village (difference significant at 5% level). There were no effects on *P. vivax* or on *P. falciparum* in any other age group or on the vector population.

A more intensive entomological study[73] in a village in which *An. farauti* was the main vector showed that permethrin impregnation of nets caused very marked reductions in (1) the number of mosquitoes caught on humans, (2) the daily survival of the mosquitoes (as estimated by two independent methods), (3) the estimated total vector population of the village, (4) the numbers resting in houses, (5) the proportion blood fed, and (6) the proportion of these meals which were from humans. Thus, a large number of mosquitoes appeared to have been killed by the nets, and most of the remainder were diverted to animal hosts. The peak time of biting was found to be shifted to early in the night. This was apparently because mosquitoes returning after oviposition late in the night could not easily feed before dawn owing to the presence of the impregnated nets, and they therefore delayed their feed until early the following night.

To study the effect of net impregnation on malaria, pairs of comparable small villages were chosen, and the existing nets in one member of each pair were impregnated with 0.4 g permethrin per square meter.[74] Over 98% of the population slept under nets. Following an initial survey children in all villages were given chloroquine and Fansidar to clear their parasitemias. Thus, the majority of children were slide negative at a 4-week survey. The incidence of *P. falciparum* indicated by a 10-week survey was 8.7% in the villages with treated nets compared with 28.9% in the controls with untreated nets. There was no significant effect on *P. falciparum* incidence in 5 to 10 year olds, perhaps because they went to bed later and were more likely to be bitten before getting under their nets. In neither age group was there a significant effect on *P. vivax,* perhaps because of relapses of preexisiting infections not eliminated by the drug treatment. The sporozoite rate was reduced five-fold in *An. farauti* and *An. koliensis* in two of the villages with impregnated nets. However, there was no such effect in *An. punctulatus* despite the fact that, since it rests indoors, a greater effect of impregnated nets might have been expected on this species than on *An. farauti.*

In parts of Papua New Guinea *An. punctulatus* transmits filariasis. Less reduction in vectorial capacity is required for filariasis control than for malaria control,[75] and the prospects for filariasis control with impregnated nets appear to be good.[76]

E. SABAH, MALAYSIA, AND SOLOMON ISLANDS

Anopheles balabacensis (a close relation of *An. dirus* — see Section F.2) is the main malaria vector in Sabah, Malaysia and is a difficult species to control because of its tendency to exit from houses immediately after feeding.

* J. D. Charlwood and P. M. Graves

In a trial[12] in six villages in Sabah, Fansidar was given to all the inhabitants and permethrin impregnated nets were provided in five of the villages, the other village being kept as a control. Much of the permethrin applied to the nets was absorbed by their thick cotton borders and the netting was estimated to have received a dose of only 0.062 g/m^2. Nevertheless this almost completely prevented mosquitoes remaining inside the nets (Table 1). The mass drug administration reduced the *P. falciparum* parasitemia in the children in all the villages. For 2 months after distribution of the nets the children in the villages with nets had a lower average parasitemia than the controls. However, after that the effect of the nets was lost. Apart from the low dose of permethrin on the netting, the rather disappointing result was attributed to poor usage and acceptance of nets by the villagers. They did not like sleeping under nets and stayed up late watching battery-operated televisions (see Chapter 12) and drinking tapioca beer.

Trials were carried out in two areas in the Solomon Islands with nets impregnated with 0.5 g permethrin per square meter and with mass antimalarial drug administration.[77] In an area where *An. farauti* was the vector and where there was considerable human population movement, the malaria parasite rate in children was reduced from 33 to 10%. In the other area, where *An. punctulatus* and *An. koliensis* were the vectors and the village was more isolated, the *P. falciparum* parasite rate and the human biting density of the vectors both dropped to undetectable levels from initial values of 35% and 28 per man hour, respectively.

The Solomon Islands malaria control program is now involved in the treatment of 20,000 nets in villages and their distribution to a population of 50,000.

F. FIELD TRIALS AND USE OF PYRETHROID-TREATED MOSQUITO NETS FOR MALARIA CONTROL IN CHINA*

Pyrethroid-impregnated nets have been much more extensively field tested and used in China than anywhere else. Most of this work has been in Sichuan, Guangdong, Henan, and Jiangsu provinces, where *Anopheles anthropophagus* and *An. sinensis* are the main malaria vectors. There has also been an important trial against *An. dirus* in Hainan province.

1. Trials against *An. anthropophagus and An. sinensis*

An. anthropophagus and *An. sinensis* are two closely related species belonging to the *An. hyrcanus* group. *An. sinensis* is the most common and the most widely distributed anopheline mosquito in China.[78] It breeds extensively in rice fields, as well as natural water bodies. Though this species is mainly animal biting (zoophilic), it plays an important role in the maintenance of a low malaria endemicity in the plains of China and is the main or sole vector in the area north of 34°N. Owing to its large population size and particularly its outdoor resting (exophilic) habits, it has hitherto not been satisfactorily controlled in China.

On the other hand, *An. anthropophagus* is an anthropophilic and endophilic species. It is found in an area between approximately 21 to 34° N and is patchy in distribution. It is an efficient vector of malaria and Brugian filariasis throughout its range, but can be successfully controlled by indoor residual sprays with DDT or other insecticides.[79,80]

An. anthropophagus is generally found together with *An. sinensis*, so the control measures described below were actually concurrently against both species.

a. Field Trials and Applications of Nets in Guangdong Province

In 1985 and 1986, a field trial was made in a hilly region of Buji District, Bao An County. It covered a population of about 4000, 72.3% of whom were immigrants from other areas. The sudden increase of the immigrant population had been followed by an outbreak of malaria. In 1985, 1855 nets were impregnated with a 2.5% wettable powder of deltamethrin

* Li Zuzi and Lu Baolin

TABLE 19
Effect of Deltamethrin Treatment of Mosquito Nets on Numbers of *Anopheles anthropophagus* and *An. sinensis* Caught at Buji, Guangdong, China[81]

		Average density		
		Treated area	Control area	Percent change[a]
Pretreatment	On nets[b]	22.5	18.6	
(May—June 1985)	Biting[c]	23.76	11.15	
After 1st treatment	On nets[b]	0.45	5.62	−93.4
(June—Nov 1985)	Biting[c]	1.42	5.08	−86.9
After 2nd treatment	On nets[b]	0.42	3.76	−90.8
(Apr—Nov 1986)	Biting[c]	3.23	6.40	−76.4

[a] Corrected for changes in the control area, i.e., $[100(T_2 \times C_1)/(T_1 \times C_2)] - 100$, where 1 refers to pretreatment and 2 to posttreatment and C and T refer to the control and treatment areas, respectively.
[b] Average number of mosquitoes caught between 2200 and 2330 h resting on 20 nets.
[c] Average number of mosquitoes caught on humans between 2200 and 2330 h.

at a dosage of 15 mg of active ingredient (a.i.) per square meter. These nets were used by 86.7% of the population. The second impregnation took place in April 1986; 2467 nets were impregnated by the same methods as before, protecting 87.5% of the population of 3744. Also in 1986, the treatment was extended to the whole subregion, and 24,665 nets were impregnated, protecting 83.6% of the total population of 32,367.

After treatment, the numbers of *An. sinensis* and *An. anthropophagus* caught resting on the nets and in outdoor biting catches were notably reduced, with a reduction of 93.4 and 86.9%, respectively, in 1985 and of 90.8 and 76.4%, respectively, in 1986 (Table 19).[81] The reductions in the numbers caught on the nets are presumably mainly due to the excito-repellency causing the mosquitoes only to rest fleetingly on the nets, whereas the reduction in the biting population was presumably because of the killing of mosquitoes on the nets. A more detailed analysis of the effectiveness of the treatment on the two species in 1986 showed that *An. anthropophagus* was more affected than was *An. sinensis*.[40]

The monthly malaria incidence decreased very significantly as shown in Figure 13. During the period July to December 1985, the average monthly incidence was 4.6 per thousand, as compared with 11.6 during the same period in 1984, and after the second treatment in 1986, the average monthly incidence was further decreased to 1.09 per thousand over the whole year, i.e., a reduction of 90% in comparison with the incidence in 1984.

In 1986, the same method was adopted in an area with a population of 236,701 in Shenzhen District. The number of mosquito nets impregnated with deltamethrin was 160,476, covering 91.3% of the total population. The malaria incidence from April to October 1986 was reduced by 59.9% after treatment in comparison with that of the same period in 1985.

b. Impregnation of Mosquito Nets in Jiangsu Province

Following the successful trials in Guangdong Province, the impregnation of mosquito nets with deltamethrin and permethrin was adopted for malaria control in two villages in Yuyi County, Jiangsu Province.[82] In one village, with a population of 22,688 who owned 13,725 nets, 93.1% of the nets were impregnated with deltamethrin emulsifiable concentrate (EC) at a dosage of 15 mg a.i. per square meter. In another village with a population of 25,993 who owned 13,486 nets, 82.2% of these nets were impregnated with permethrin, at a dosage of 200 mg a.i. per square meter. The impregnation was carried out by the villagers themselves under the instruction of technicians (Figure 1) and, thus, the task was accomplished within 2 d. Following the impregnation the number of mosquitoes caught resting on 30 nets at night was greatly reduced (Table 20).

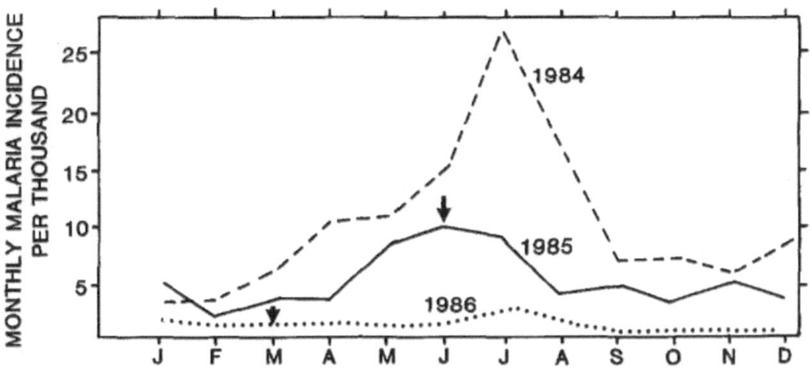

FIGURE 13. Monthly malaria incidence per thousand of the population in Buji county, Guangdong Province, China, before and after deltamethrin treatment of mosquito nets in June 1985 and March 1986.[81] Arrows indicate times of impregnation of nets in 1985 and 1986.

TABLE 20
Number of *Anopheles sinensis* and *Culex pipiens pallens* Caught on Pyrethroid-Treated and Control Mosquito Nets at Night (Yuyi County, Jiangsu Province, China, July—October 1986)[82]

	Control village[a]	Deltamethrin-treated village[a]	Permethrin-treated village[a]
Before treatment	200.5	151.5	88
Days after treatment			
2	337	0	24
12	95	10	4
22	69	15	8
33	34	0	1
43	32	3	0
53	11	0	0
64	26	1	0
74	44	0	0
84	9	0	2
94	21	0	1
104	4	0	0

[a] On 30 nets in each village.

At the same time, indoor residual spraying with DDT was carried out in the neighboring six villages with a combined population of 109,701. The dosage applied was approximately 2 g/m^2.

Malaria incidence was reduced by 87.4% in the deltamethrin-treated village, 85.7% in the permethrin-treated village and by 20.5 to 92.0% in the DDT sprayed villages in comparison with those during the same period of 1985 (Table 21).[82] A reduction of 40% was also observed in the control village, though this was not statistically significant. The percentage reductions in incidence in both the villages with nets were only exceeded by that in one of the six DDT-sprayed villages, suggesting, but not proving, that the use of impregnated nets is more effective than house spraying against malaria transmitted by *An. sinensis* and *An. anthropophagus*.

TABLE 21
Effects of DDT Spraying and Pyrethroid Impregnation of Mosquito Nets on Malaria Incidence in Yuyi County, Jiangsu Province, China[82]

	June—December 1985				July—December 1986			
Village	Population	No. cases	Incidence per 1000	Treatment (June 1986)	Population	No. cases	Incidence per 1000	Reduction (%)
A	18,022	75	4.16	DDT[a]	17,956	28	1.56	62.5
B	15,244	67	4.39	DDT[a]	15,185	53	3.49	20.5
C	16,423	111	6.76	DDT[a]	16,369	43	2.62	61.2
D	16,499	37	2.24	DDT[a]	16,472	18	1.09	51.3
E	26,739	234	8.75	DDT[a]	27,030	19	0.70	92.0
F	16,774	117	6.97	DDT[a]	16,796	17	1.01	85.5
G	22,600	236	10.44	Delta[b]	22,688	30	1.32	87.4
H	27,191	257	9.45	Perm[c]	25,993	35	1.35	85.7
I	15,045	32	2.12	None	15,898	20	1.26	40.6

[a] House spraying.
[b] Deltamethrin-impregnated nets.
[c] Permethrin-impregnated nets.

The area in which deltamethrin-impregnated nets were used for malaria control in Jiangsu Province was doubled in 1987 and will be further expanded in 1988.

c. Deltamethrin Treatment of Nets in Sichuan Province

In 1985, a pilot study was made in a small village with a population of 1335 in Yiping County, Sichuan Province, by the Yiping Hygiene and Anti-Epidemic Station.[83] There were 893 nets sprayed with deltamethrin at a dosage of 11.7 mg a.i. per square meter. Good results were obtained both in reduction of mosquito density and malaria incidence. In 1986, the method was extended to an area with a population of 30,374. There were 18,837 nets (i.e., 96.9% of the nets in the area) sprayed with deltamethrin at a dosage of 9.6 mg a.i. per square meter.

The number of mosquitoes caught resting on and inside the nets at night was reduced after the treatment by 99% for *An. anthropophagus* and 94% for *An. sinensis* (Table 22). As in Table 15 these calculations take into account the fact that before treatment the numbers caught in the intervention area were somewhat greater than in the control area. Similarly, the outdoor biting rate was reduced by 98% for *An. anthropophagus* and 91% for *An. sinensis* (Table 22). The latter result shows that the killing of mosquitoes on the impregnated nets was having a significant impact on the population in spite of the predominantly zoophilic habits of *An. sinensis*. This was also reflected in a significant lowering of the percentage of mosquitoes of both species which were parous, as also reported above from Burkina Faso (Table 15) and Tanzania (Table 18). The malaria incidence in the treated area was significantly reduced as shown in Figure 14.

In 1987, this control measure was extended to 34 counties, with a total population of 2.48 million, and about 1.45 million mosquito nets were treated.

2. Deltamethrin Against *An. dirus* in Hainan Province

An. dirus is an important malaria vector in the jungle and hilly regions in Hainan Province, China, as well as in similar habitats in some Southeast Asian countries. Owing to its strongly exophilic habit, attempts to control it by indoor residual spraying with DDT or other insecticides have not been successful. The satisfactory results obtained in the control of *An. sinensis* with treated nets encouraged the adoption of the new treatment method against *An. dirus* and a field trial was carried out in Wanning County, Hainan Province.

TABLE 22
Numbers of *Anopheles anthropophagus* and *An. sinensis* in Areas of Yiping County, Sichuan Province, China with Deltamethrin-Impregnated and Untreated Control Nets

	No. *An. anthropophagus* caught				No. *An. sinensis* caught			
	Control area		Impreg. area		Control area		Impreg. area	
Time	Biting	On nets	Biting	On nets	Biting	On nets	Biting	On nets
Before impreg.[a]	14.0	31.3	29.3	44.3	5.7	13.3	12.3	26.7
After impreg.								
June								
Late	41	40	1	0	16	12	1	0
July								
Early	50	50	1	1	11	11	3	0
Mid	56	54	2	0	17	12	4	0
Late	48	56	1	0	19	13	3	0
Aug.								
Early	50	59	3	1	21	21	4	3
Mid	44	50	2	2	19	20	5	4
Late	31	41	3	2	12	14	3	3
Sept								
Early	26	40	2	1	11	12	3	3
Mid	22	29	2	0	13	13	2	3
Late	21	31	1	0	11	11	3	2
Oct.								
Early	23	20	1	0	8	0	2	0
Mid	10	10	0	0	7	5	0	0
Late	10	8	1	0	5	3	1	0
Av. after impreg.	32.5	37.5	1.5	0.5	13.1	12.3	2.6	1.4

Note: Data are shown from collections of mosquitoes resting on or in nets and from those biting humans outside houses.[83]

[a] Data from average of 3 catches.

By Chinese standards the malaria incidence was rather high in this area, with an annual incidence rate of 40 to 110 per thousand, in spite of DDT residual spraying. The treatment area had a population of 6407 about two thirds of whom had nets. There were 3095 nets, and 92.9% of these were impregnated with deltamethrin (EC) at a dosage of 25 mg a.i. per square meter in April 1986 and again in April 1987. In 1986 deltamethrin residual spray was applied to the eaves of houses, though subsequent studies suggest that this measure would have had little impact and it was not repeated in 1987.[40]

After the treatment, the number of mosquitoes caught resting on the nets and the outdoor biting rate were greatly reduced. The reduction in the outdoor biting indicates that a large proportion of the mosquitoes were attracted to the treated nets and killed as a result of contact with these nets.

The monthly malaria incidences (detected by a combination of active and passive case detection) were markedly reduced in the treatment area in 1986 and 1987, compared with the corresponding months in 1985. In the control area the monthly malaria incidences showed progressive increases in 1986 and 1987 compared with the corresponding months in 1985 (Figure 15).[84]

A more detailed analysis of the individual protective effect of the treated nets is shown in Table 23(a). The malaria incidence in the people protected with deltamethrin-treated mosquito nets was significantly lower than those with untreated nets or without nets.

Surveys of the blood of about 1,000 people, regardless of malaria symptoms, in November 1985, 1986, and 1987 showed a progressive decline in the parasite rate over the two years after net treatment, and a disappearance of *P. falciparum* (Table 23 (b)).

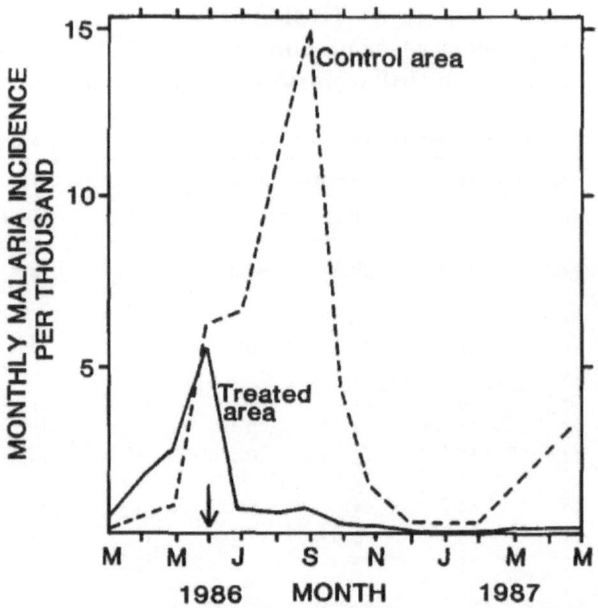

FIGURE 14. Monthly malaria incidence per thousand of the population in areas with and without deltamethrin-treated nets in Yiping County, Sichuan Province, China.[83] Deltamethrin treatment was applied in June 1986, as indicated by the arrow.

Impregnated nets are now provided in some of the temporary shelters on the mountainsides in Hainan for the use of anyone who is sleeping there (Figure 16).

V. PROSPECTS FOR THE APPLICATION OF PYRETHROID-IMPREGNATED NETS OR CURTAINS TO LARGE-SCALE MALARIA CONTROL

A. TIME PEOPLE SPEND UNDER NETS AND TIME OF MOSQUITO BITING

Many malaria vectors bite in the middle of the night when most people are in their houses and in bed. Figure 17 gives data on biting hour by hour during the night by *An. gambiae* s.l. in The Gambia.[62] It also gives data from observations on the proportion of the time people spend out of their houses in each hour of the night. These proportions are multiplied by the amount of biting to give the black sectors in Figure 17. Thus, the black (10% of the whole) and white (90% of the whole) sectors might be expected to indicate, respectively, the remaining risk and the degree of protection from infective biting when someone has a net. However, several complications need to be considered:

1. When most people are under nets mosquitoes may "wait their opportunity" and attack in force when someone leaves the protection of their net.
2. The "mass effect" on the sporozoite inoculation rate due to killing of mosquitoes by impregnated nets will tend to reduce the danger even when some people are not under their nets (Tables 15 and 18).
3. Mosquitoes may change their biting habits because of the presence of insecticide so as to bite out of doors and in the evening or morning rather than the middle of the night; such changes were seen in *An. farauti* as a result of DDT spraying in Irian Jaya[85] and the Solomon Islands[86] and of permethrin treatment of nets in Papua New Guinea (see Section IV.D),[73] but were not seen in *An. gambiae* s.s. in a Tanzanian

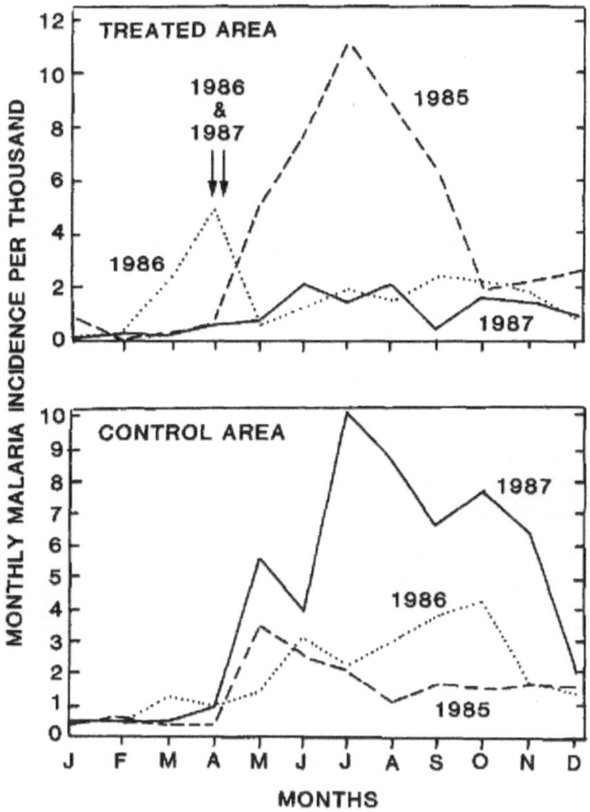

FIGURE 15. Monthly malaria incidence per thousand of the population in two areas of Hainan Province, China, 1985—1987.[84] Nets were treated with deltamethrin in one of the areas in April 1986 and 1987, as indicated by the arrows.

TABLE 23
Effectiveness of Deltamethrin-Treated Mosquito Nets on Incidence of Malaria Cases and Parasite Prevalence in the Net Treatment Area in Hainan Province, China[84]

(a) Incidence in 1986 Among People with or without Treated Nets

	No. of people	Malaria cases	Incidence per thousand
With treated nets	4,013	5	1.2[a]
With untreated nets	300	10	33.3[b]
Without nets	2,094	70	33.4[b]

Note: Different letters indicate differences significant at the 1% level.

(b) Malaria Parasite Rates in Mass Blood Surveys in 1985 (Pretreatment) and 1986 and 1987 (Post-treatment)

Date	No. examined	No. positive *P. vivax*	No. positive *P. falciparum*
Nov. 1985	964	27	58
Nov. 1986	1,001	14	6
Nov. 1987	1,025	2	0

FIGURE 16. Impregnated net provided in a tempory shelter in Hainan, China. The message on the net may be translated "Insecticide impregnated mosquito net. If you spend the night in this hut please sleep under the net which can protect you from malaria." (Photograph by Cai Xianzheng.)

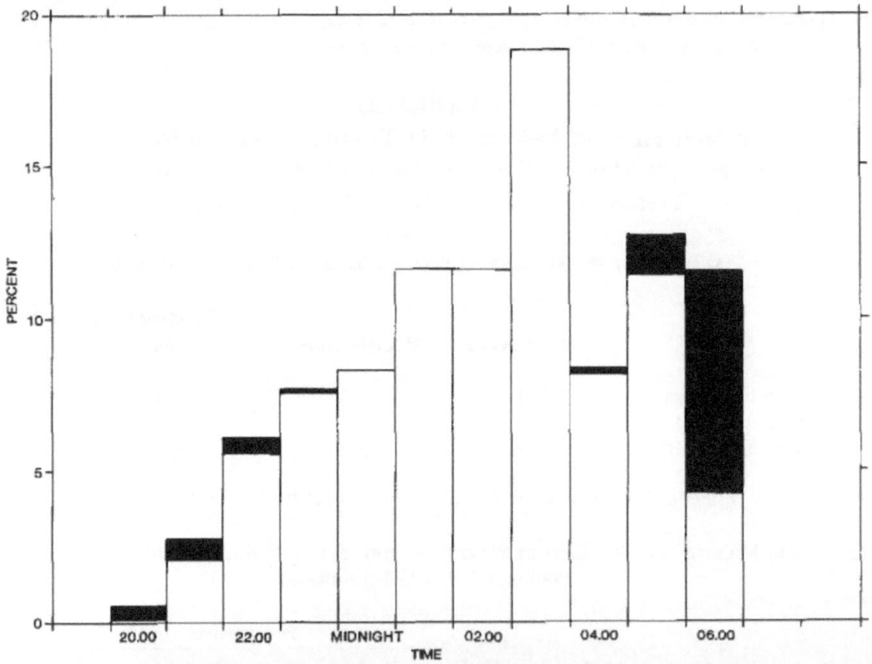

FIGURE 17. Number of bites in each hour of the night by *Anopheles gambiae* s.l. in The Gambia.[62] The black sectors indicate the proportion of each hour that people spend outside their houses multiplied by the amount of biting in each hour. Thus, the white sectors indicate the amount of protection which might be expected from having a net. (From Snow, R. W., et al., *Trans. R. Soc. Trop. Med. Hygiene*, 82, 212, 1988. With permission.)

TABLE 24
Questionnaire on Side Effects of Use of Permethrin Nets in The Gambia[65]

	Permethrin treated (n = 209)	Placebo treated (n = 240)
Coughing	1.1%	2.7%
Headaches	1.6%*	7.1%*
Itching	0.6%	0.4%
Sweating	0.0%*	8.9%*

Note: * = difference significant at 5% level.

village with permethrin-treated nets;[70] such changes could be due to the excito-repellency of the insecticide and/or the long-term selective effect of insecticidal killing.

B. SAFETY AND ACCEPTABIILTY

Pyrethroids belong to the safest category of insecticide.[87] A WHO Expert Committee[88] stated the following about the safety of permethrin impregnated nets:

Properly treated bednets should pose no hazard to those who use them. The acute oral toxicity of permethrin in aqueous solution for rats is very low, the LD_{50} being over 4000 mg/kg. Dermal toxicity is so low that it could not be demonstrated. Permethrin has a very low vapour pressure and hardly evaporates.

The LD_{50} for rats of deltamethrin is similar to that of permethrin.[87] However, deltamethrin is an alpha cyano pyrethroid, and, like other members of this group, deltamethrin can cause parasthesia — a tingling sensation in the skin which is well known to farmers who use this class of insecticide.[89] During the dipping of nets in deltamethrin emulsion without the use of rubber gloves in China (Figure 1)[40] parasthesia has been noticed, but there have been no reports of adverse side effects once the nets are dry and in use in China, where about 1.6 million nets have been impregnated.

Pyrethroids are toxic to fish, and surplus pyrethroid emulsion should not be disposed of into streams, rivers, or ponds. However, the degradation, adsorption, and leaching of pyrethroids in soil has been extensively studied, and the data clearly indicate that little if any unchanged pyrethroid would reach aquatic environments via leaching from soil.[90]

The side effects of permethrin impregnation of bed nets have been investigated by questionnaire in The Gambia (Table 24).[65] The only significant differences were in favor of permethrin impregnation against the placebo. These reductions in reported cases of headache and excessive sweating may have been partly via the effect of permethrin impregnation on malaria, but perhaps more likely were because the children slept better under permethrin-treated nets, because of reduced insect biting nuisance. The insects known to be affected include not only mosquitoes, but also bedbugs (suppression/eradication reported from villages in The Gambia,[91] Papua New Guinea,[92] and Tanzania),[70] headlice (The Gambia,[91] Papua New Guinea),[92] chicken ticks (The Gambia),[91] and houseflies (reduced by permethrin-impregnated curtains in Burkina Faso and permethrin-impregnated nets in The Gambia).[47,91] Such effects may be much stronger "selling points" for impregnated nets than their effect on malaria.

Most of the recent field studies have found that villagers are delighted to be provided with impregnated nets and use them readily. An exception was in Sabah[12] (see Section IV.E), and in other areas there are inevitably some people who neglect to use their nets.[67] Where people object to the lack of ventilation under bed nets and/or where vectors bite indoors but before people go to bed, impregnated curtains would be more appropriate than bed nets (see Section III).[47,57]

TABLE 25
Cost in China of Impregnation of Nets and House Spraying[40]

	Deltamethrin on nets	DDT house spraying
Area	15 m^2/net	150 m^2/house
People	1.5/net	5/house
a.i./m^2/yr	0.015 g	2 g
a.i./person/yr	0.15 g	60 g
Cost of a.i./kg	$433	$2
a.i./person/yr	$0.065	$0.12
Labor cost	Nil (home dipping)	$0.03
Total/person/yr	$0.065	$0.15

Note: 85% of people have nets costing $2/person; a.i. = active ingredient of insecticide; costs converted to U.S. dollars at $1 = 3.6 yuan.

In countries such as China and India where house spraying has been used for a long time, this procedure is now unpopular with householders. However, in China impregnation of nets is much more acceptable and people are willing to participate in doing the impregnation (Figure 1),[40] as found also in Tanzania (Figure 2).[70] The primary health care system in many countries should be well suited to the organization of such operations.

C. COST

Pyrethroids are far more expensive per unit weight than DDT. However, far less has to be used for pyrethroid impregnation of nets than for DDT house spraying, because of both the lower dosages used per unit area and the smaller area needing to be treated. Treatment of nets or curtains may be considered as the most rational form of selective insecticidal treatment in the sense that the treated surfaces are those which blood-seeking mosquitoes are bound to encounter and will be attracted to in their efforts to approach humans.[35] According to costings in China the very low dosages of deltamethrin on nets which are known to be effective work out cheaper than DDT spraying to protect the same human population (Table 25).[40] This table includes a very low estimate, compared with other parts of the world, for the cost of employing trained teams for house spraying. In contrast, no labor cost is ascribed to net impregnation because it can be done by householders, although some supervision is necessary.

In countries such as China and The Gambia bed nets are already widely used, but in many other malarious countries this is not so. In The Gambia the preferred type of opaque net (or bed curtain) costs the equivalent of about $10 and lasts for about 6 years.[93] On the international market the price of less durable nets made from nylon netting is only about $2.50, but in some countries imported goods acquire inflated prices, e.g., in Tanzania the cost of a net is almost as much as the legal minimum monthly wage[70] and in Burkina Faso their price is the equivalent of $15.[67] The situation might be improved if governments taxed nets as medical appliances rather than luxury items or if they were made locally by village tailors (see Chapter 13).[93] Another possibility is the selling of nets at subsidized prices. This was tried in seven villages in Papua New Guinea, the selling price being set at one third of the market rate.[94] Net usage was initially 86%, and after 3 months of subsidized sales it had risen to 99%. However, a considerable number of the sales were to people who already used nets and who kept their new nets for the use of visitors. A survey of one village more than 2 years after the sale showed that many of the nets (which were made of cotton) were still in use and in good condition.

Many governments would not be able or willing to subsidize net sales on a large scale and in such cases "curtains" made of locally available material (Section III) may be the only affordable substrate for pyrethroid impregnation.

D. RESISTANCE

There have been a number of reports of pyrethroid resistance in mosquitoes,[95] but most seem to derive from low levels of enhanced tolerance which would be unlikely to interfere with practical use of impregnated nets.[96] However, in the Horn Fly, *Haematobia irritans*, pyrethroid resistance spread across the U.S. within a few years of the introduction of permethrin-impregnated cattle ear tags.[97] The gene concerned confers resistance to all the pyrethroids tested. The widespread introduction of pyrethroid-impregnated nets would be expected to cause selection pressure for resistance among the female mosquitoes which are attracted to occupied bednets. Therefore, at an early stage in the introduction of these methods it seems highly desirable to instigate programs of testing for resistance. Furthermore, methods proposed for managing resistance[98] appear to have more chance of success if introduced when resistance genes are still rare rather than if one waits until a resistance crisis has already arisen. Perhaps the most hopeful of these proposals is the use of a mixture of unrelated insecticides. The principle of this idea is that insects resistant to compound A will be killed by B, those resistant to B will be killed by A, and double resistants will be exceedingly rare if one starts early when each resistance is still at a low frequency. It seems a matter of urgency that suitable mixtures are investigated now to try to avoid the promise of pyrethroid-impregnated netting being vitiated by the early appearance of resistance.

REFERENCES

1. **Ghana Health Assessment Project Team,** A quantitative method of assessing the health impact of different diseases in less developed countries, *Int. J. Epidemiol.*, 10, 73, 1981.
2. **MacDonald, G.,** *The Epidemiology and Control of Malaria*, Oxford University Press, London, 1957.
3. Chemical Methods for the Control of Arthropod Vectors and Pests of Public Health Importance, World Health Organization, Geneva, 1984.
4. **MacCormack, C. and Snow, R. W.,** Gambian cultural preferences in the use of insecticide treated bed nets, *J. Trop. Med. Hyg.*, 89, 295, 1986.
5. **Ross, R.,** *The Prevention of Malaria*, John Murray, London, 1910.
6. **Lindsay, S. W. and Gibson, M.,** Bednets revisited: an old idea, a new angle, *Parasitol. Today*, 4, 237, 1988.
7. **Liu, Y., Wu, K., Jia, J., and Jiang, W.,** [Integrated approach to malaria control including environmental management to reduce man-mosquito contact and reduction of the source of infection in Huanghai plain] (in Chinese with English summary), *J. Parasitol. Parasit. Dis.*, 4, 246, 1986.
8. **Nevill, C. G., Watkins, W. M., Carter, J. Y., and Munafu, C. G.,** Comparison of mosquito nets, proguanil hydrochloride, and placebo to prevent malaria, *Br. Med. J.*, 297, 401, 1988.
9. **Molineaux, L. and Gramiccia, G.,** *The Garki Project*, World Health Organization, Geneva, 1980.
10. **Rozendaal, J. A., Voorham, J., Van Hoof, J. P. M., and Oostburg, B. F. J.,** The efficacy of local mosquito nets treated with permethrin in Suriname, *Med. Vet. Entomol.*, in press.
11. **Bryan, J. H.,** *Anopheles gambiae* and *Anopheles melas* at Brefet, The Gambia and their role in malaria transmission, *Ann. Trop. Med. Parasitol.*, 77, 1, 1983.
12. **Hii, J., Vun, Y. S., Chin, K. F., Chua, R., Tambakau, S., Binisol, E. S., Fernandez, E., Singh, N., and Chan, M. K. C.,** The influence of permethrin-impregnated bednets and mass drug administration on the incidence of *Plasmodium falciparum* malaria in children in Sabah, Malaysia, *Med. Vet. Entomol.*, 1, 393, 1987.
13. **Snow, R. W., Jawara, M., and Curtis, C. F.,** Observations on *Anopheles gambiae s.l.* made during a trial of permethrin treated bed nets in The Gambia, *Bull. Entomol. Res.*, 77, 279, 1987.
14. **Lindsay, S. W., Snow, R. W., Bloomfield, G., Semega Janneh, M., Wirtz, R. A., and Greenwood, B. M.,** The effect of permethrin-treated bednets (mosquito nets) on malaria transmission by members of the *Anopheles gambiae* complex in The Gambia, *Med. Vet. Entomol.*, 3(3), 266, 1989.
15. **Port, G. R. and Boreham, P. F. L.,** The effect of bed nets on feeding by *Anopheles gambiae*, *Bull. Entomol. Res.*, 72, 483, 1982.
16. **Charlwood, J. D.,** A differential response to mosquito nets by *Anopheles* and *Culex* mosquitoes from Papua New Guinea, *Trans. R. Soc. Trop. Med. Hyg.*, 80, 958, 1986.

17. **Lines, J. D., Myamba, J., and Curtis, C. F.,** Experimental hut trials of permethrin-impregnated mosquito nets and eave curtains against malaria vectors in Tanzania, *Med. Vet. Entomol.*, 1, 37, 1987.
18. **Herodotus,** [*The History*] Book 2, (in Greek with an English translation by A. D. Godley), Harvard University Press, Cambridge, MA, and William Heinemann, London revised ed., 1981.
19. **Pavlovsky, E.,** [*Protective Measures against Blood-sucking Insects*] (in Russian), U.S.S.R. Academy of Sciences, Moscow, 1941.
20. **Blagoveschensky, D., Bregetova, N., and Monchadsky, A.,** An investigation on new repellents for the protection of man against mosquito attacks, *Trans. R. Soc. Trop. Med. Hyg.*, 39, 147, 1945.
21. **Nabokov, V. A.,** [*Guide to Controlling Malaria Mosquitoes*] (in Russian), Medgiz, Moscow, 1952.
22. **Harper, P. A., Lisansky, E. T., and Sasse, B. E.,** Malaria and other insect-borne diseases in the South Pacific campaign, 1942—1945, *Am. J. Trop. Med.*, 27 (Suppl.), 1, 1947.
23. **Nauck, E. G., Reichenow, E., Vogel, J., Westphal, A., and Weyer, F.,** Tropenkrankheiten des Menschen, in *Tropical Medicine and Parasitology,* Nauck, E. G., senior author, FIAT review of German Science 1939—1946, Field Information Agencies, Technical, Office of Military Government for Germany, Wiesbaden, West Germany, 1948; *Rev. Appl. Entomol.*, 36 (Abstr.), 158, 1948.
24. **Brun, L.-O. and Sales, S.,** Stage IV evaluation of four organophosphorus insecticides, OMS-43, OMS-1155, OMS-1197 and OMS-1424 applied at 0.2 gm/m^2 to cotton mosquito nets, WHO mimeographed doc., WHO/VBC 76.630, World Health Organization, Geneva, 1976.
25. **Elliot, M., Farnham, A. W., Janes, N. F., Needham, P. H., Pulman, D. A., and Stevenson, J. H.,** A photostable pyrethroid, *Nature (London),* 246, 169, 1973.
26. **Lindsay, I. S. and McAndless, J. M.,** Permethrin-treated jackets versus repellent treated jackets and hoods for personal protection against blackflies and mosquitoes, *Mosq. News,* 38, 350, 1978.
27. **Schreck, C. E., Smith, N., Weidhaas, D., Posey, K., and Smith, D.,** Repellents vs toxicants as clothing treatments for protection from mosquitoes and other biting flies, *J. Econ. Entomol.*, 71, 919, 1978.
28. **Schreck, C. E., Mount, G. A., and Carlson, D. A.,** Wear and wash persistence of permethrin used as a clothing treatment for personal protection against Lone Star Tick, *J. Med. Entomol.*, 19, 143, 1982.
29. **Hervy, J. P. and Sales, S.,** Evaluation de la rémanence de deux pyrethrinoïdes de synthèse — OMS-1821 et OMS-1998, Doc. multigr., Centre Muraz, Bobo-Dioulasso, No. 7353/80/Doc Tech. OCCGE No. 04/ENT/80, 1980.
30. **Chao, X.,** [Field trials of bed nets impregnated with permethrin for control of mosquitoes] (in Chinese), *Chinese J. Prevent. Med.*, 1, 89, 1984.
31. Guangdong Institute of Parasitic Disease Control, [Studies on the control of *Anopheles sinensis* and *Anopheles dirus* by bed-nets impregnated with deltamethrin] (in Chinese), *Ann. Bull. Soc. Parasitol. Guangdong Province*, 7, 180, 1984.
32. **Li, M. and Liang, Z.,** [Experiments on the toxic effects of deltamethrin to *Anopheles dirus*] (in Chinese), *Ann. Bull. Soc. Parasitol. Guangdong Province,* 7, 184, 1984.
33. Integrated Vector Control, *W.H.O. Tech. Rep. Ser.*, No. 688, World Health Organization, Geneva, 1983.
34. **Curtis, C. F. and Lines, J. D.,** Impregnated fabrics against malaria mosquitoes, *Parasitol. Today,* 1, 147, 1985.
35. **Mouchet, J.,** Deltamethrin impregnated bednets, an alternative for mosquito and malaria control, in *VIII Congreso Latinoamericano de Parasitologia, 1 Congreso Guatemalteco de Parasitologia y Medicina Tropical, Guatemala,* 1987, 201.
36. **Schreck, C. E. and Self, L. S.,** Treating mosquito nets for better protection from bites and mosquito borne diseases, WHO mimeographed doc., WHO/VBC/85, 914, World Health Organization, Geneva, 1985.
37. **Hossain, M. I., Curtis, C. F., and Heekin, J. P.,** Assays of permethrin impregnated fabrics and bioassays with mosquitoes *(Diptera, Culicidae), Bull. Entomol. Res.*, 79, 299, 1988.
38. **Snow, R. W., Philips, A., Lindsay, S. W., and Greenwood, B. M.,** How best to treat bednets with insecticide in the field, *Trans. R. Soc. Trop. Med. Hyg.*, 82, 647, 1988.
39. **Li, Z., Xu, J., Li, B., Zhu, T., and Li, M.,** Mosquito nets impregnated with deltamethrin against malaria vectors in China, WHO mimeographed doc., WHO/VBC/87, 939, World Health Organization, Geneva, 1987.
40. **Li, Z.,** unpublished data, 1987.
41. **Lindsay, S. W., Hossain, M. I., Bennett, S., and Curtis, C. F.,** The insecticide activity and wash fastness of 12 synthetic pyrethroid treatments into bednetting assayed against mosquitoes, *Pestic. Sci.*, in press.
42. **Bry, R. E., Simonaitis, R. A., Lang, J. H., and Boatright, R. E.,** Durability of FMC33297 in moth proofing, *Soap, Cosmet., Chem. Spec.*, 52, 31, 1976.
43. **Wu, N., Xiao, Y., and Huang, F.,** [Studies on the toxic effects of mosquito nets impregnated with deltamethrin and permethrin to mosquitoes], (in Chinese), unpublished document, 1987.
44. **Schreck, C. E., Posey, K., and Smith, D.,** Durability of permethrin as a potential clothing treatment to protect against blood-feeding arthropods, *J. Econ. Entomol.*, 77, 397, 1978.
45. **Njunwa, K. J., Wilkes, T. J., Myamba, J., and Curtis, C. F.,** unpublished data.

46. **Darriet, F., Robert, V., Tho Vien, N., and Carnevale, P.,** Evaluation of the efficacy of permethrin-impregnated intact and perforated mosquito nets against vectors of malaria, WHO mimeographed doc., WHO/VBC/84.899, World Health Organization, Geneva, 1984.
47. **Majori, G., Sabatinelli, G., and Coluzzi, M.,** Efficacy of permethrin impregnated curtains for malaria vector control, *Med. Vet. Entomol.*, 1, 185, 1987.
48. **Kurihara, T., Fujita, K., and Suzuki, T.,** [Insecticide treatment of wide-mesh net curtain for vector control and the effect upon behavioral response of adult mosquitoes] (in Japanese with English summary), *Jpn. J. Sanit. Zool.*, 36, 25, 1980.
49. **Kurihara, T. and Umino, T.,** The effects of insecticides on the behavior of female mosquitoes under laboratory conditions, *Jpn. J. Sanit. Zool.*, 38, 25, 1987.
50. **Hossain, M. I. and Curtis, C. F.,** Laboratory evaluation of deet and permethrin impregnated wide-mesh netting against mosquitoes, *Entomol. Exp. Appl.*, in press.
51. **Itoh, T., Shinjo, G., and Kurihara, T.,** Studies on wide-mesh netting impregnated with insecticides against *Culex* mosquitoes, *J. Am. Mosq. Control Assoc.*, 2, 503, 1986.
52. **Hossain, M. I. and Curtis, C. F.,** Behavioural and killing effects of permethrin impregnated bednets — a laboratory evaluation, *Med. Vet. Entomol.*, in press.
53. **Service, M. W.,** *Mosquito Ecology*, Applied Science Publishers, London, 1976.
54. **Curtis, C. F. and Lines, J. D.,** Insecticidal control of insect vectors of diseases, *Insect. Sci. Appl.*, 8, 709, 1988.
55. **Curtis, C. F. and Myamba, J.,** unpublished data.
56. **Xu, R., Lu, B., Chen, X., Jiang, Z., and Nie, Z.,** [Experimental hut studies on the effect of mosquito nets impregnated with deltamethrin and permethrin to *An. sinensis* and *Cx. tritaeniorhynchus*] (in Chinese), unpublished document, 1987.
57. **Kurihara, T., Kamimura, K., and Arakawa, R.,** Phenothrin impregnation of wide-mesh net for protection from biting mosquitoes, *Jpn. J. Sanit. Zool.*, 37, 261, 1986.
58. **Lines, J. D., Curtis, C. F., Myamba, J., and Njau, R.,** Tests of repellent or insecticide on curtains, bednets and anklets against malaria vectors in Tanzania, WHO mimeographed doc., WHO/VBC 85.920, World Health Organization, Geneva, 1985.
59. **Lines, J. D., Nicholson, E. A., Curtis, C. F., and Myamba, J.,** unpublished data, 1987.
60. **Bradley, A. K., Greenwood, B. M., Greenwood, A. M., Marsh, K., Byass, P., Tullock, S., and Hayes, R.,** Bed nets (mosquito nets) and morbidity from malaria, *Lancet*, 2, 204, 1986.
61. **Campbell, H., Byass, P., and Greenwood, B. M.,** Bed nets and malaria suppression, *Lancet*, 1, 859, 1987.
62. **Snow, R. W., Rowan, K. M., Lindsay, S. W., and Greenwood, B. M.,** A trial of bed nets (mosquito nets) as a malaria control strategy in a rural area of The Gambia, West Africa, *Trans. R. Soc. Trop. Med. Hyg.*, 82, 212, 1988.
63. **Lindsay, S. W., Shenton, F. C., Snow, R. W., and Greenwood, B. M.,** The effects of untreated bednets on the population dynamics of the *Anopheles gambiae* complex in The Gambia, *Med. Vet. Entomol.*, in press.
64. **Snow, R. W., Rowan, K. M., and Greenwood, B. M.,** A trial of permethrin treated bed nets in the prevention of malaria in Gambian children, *Trans. R. Soc. Trop. Med. Hyg.*, 81, 563, 1987.
65. **Snow, R. W., Lindsay, S. W., Hayes, R. J., and Greenwood, B. M.,** Permethrin treated bednets (mosquito nets) prevent malaria in Gambian children, *Trans. R. Soc. Trop. Med. Hyg.*, 82, 838, 1988.
66. **Ranque, P., Touré, T. Y., Soula, G., Le Du, Diallo, Y., Traoré, O., Duflo, B., and Balique, H.,** Etude expérimentale sur l'utilisation de moustiquaires imprégnées de deltaméthrine dans la lutte contre le paludisme, *Parassitologia (Rome)*, 26, 261, 1984.
67. **Carnevale, P., Robert, V., Boudin, C., Halna, J.-M., Pazart, L., Gazin, P., Richard, A., and Mouchet, J.,** La lutte contre le paludisme par des moustiquaires imprégnées de pyrethinoides au Burkina Faso, *Bull. Soc. Pathol. Exot.* 81, 832, 1988.
68. **Robert, V., Petrarca, V., Carnevale, P., and Coluzzi, M.,** Analyses cytogénétiques du complexe *Anopheles gambiae* dans la région et la ville de Bobo Dioulasso (Burkina Faso), in *3ème Conf. Int. sur le Paludisme et les Babésioses, Annecy, France*, 1987, 167.
69. **Detinova, T. S.,** Age Grouping Methods in Diptera of Medical Importance, Monogr. Ser. No. 47, World Health Organization, Geneva, 1962.
70. **Njunwa, K., Wilkes, T., Lyimo, E., Curtis, C. F., and Lines, J. D.,** unpublished data.
71. **Curtis, C. F. and Lines, J. D.,** unpublished data.
72. **Millen, B. D.,** A strategy for personal and community protection against the vectors of malaria in Papua New Guinea with emphasis on the evaluation of bednets impregnated with permethrin, *Prof. Pest Manage. Assoc. B. C.*, 19, 1986.
73. **Charlwood, J. D. and Graves, P. M.,** The effects of permethrin impregnated bednets on the population of *Anopheles farauti* in coastal Papua New Guinea, *Med. Vet. Entomol.*, 1, 319, 1987.

74. **Graves, P. M., Brabin, B. J., Charlwood, J. D., Burkot, T. R., Cattani, J. A., Ginny, M., Paina, J., Gibson, F. D., and Alpers, M.,** Reduction in incidence and prevalence of *Plasmodium falciparum* in under-5-year-old children by permethrin impregnation of mosquito nets, *Bull. W.H.O.*, 65, 869, 1987.
75. **Webber, R. H.,** The natural decline of *Wuchereria bancrofti* infection in a vector control situation in the Solomon Islands, *Trans. R. Soc. Trop. Med. Hyg.*, 71, 396, 1977.
76. **Charlwood, J. D. and Dagaro, H.,** Impregnated bed-nets for the control if filariasis transmitted by *Anopheles punctulatus* in rural Papua New Guinea, *Papua New Guinea Med. J.*, 30, 199, 1987.
77. **Self, L. S.,** Malaria vector control based on personal protection, W.H.O. mimeographed document for 10th Southwest Pacific Malaria Meeting, Honiara, Solomon Islands, July 1988, World Health Organization, Regional Office for Western Pacific, Manila, Philippines.
78. **Lu, B.,** The geographical distribution of the *An. sinensis* gr., in China, in Sino-American Symposium on *An. sinensis* and *An. balabacensis*, Shanghai, 1981.
79. **Lu, B.,** [Integrated control of mosquito vectors] (in Chinese), Institute of Zoology, Academica Sinica, Ed., Science Press, Beijing, 1979, 441.
80. **Lu, B.,** [The malaria vectors and their control in China] (in Chinese), *Chin. J. Epidemiol.*, 3, 379, 1982.
81. **Li, Z., Zhang, M., Wu, Y., Zhong, B., Lin G., and Huang, H.,** Trial of deltamethrin impregnated bed nets for the control of malaria transmitted by *Anopheles sinensis* and *Anopheles anthropophagus*, *Am. J. Trop. Med. Hyg.*, 40, 356, 1989.
82. Yuyi Hygiene and Anti-Epidemic Station, unpublished data, 1987.
83. **Yang, J., Liu, G., Wu, L., Yang, X., and Zhu, W.,** [The effect of malaria control by the application of mosquito nets impregnated with deltamethrin] (in Chinese with English summary), *J. Parasitol. Parasit. Dis.*, Shanghai, in press.
84. **Li, Z., Zhang, M., Shen, M., Li, M., and Zhang, L.,** Field trials of deltamethrin impregnated mosquito nets in Hainan Island, in *The Studies of Bed Nets Impregnated with Deltamethrin for the Control of Vectors and Malaria*, Symposium Proceedings, Li Zuzi, Lu Baolin, and Xie Wanling, Eds., Anti-Parasitic Diseases Institute, Guangzhou, China, 1988, 35.
85. **Sloof, R.,** *Observations on the Effect of Residual DDT House Spraying on Behaviour and Mortality in Species of the Anopheles punctulatus Group*, A. W. Sythoff, Leiden, The Netherlands, 1964.
86. **Taylor, B.,** Changes in the feeding behaviour of a malaria vector, *Anopheles farauti* Lav., following use of DDT as a residual spray in houses in the British Solomon Islands Protectorate, *Trans. R. Entomol. Soc. London*, 127, 277, 1975.
87. The WHO recommended classification of pesticides by hazard and guidelines to classification 1986-87, WHO mimeographed doc., VBC/86.1, World Health Organization, Geneva, 1986.
88. Safe Use of Pesticides, W.H.O. Tech. Rep. Ser., No. 720, World Health Organization, Geneva, 1988.
89. **Litchfield, M. H.,** Toxicity to mammals, in *The Pyrethroid Insecticides*, Leahey, J. P., Ed., Taylor & Francis, London, 1985, 99.
90. **Hill, I. R.,** Effects on non-target organisms in terrestrial and aquatic environments, in *The Pyrethroid Insecticides*, Leahey, J. P., Ed., Taylor & Francis, London, 1985, 151.
91. **Lindsay, S. W., Snow, R. W., Armstrong, J. R. M., and Greenwood, B. M.,** Permethrin-impregnated bednets (mosquito nets) reduce nuisance arthropods in Gambian houses, *Med. Vet. Entomol.*, in press.
92. **Charlwood, J. D. and Dagaro, H.,** Collateral effects of bed-nets impregnated with permethrin in controlling bed bugs (Cimicidae) in Papua New Guinea, *Trans. R. Soc. Trop. Med. Hyg.*, 83, 261, 1989.
93. **MacCormack, C. P., Snow, R. W., and Greenwood, B. M.,** Economic aspects of impregnated bed nets in Gambian primary health care, *Bull. W.H.O.*, in press.
94. **Charlwood, J. D., Graves, P. M., and Birley, M. H.,** unpublished data, 1986.
95. Resistance of Vectors and Reservoirs of Disease to Pesticides, W.H.O. Tech. Rep. Ser., No. 737, World Health Organization, Geneva, 1986.
96. **Malcolm, C. M.,** Current status of pyrethroid resistance in anopheline mosquitoes, *Parasitol. Today*, 4, S13, 1988.
97. **Byford, R. L. and Sparks, T. C.,** Chemical approaches to the management of resistant horn fly, *Haematobia irritans*, populations, in *Combating Resistance to Xenobiotics*, Ford, M. G., Holloman, D. W., Khambay, B. P. S., and Sawicki, R. M., Eds., Ellis Horwood, Chichester, 1987, 178.
98. **Curtis, C. F.,** Genetic aspects of selection for resistance, in *Combating Resistance to Xenobiotics*, Ford, M. G., Holloman, D. W., Khambay, B. P. S., and Sawicki, R. M., Eds., Ellis Horwood, Chichester, 1987, 150.

Chapter 3

BAIT METHODS FOR TSETSE CONTROL

C. Laveissière, G. A. Vale, and J.-P. Gouteux

TABLE OF CONTENTS

I. TSETSE BIOLOGY AND BAIT METHODS*

The tsetse fly is often regarded as a despicable parasite, but there are some features of its life-style that make it appear a more majestic beast — a little lion on wings. These features make it feasible to control tsetse using natural baits, such as cattle sprayed with insecticide, or artificial baits utilizing synthetic host odor. The latter could either be traps or devices, known variously as screens or targets, coated with insecticides.

Like the big cats the tsetse fly is free living; males and females feed exclusively on the larger vertebrates, hunting their prey over extensive areas to take a blood meal every few days. This regular supply of protein-rich food fits well with the reproductive strategy of tsetse, which is distinct from that of most insects and somewhat like that of mammals. The female of many insect species produces large numbers of eggs and then abandons them to their fate in which the developing larvae must find their own food. However, the female tsetse matures one egg every 9 d, retaining it in her uterus where it develops into a fully grown larva by feeding on a milk-like secretion from the mother. The larva is then deposited in a sandy place where it burrows into the soil to pupate. It emerges as an adult about a month later, having spent a remarkably sheltered existence for all of its early life. This slow but sure method of reproduction, coupled with a constant availability of food, guarantees that the abundance of tsetse flies is normally fairly stable from season to season and year to year, being free of the marked fluctuations that characterize many insect populations.

In some ways the life-style of tsetse flies makes them promising candidates for bait methods of control. The slow reproductive capacity of the flies means that baits need to attract no more than a small proportion of the population per day in order to ensure that the population declines steadily at an acceptable rate of, say, a halving each month. Chapter 9 discusses the relative merits of killing or sterilizing the flies which are caught.

The fact that the whole of the adult population of tsetse succeeds in finding thinly spread hosts every few days indicates that a few sparsely distributed baits, offering host-type stimuli, must be able to attract the required daily percentages of the tsetse population.

Against the favorable indications for the applicability of bait techniques, there is the problem that the hunting life-style of tsetse requires them to be highly mobile within their habitat, so that if attempts are made to clear tsetse from only part of an extensive infestation, there is continual invasion pressure. The *palpalis* and *fusca* groups of species are mostly confined to riverine fringes and forests, venturing only occasionally into more open situations. With these flies, baits placed in the limited areas of dense vegetation that occur within settled regions can be an economical means of protecting large numbers of people and cattle against tsetse, even if many baits are needed per unit area. On the other hand, for those tsetse species that occupy vast areas of savanna, mostly flies of the *morsitans* group, the mobility of tsetse is a more difficult problem. The main savanna species, *Glossina morsitans* and *G. pallidipes,* can move with an average displacement of up to 1 km/d.[1] This displacement is in random steps from day to day so that the overall displacement in, say, 100 d is not 100 km, rather it is closer to the square root of this figure, that is 10 km. Nevertheless, if savanna settlements are to be held free of tsetse, it is necessary to remove the flies from infested areas nearby, so eliminating the source of invasion. Alternatively, the edge of the settlement must be covered by a dense placement of baits to form an invasion barrier. In any event, if the baits for the control of savanna tsetse are to be economical, they must be cheaper or required in lower density than those for the tsetse of restricted habitats.

The distinction between the riverine and savanna tsetse in the context of control is further enhanced by their role in disease transmission. The riverine flies are especially important in the transmission of sleeping sickness due to *Trypanosoma gambiense* in West Africa,

* G. A. Vale

largely by a man-fly-man contact. In these circumstances it may not be necessary to eliminate the flies completely; the mere reduction of tsetse densities, coupled with the curing of patients, may be sufficient to eliminate the disease. On the other hand, the savanna flies are important mainly as the vectors of sleeping sickness of man due to *T. rhodesiense,* and especially of nagana to domestic stock. A permanent reservoir of the trypanosomes that produce these diseases exists in the populations of wild animals, so that even sparse populations of tsetse will eventually transmit the trypanosomes from the wild animals to men and domestic animals. Under such conditions the complete elimination of the disease risk requires the complete removal of the flies.

In some circumstances where the invasion pressure of tsetse is particularly strong, say on a small ranch in the center of a dense infestation of tsetse, it is unlikely that bait techniques, or any other method of tsetse control, will be able to eliminate tsetse completely. Here the aim might be to reduce the tsetse population and the risk of disease to the point where a costly program of protecting cattle by prophylactic drugs can be replaced by a cheaper program in which curative drugs are administered on the infrequent occasions when the cattle become infected. Such an aim might be justified if there is resistance to prophylactic drugs.

II. EARLY DEVELOPMENTS IN BAIT TECHNOLOGY*

The first trials of bait methods of tsetse control used human baits against tsetse of the *palpalis* group, the attracted flies being caught by hand nets or by sticky screens worn on men's backs as in the eradication campaign (also involving wild pig eradication and bush clearance) organized by Maldonado[2,3] early in this century on the island of Principe. Although this trial was highly successful, the sticky-screen technique would only be applicable where tsetse are strongly attracted to people and where it is practicable and economical to enroll large numbers of willing human baits for extended periods. Such conditions are seldom met in modern Africa. Moreover, for *G. morsitans* and *G. pallidipes,* humans can be repellent. Thus, it seems more feasible to use artificial baits or cattle.

Harris[4] was the first to appreciate the potential of artificial baits. In the late 1920s he used traps designed to simulate the appearance of large ungulates, to supplement other control measures against *G. pallidipes* over several hundred square kilometers of Zululand. The tsetse population was reduced by about 99.99% over 7 years, and although it was not clear to what extent this effect was due to trapping, the results stimulated much study of traps in other parts of Africa and with other tsetse species.[5-9] However, the efficacy of each of a variety of traps was disappointing for control purposes,[10,11] and they were studied mainly as a monitoring technique which might be less biased than conventional methods involving human or animal baits.[12-14]

The first large field trial with cattle as baits was conducted in Tanzania against *G. pallidipes* in the 1940s. The cattle were sprayed with DDT to kill tsetse alighting on them. This produced a 95% reduction in the abundance of tsetse over a 5-month period, but there were problems with the technique: the insecticide deposit was lost after a few days and the deployment of the cattle was costly. In Zimbabwe, where cattle are regularly dipped with preparations against ticks, the use of a dip that could deal with both tsetse and ticks seemed economically attractive. However, tests of the many dips available in the 1960s failed to identify one that would be suitable against tsetse, and interest in cattle as baits declined. This decline of interest was reinforced by the evidence of hand-net catches of tsetse from cattle, which were interpreted at the time as indicating that cattle were poor baits for female tsetse.

* G. A. Vale and J.-P. Gouteux

Since the 1970s the idea of using baits to control or suppress tsetse populations has been revived. On the one hand, there has been development, based in Zimbabwe, of odor-baited, pyrethroid-treated traps and screens and pyrethroid treatment of cattle for control of the *morsitans* group of tsetse of the open savanna. On the other hand, there has been development in West and Central Africa of traps and screens which are visually attractive to the *palpalis* and *fusca* groups of riverine and forest-dwelling tsetse.

III. BAIT METHODS FOR FLIES OF THE OPEN SAVANNA*

A. RECENT DEVELOPMENTS

1. Studies with Electrocuting Devices

In Zimbabwe in the late 1960s, a variety of electrocuting devices[15] to trap tsetse flies allowed a more precise and objective indication of the responses of *G. morsitans* and *G. pallidipes* to traps and cattle, and several important facts emerged. First, men are repellent to these tsetse, especially to females; in the absence of the men usually employed with cattle baits the numbers of flies visiting the baits increased severalfold.[16] Second, most of the tsetse that visit stationary cattle do so in response to the host odor, and massive doses of such odor, from several tons of cattle, attract huge numbers of tsetse;[17,18] traps baited with such odor would be needed at densities of only one per 10 km^2 to achieve effective control.[19] Third, it was shown that about 90% of the tsetse that visit conventional traps failed to enter the traps and be retained, so that there was ample scope to improve their efficacy.[20,21] Such discoveries helped to rekindle interest in cattle and artificial baits for tsetse control.

2. Artificial Baits

For the economic use of odor attractants with traps, it is necessary to identify the effective components of natural host odor and to dispense them artificially. Several components have been identified so far, namely carbon dioxide, acetone, butanone, 1-octen-3-ol (henceforth termed octenol), and various phenols, especially the 4-methyl and 3-*n*-propyl derivatives.[21,22] Unfortunately, carbon dioxide cannot be dispensed economically and conveniently on a routine basis in large-scale operations, and the phenolic attractants are much more effective for *G. pallidipes* than for *G. morsitans*. Moreover, not all of the effective components of host odor have been identified as yet.[23] Nevertheless, by using octenol and phenols, with either acetone or butanone, it is possible to increase trap catches by 5 to 20 times at little cost.

So far it has proved difficult to increase the proportion of tsetse that enter a trap after arriving near it. Even with the better traps the proportion of flies that enter is only about 20% for *G. morsitans* and 50% for *G. pallidipes*.[24] Thus, it seems more effective to employ baits that do not need to be entered in order to be effective. Such considerations led to the development of simple visual targets, consisting of a screen of black cloth about 1 m^2, coated with insecticide to kill the flies instead of trapping them.[25,26] The screens have the advantage of being cheaper and simpler than traps, and whereas traps can rapidly become ineffective due, for example, to small rips in the netting components, screens can remain effective even when they are partially damaged. On the other hand, screens are disadvantageous in that they do not retain the flies with which they deal. Hence, to monitor the changes in the populations of tsetse, special traps or other sampling systems must be deployed. Moreover, if the bait campaign is to rely on community motivation, a trap that is seen to be full of flies is of great public relations value.

The technical development of screens has been attended by its own special difficulties. One problem was to find a suitably effective insecticide that would not be destroyed by

* G. A. Vale

sunlight or washed off the target by rain. Such an insecticide has been identified, namely, deltamethrin suspension concentrate used with an ultraviolet absorber.[27] The formulation is called Glossinex (Coopers Ltd.). For the first application to a screen the insecticide is diluted in water to a concentration of 0.1% of active ingredient (a.i.). For resprays the insecticide concentration should be reduced to 0.05%. The spray should be applied from a knapsack sprayer, at low (2 bar) pressure, via a 90° fan nozzle held obliquely to the screen. Each screen requires 400 ml of spray mixture.

Another problem is that the black dye of the target fades, making the target less effective. This has been overcome by adding black dye (0.5% Colanyl Black) to the insecticide spray so that the fading is corrected each time the target is resprayed, which should be every 2 to 3 months in the wet season and every 4 to 5 months in the dry season.

Yet another difficulty is that many tsetse that visit the screen fail to alight on it and so do not contact the insecticide. This problem has been partially overcome by fixing sheets of fine black netting beside the screen, so that tsetse flying around the screen collide with the netting and contact insecticide deposited there. Subsequent research has shown that a screen in which the netting sheets are replaced by cloth, forming a large black screen 1 m tall and 1.7 m long, can be as effective as the screens made of netting and cloth. The extra visual stimulation of the large screen encourages alighting responses, but with all of the screens so far available, no more than about 50% of the attracted flies contact the insecticide deposits.[27]

Trials using four screens per square kilometer, baited with acetone and octenol, have been conducted in areas of up to 1500 km^2 in Zimbabwe, giving within a year a reduction of at least 99.99% in the initially dense populations of *G.m. morsitans* and *G. pallidipes*.[28,29] Moreover, bands of bush about 1 km wide, provided with 30 screens per linear kilometer, show promise as barriers to prevent or retard invasion by tsetse. Subsequently, a number of small field trials of odor-baited traps and screens have been conducted with promising results against *G. morsitans centralis* in Zambia,[30] *G.m. submorsitans* in Ethiopia[31] and Burkina Faso,[32] and against *G. pallidipes* in Kenya.[33] There are preliminary indications that the technology is effective against *G. longipennis*[34] in the savanna and also *G. tachinoides*,[35,36] *G. brevipalpis*,[37] and *G. austeni*[38] in more wooded situations, but so far the subspecies of *G. palpalis* have shown no clear response to odor attractants other than carbon dioxide.[32,39]

3. Cattle as Baits

Following the demonstrations that cattle without men in close attendance are excellent baits for tsetse,[16,18] and that the deltamethrin formulation used on screens is also effective on cattle, several field trials with cattle dipped in this insecticide have been performed with encouraging results against ticks and tsetse flies in Zimbabwe and Zambia.[40,41] With cattle at average densities of two to ten per square kilometer, the dipping regime has drastically reduced the incidence of nagana, to the point where the drug program can be switched from blanket prophylaxis to curative treatment.

The use of cattle as baits for tsetse control now seems feasible, provided cattle occur in sufficient density, are fairly evenly distributed in the tsetse habitat, and are already on a regular regime of dipping against ticks. The requirement then is simply to substitute deltamethrin for the normal tick dip, at little extra cost and no extra effort. This is likely to constitute the most advantageous method of tsetse control in all areas where it is possible. Unfortunately, cattle are not sufficiently numerous in all areas where tsetse control is required, and in many regions the cattle may not be dipped against ticks for fear of destroying their natural immunity to tick-borne diseases. Moreover, a serious limitation to the concept of treating cattle with insecticide is that, in order for the tsetse to be killed, they must contact the cattle and hence would be able to transmit disease. This would be important if the area where the cattle occur is subject to a permanent invasion pressure of tsetse; the cattle on the edge of the area would be at constant risk of a severity that may well require prophylactic drugs unless the bait cattle there are supplemented by artificial baits. It is only in the center of the area, where tsetse invasion is restricted by the peripheral baits, that the tsetse may be expected to decline sufficiently for the prophylactic regime to be lifted without the provision of extra baits.

The current recommended cattle dip is Decatix (Coopers Ltd), a suspension concentrate of deltamethrin. This should be diluted to a concentration of 0.00375% a.i. in the dip tank. Cattle should be dipped at 2-week intervals in the rainy season and monthly intervals in the dry season.

In the absence of dip tanks, one can treat the sides of cattle with a pour-on formulation which spreads over the body surface.

B. IMPLEMENTATION OF BAIT METHODS AGAINST TSETSE OF THE OPEN SAVANNA

It is possible to offer only provisional guidelines for the use of baits against savanna tsetse because the methods are new and are being changed in the light of ongoing research in several parts of Africa. Before incurring the costs of large-scale operations it would be wise to implement a short and inexpensive program to test the main elements of bait techniques in the area to be tackled, if this is distant from an area where the techniques have already been applied successfully, and especially if the species of tsetse present have not been studied extensively elsewhere.

Any initial investigations might be restricted to three aspects of the use of artificial baits:

1. The effectiveness of odor attractants
2. The design of the bait itself
3. The proportion of the population that could be expected to be removed per day

1. Odor Attractants

The most convenient tool for an initial study of odor attractants is a trap, being simple and cheap, and capable of being used in large numbers with and without various odor attractants. However, if screens are to be employed in the main operation, the use of traps for the initial studies may be misleading because the catches from traps depend on two distinct factors: the numbers of tsetse attracted to the bait from a distance and the proportion of the attracted flies that enter the trap after arriving near it. Both of these factors can be affected by odor attractants, but only the attraction of tsetse from a distance has a bearing on the efficacy of screens. Hence, if an odor increases the catches from a trap it does not necessarily follow that it will improve the efficiency of a screen. Nevertheless, the data produced for the effects of odor added to traps in the project area should provide a rough indication of the extent to which the tsetse there behave similarly to the tsetse in other areas where the relationship between the performance of odor-baited traps and screens has already been established.[25] It is only when considering species of tsetse that have not been studied extensively in relation to odor-baited systems that studies with electrocuting devices[15] might have to be undertaken.

As a rough guide, it is to be expected that for *G. morsitans* and *G. pallidipes* the emission of acetone vapor (at doses of 0.1 to 1.0 g/h) or butanone (0.02 to 0.2 g/h) will increase trap catches by two to three times; octenol (0.1 to 1.0 mg/h) will increase catches by about 50% more, and the further addition of a mixture of 4-methylphenol (0.5 to 5 mg/h) and 3-*n*-propylphenol (0.05 to 0.5 mg/h) will increase catches by another two to five times for *G. pallidipes* or up to another two times for *G. morsitans*.[27] That is, the overall effect of the combination of attractants should be to increase trap catches by about five times for *G.*

FIGURE 1. The Epsilon trap developed in Zimbabwe. (Photograph by J. W. Hargrove.)

morsitans and 20 times for *G. pallidipes*. Acetone or butanone are dispensed from a 500-ml bottle with a hole in the lid, the vapor of the chemicals diffusing through this hole. For screens the hole should be 2 mm in diameter, giving release rates of about 100 mg/h for acetone and 20 mg/h for butanone. With traps the diameter of the hole should be increased to 6 mm, giving about a fivefold increase in release rate.

The octenol and phenols are dispensed from sachets made of low-density polyethylene, 150 μm thick, with a surface area (both sides combined) of about 50 cm^2. Each sachet is charged with 4 ml of a 1:4:8 mixture of 3-*n*-propyl phenol:octenol:4-methyl phenol, and then heat sealed. The sachet should be placed in a cloth cover to protect it from direct sunlight; it will remain effective for about 4 months.

Phenols are caustic and must be handled with rubber gloves. The sachets containing the phenols are not dangerous to handle briefly.

2. Bait Design

If traps are to be used for the large-scale operation it would be simple and convenient to explore first a variety of trap designs for their effectiveness and convenience of operation and manufacture. Some traps that might be considered for testing are the biconical trap (see Section IV below), the F3 trap[29] developed in Zimbabwe, and the NG2B developed in Kenya.[42] The most economical and effective trap currently recommended for use in Zimbabwe is the Epsilon (Figure 1). If the trap is to be inspected at infrequent intervals, a little deltamethrin should be placed inside the cage to prevent ants from damaging or removing the catch.

A full study of the optimum screen to employ in a large-scale project would require the use of electrocuting devices[27] even more complex and expensive than those that would be

FIGURE 2. The S type of screen developed in Zimbabwe. (Photograph by J. W. Hargrove.)

necessary for studies of the effect of odors with screens. Hence, it might be easiest to accept that one of the Zimbabwean designs (Figure 2) will be used. The screen is made of black cotton drill and fine black terylene netting mounted on a frame made of 4-mm diameter galvanized wire and galvanized iron tubing.

The initial investigations with screens might be limited to using traps to gauge the likely performance of such a target. One can assume that the numbers of *G. pallidipes* caught by a trap would be no less than the number killed if a screen had been used instead. With *G. morsitans,* a species that shows a weak trap-entering response, the numbers caught by traps are likely to underestimate by at least 50% the numbers that would be killed by screens.[27]

3. Population Studies

If data exist already for the population density of tsetse in the project area, it would be possible to use the information from trap catches to assess the number of odor-baited traps or screens to give the required daily catching rate of 2 to 4% of the adult female population, from which a rough indication can be made of the cost and feasibility of a large-scale operation.

If the required population data do not exist, it may be possible to obtain them using existing information relating trap catches to population densities, but the use of mark, release, and recapture studies to assess population parameters is likely to prove as long and costly as the large-scale control operation itself. In such circumstances it would be more sensible to embark on the control operation and assess and adjust it as one goes along.

C. STRATEGY FOR THE USE OF BAITS AGAINST SAVANNA TSETSE

The strategy for the implementation of bait-based control operations depends on the aims of the operation, the type of infestations to be tackled, and the resources available. These factors vary considerably from one project to another. For example, the intention might be to attack the tsetse in a small area of, say, 500 km^2 and then hold the area free from dense infestation. Alternatively the aim might be to clear progressively much wider infestations. For some projects it is possible to use deltamethrin-dipped cattle as partial or complete substitutes for artificial baits, or to use community participation to augment core personnel. In projects where it is proposed to create a tsetse-free area in the center of a dense infestation, with a consequently strong invasion pressure on all sides, the placement of baits should be geared largely to prevent invasion and is distinct from that required for the control of sparse populations of tsetse occurring in isolated pockets or a tongue of habitat extending from the main infestation.

Despite this wide variation in circumstances, the basic strategy should remain the same: place the baits in a small area of about 100 km^2 first; if this proves successful, progressively expand the treated area, modifying the techniques and procedures in the light of experience. The opportunity to adopt this gradual and flexible approach is one of the main advantages of the bait methods of control and should be exploited fully. It may well require making a clean break with previous approaches. Thus, with conventional spraying operations, the main sampling of the tsetse population is conducted before and after the control campaign. With bait operations the emphasis is to be placed on sampling during the campaign.

The following is an example of the way an operation might progress with screens.

1. An initial trial area (termed A), about 10 × 10 km, is identified where tsetse populations are densest and, hence, where the screens will probably have the most demanding task and where the availability of the flies will allow population data to be produced most readily.
2. Surveys for tsetse and trypanosomiasis are made in area A and in adjacent areas (B,C,D . . .) into which it is proposed to extend the screens later, and in a sufficiently distant control area, say 10 km away, to be free of influence of the screens.
3. After surveys with traps have established baseline data, screens are placed at a fairly high density, say 6/km^2 of habitat for *G. morsitans* and 2/km^2 for *G. pallidipes*. This high density of screens should ensure that the tsetse population declines sharply in relation to the control population in, say, 3 months.
4. When the population in area A has started to decline convincingly, screens are deployed in area B at, say, 3 or 1/km^2 for *G. morsitans* and *G. pallidipes*, respectively.
5. Later, the screens are placed progressively in areas C,D . . . , perhaps obtaining some of them by removing half the screens in area A.
6. Once the combined area of A, B, C . . . is fairly large and compact, say, 30 × 30 km, the density of screens in the center of the combined area might be reduced further if the population there is at or near extinction. If the total area treated is then as great as intended for the time being, the screens removed from the center might be placed in a dense band on the border to form an invasion barrier.

A somewhat similar program might be adopted with cattle as baits, except that on the borders of the areas where cattle graze, one may have to use artificial baits to reduce the risk of trypanosomiasis.

IV. BAIT METHODS FOR RIVERINE VECTORS OF SLEEPING SICKNESS*

A. THE BICONICAL TRAP

Challier and Laveissière concentrated on the development of a simple, light, and effective

* C. Laveissière

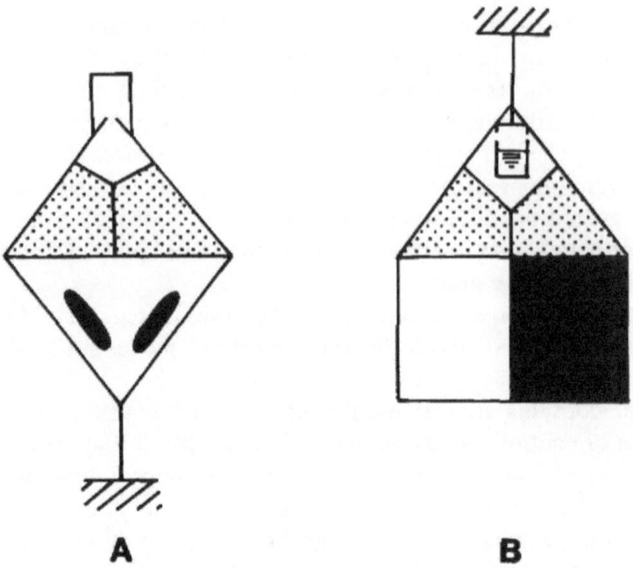

FIGURE 3. (A) Biconical trap; (B) pyramidal trap.

trap, and this work led in 1973 to the biconical trap (Figure 3A).[43] The attractiveness of this trap to flies is entirely visual, and it is well suited to the *palpalis* and *fusca* groups of tsetse in riverine and forest habitats. The biconical trap contains a dark cavity into which the flies are attracted.

In 1978 the performance of the biconical trap was improved threefold by using blue material for the lower cone.[44] This was originally introduced on the basis of field experience but has since been justified by a study of spectral sensitivity in the range of colors visible to tsetse.[45] This study showed that tsetse use information based on color, not only on contrast. However, the use of a rotating cylindrical variant of the biconical trap showed that this type of movement had no appreciable attractive effect.[46]

Biconical traps have been tested and found effective in many different parts of Africa.[14,20,47-54] The traps are more effective for control purposes if they are impregnated with deltamethrin.

B. TRAPPING OF RIVERINE TSETSE IN CÔTE D'IVOIRE

Extensive trials have been carried out against the riverine tsetse of the savanna zone. Biconical traps are impregnated with 400 mg of deltamethrin, and they are installed along riverine forest galleries as near as possible to river banks and in sunlit places to increase their attractiveness. The traps may be hung from supports driven into the riverbanks, or they may simply be hung from low branches. In any case the base of the trap should be about 10 cm from the ground or water surface. A control campaign with traps should begin after the floods have receded when the population of riverine tsetse is aging and beginning to decline, and the campaign should continue until the beginning of the wet season just before the first floods, when the traps should be collected up to avoid losses.

Traps can be quickly installed. Two men in a motorized canoe (supported by a four-wheel-drive vehicle carrying the traps) can install traps in 15 linear kilometers of riverine forest per day.[55] The method is highly effective. For example, in 62 linear kilometers of gallery forest treated with biconical traps impregnated with deltamethrin, 99% reductions in populations of *G. tachinoides* and *G. palpalis* were achieved in 1 to 2 months, and these reductions were maintained as long as the traps were left in place (Figure 4).[56] Eradication can be achieved in the middle of a treated area, but a few tsetse continue to be found toward the edges.

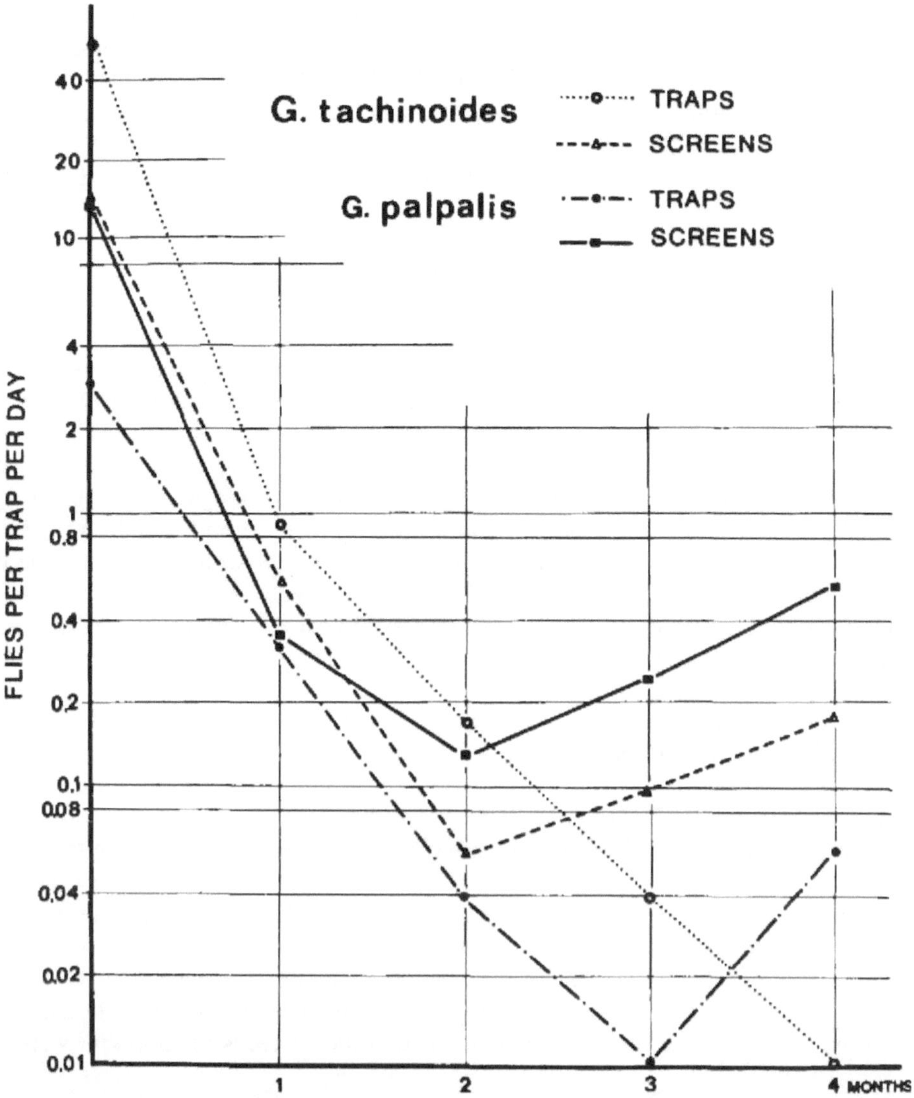

FIGURE 4. Comparative effect of traps and screens impregnated with deltamethrin on the riverine species of tsetse.

Even though biconical traps are each rather expensive, the overall cost of a program is highly competitive. The most recent estimate of the cost of spraying was the equivalent of \$600 per linear kilometer,[57] trapping costs \$74 per kilometer, and this cost is halved in the second year because most traps can be reused.[43] Recent experiments suggest that the method remains fully effective with only one trap every 300 m[58] with a substantial saving in labor and cost of material.

In an attempt to reduce the cost still further, traps have been replaced by simple squares of blue material, impregnated with 150 mg deltamethrin per square meter and suspended from a metal "gallows" and placed every 100 m along forest galleries. The results (Figure 4) are good, but are somewhat inferior to those achieved by trapping, the difference being more marked for *G. palpalis* than *G. tachinoides*. However, the cost was reduced to about \$53/km in the first year and \$13 in the second.[59-61]

C. TRAP BARRIERS ON RIVERS

Reinvasion of areas cleared of tsetse is a major problem, since tsetse can fly substantial distances across unfavorable territory. However, several studies have shown that an area is effectively protected from reinvasion by carefully placed traps or screens forming a barrier less than 5 km wide, i.e., not more than 50 traps.[55,59,62,63] If a complete elimination of tsetse is achieved in the middle of a given area by trapping, alone or in combination with some other method such as sterile male releases,[61,64] the traps are no longer required there and should be moved somewhere else, thus increasing the area which can be treated with the money invested.

V. VECTOR CONTROL IN THE FOREST ZONE*

A. INSECTICIDE SPRAYING

The ecological complexity of the forest zone makes it difficult to control tsetse there. Spraying has achieved variable results depending on the insecticide used: DDT has consistently given worse long-term results than pyrethroids.[65] Deltamethrin and alpha cypermethrin are particularly effective and long lasting, and if they are sprayed selectively on *G. palpalis* resting places around the fringes of plantations and villages, they quickly give drastic and long-term population suppression. However, the method is not cost effective because the fringing areas are not accessible to vehicles and the drums of insecticide have to be carried on foot. It has been estimated that it takes 3 weeks to treat 4 km^2 with a team of 2 supervisors, 12 spraymen, 24 drum carriers, and 2 vehicles with drivers.[66] The cost of treating a sleeping sickness focus in this way is excessive, not only for labor, but for insecticide.

On the other hand, aerial spraying is impracticable for several reasons, notably the danger of tall trees to low-flying aircraft.

B. USE OF IMPREGNATED SCREENS

Following the successes of traps in the savanna zone it seemed appropriate to test them in the forest. It was out of the question to treat a whole area of forest with the relatively expensive biconical traps. A simpler device was needed. Medium-scale trials with blue screens showed that they are cheaper and more effective than spraying.[66] However, it is essential to reimpregnate the screens regularly and to treat a large area at once: 1000 screens in 4 km^2 had a definite effect, but 16,000 screens, reimpregnated every 3 months and spread over 86 km^2, had a better effect (Figure 5A),[67] and 36,000 screens in 1300 km^2 were even better (Figure 5B).

In some situations screens are ineffective:

1. Along roadsides screens become covered in dust thrown up by vehicles, and this changes their color and masks the insecticide.
2. In the middle of islands of forest and in areas of secondary growth, screens are almost invisible to the flies.
3. In the early trials when the research team itself installed screens, which were not then kept under supervision, many were stolen, destroyed by fire, or overgrown.[66]

However, not all sites are of equal epidemiological importance, and this allows one some degree of selectivity as to where screens need to be placed.

A control campaign should include barrier zones to limit reinvasion into the land holdings of all the villages at risk. Hervouet[68] has shown that the landholdings of the residents of different villages overlap and intermingle, and since sleeping sickness is generally transmitted while farmers are at work, one might think that the area needing treatment would extend

* C. Laveissière

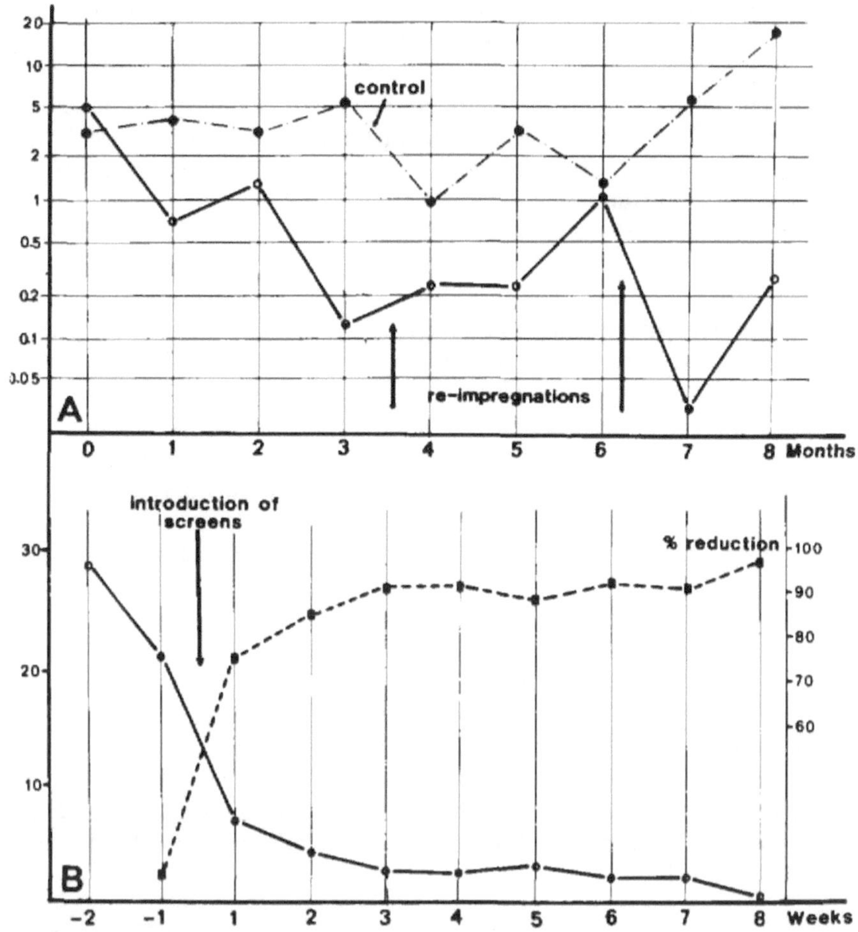

FIGURE 5. Results obtained in the forest zone with screens impregnated with deltamethrin against *G. palpalis*. (A) Campaign in 86 km² with simple blue screens; (B) beginning of a campaign in 1300 km² with improved screens (see Figure 10).

all the way from the ocean in the south to the savannas in the north. However, given the above limitations on where screens are effective, it is appropriate to concentrate one's efforts on high-risk sites such as places where people go for water, the fringes of plantations, riverine forest galleries (which are the principal tsetse flight lines), human settlements, footpaths and tracks adjacent to plantations, and perimeters of villages. Everywhere else — fields, fallow ground, clearings, and even savanna, which tsetse cross or occasionally hunt in — will be cleared of fly as an indirect consequence of placing screens in the principal habitats.

Despite these limitations on where screens are needed, a vast area still remains, and one has to consider how the job can be done.

C. UTILIZATION OF THE HUMAN POTENTIAL OF THE AREA

A farmer knows his land better than anyone else, and it is appropriate that he should install the screens on his own land. This approach was used in a medium sized trial in Côte d'Ivoire in 1983—1984[69] and on a large scale in 1987 to 1990.[39]

The mobilization of rural communities requires that they are well informed about:

1. Sleeping sickness and the importance of vector control for its prevention
2. The basic facts about the use of screens and their maintenance

In areas of high endemicity every family has had some victims of the disease and even some deaths from it. However, outside these foci people are much less aware of the disease and think of it as "someone else's problem". In general tsetse are thought of more as a nuisance than as vectors of disease.

Nontechnical but detailed explanation is required to convince people of the true importance of tsetse, and only when this has been done can the screens be distributed. It is important to emphasize collective responsibility and mutual assistance in the vector control operations. This is essential, not only to ensure the even distribution of the screens, but also for their maintenance and reimpregnation. The timing of the latter should be decided not only on the basis of the persistence of the insecticide, but also on the fluctuations of the tsetse population and to avoid the seasons when people are occupied with cultivation, traditional ceremonies, and annual migrations. To avoid the need for reimpregnation work during these periods a system has been developed which requires only two treatments per year.[70]

The involvement of well-informed and well-motivated rural communities gives the advantages of:

1. Speed of treatment. In 1983 during the pilot trial in Vavoua in less than 6 d 363 farmers installed 16,000 screens in 451 plots with a total area of 86 km^2; in 2 weeks in 1987 3800 farmers installed 36,000 screens in 5000 plots.
2. Completeness of coverage. Almost all plantations, except for a few abandoned areas, were treated, and in many cases farmers treated the plots of absent neighbors.
3. Maintenance. In 1983—1984 more than 95% of farmers took part in the first reimpregnation of the screens, and 90% did so in the second, 94 and 88% of screens being treated, respectively.

No specialist team without community participation could possibly have covered so quickly such a large area nor rediscovered so many screens for retreatment.

Someone must also treat roads, villages, and strips of riverine forest which belong to everyone, but to no one in particular. Experience shows that these will only be attended to if there is a team appointed for the purpose. The use of biconical traps for the riverine forest allows effective control without the need for such frequent reimpregnation as with screens. Along roads and around edges of villages one can selectively spray vegetation up to 1.2 m in height. This is effective, but only for a short time, because of deposition of dust, fire, and periodic cutting down of the vegetation.

A new system was tried in the Vavoua area in 1987. A village appointed two young people to be in charge of the positioning and maintenance of simplified impregnated traps ("Vavoua" traps) and to keep a continuous record of fly densities in their villages. This gave good results in rapid treatment of a large number of communities, accurate evaluation, and reduction in the work load of the team in charge of treatment. It is important that everyone in the neighborhood is made aware of the purpose of the traps so that they are not stolen.

D. TIMING OF TREATMENT

As already stated the timing of treatment must be adapted to the needs of the farmer and the community. The visibility of the screens to the flies is greatest at the beginning of the dry season when the coffee and other plantations are cleared of undergrowth in preparation for the harvest. At this time all the farmers are to be found on their plantations, which increases the coverage.

There are several arguments against delaying the start of a campaign until the middle of the dry season. At this time, most farmers have finished harvesting their cash crops and have left the settlements near the plantations and returned to their home villages to plant their subsistence crops. The few people remaining at the plantations are tired after the harvest and so are not inclined to take part in tsetse control operations. Furthermore, as soon as the rains come, undergrowth springs up and masks the screens. Thus, treatment in the middle of the dry season would only give 1 or 2 months of effectiveness at a time when the tsetse population is difficult to control.[71] In the hot season pupal development is accelerated, leading to a surge in the number of teneral flies, which are less readily attracted to traps or screens than are older flies.[56]

It is, therefore, preferable to attack the population of older flies at the end of the wet season. The fact that the population density is then high makes a 90% reduction in tsetse nuisance all the more spectacular. This increases the willingness of the farmers to take part in the operation and to present themselves for medical examination.

E. MEDICAL EXAMINATION FOR SLEEPING SICKNESS

From the point of view of trypanosomiasis control one of the most significant effects of the involvement of farmers in the program has been on the percentage of the population coming forward for examination by the mobile medical teams.[72] This percentage averages 88%, with marked variation between villages depending on the endemicity of the disease and the number of farmers involved in the vector control program. In villages seriously affected by sleeping sickness and where all the farmers had received screens, more than 98.2% of the population presented themselves for examination.

Thus, by involvement of the community, not only was a high degree of vector control attained, but also almost all the remaining infected individuals could be identified and treated. Conversely, if medical coverage is poor, the disease reappears apparently inexplicably, and people lose faith in the vector control operations.

F. COST OF CONTROL IN THE FOREST ZONE

In calculating costs one should take into account not only the primary habitats in which baits are actually installed, but also the secondary habitats from which *G. palpalis* disappears as an indirect consequence of the program, i.e., one should consider the overall area protected by the program.

At Vavoua in 1983-1984 the cost of a screen was $2.50 and the total cost of the screens, their installation, and three reimpregnations was $580/km^2 protected (or about $6 per inhabitant). In subsequent years one only needs to count the cost of maintenance and replacement of about 10% of the fabric, i.e., about $100/km^2. In 1987 costs were further reduced by using only 36,000 screens for 1300 km^2 and by making the villagers responsible for treating the publicly owned land.

It is to be hoped that farmers will one day be able to take full charge of tsetse control and assign it the same importance that they now do to control of agricultural pests. At Vavoua in 1983 the number of screens taken by each farmer averaged 43, with marked variation between different ethnic groups. The cost of these 43 screens was $108, and the subsequent maintenance cost was $24 per year.[73] The costs in 1987 were considerably reduced to ten screens per farmer, but unless the farmers are convinced of the pressing need for tsetse control, it is unlikely that they would be willing to invest in screens. If the screens were provided free, perhaps the farmers could be induced to pay for their maintenance.

G. PLANNING FOR THE DISTRIBUTION OF SCREENS AND TRAPS

Aerial photographs are helpful to get an idea of the road network, the proportion of the area which is cultivated, and how much riverine forest there is. The latter two pieces of information allow one to estimate the number of screens and traps that will be required.

The screens, traps, and insecticides are delivered to the farmers in their villages or by prearranged meetings with them at the roadside. At that time the farmer is asked to provide information from which the number of screens that he will require can be estimated, i.e., the nature and area of his plantation, the presence of settlements or water collection places, and the existence and approximate length of forest edges, tracks, or footpaths. During this meeting advice about installation of the screens or traps is given.

Experience has shown that few screens are needed in the middle of plantations, and one should concentrate on the edges and on tracks and footpaths (which need one screen every 100 m), at settlements, water collection points, and where crops are dried or decorticated (1 to 2 screens at each of these sites). One should make clear that the screens become the farmer's property, and it is up to him to maintain them, i.e., to prevent vegetation overgrowing them and to reerect any which fall over. Though the entomologist can give the farmer some ideas about where to place screens, the farmer knows much better where tsetse are concentrated because he is attacked by them every day.

H. CHOICE OF INSECTICIDE

Although flies which enter traps are killed by exposure to the sun, their effectiveness is increased by insecticide impregnation which kills the large numbers (about two thirds) which rest on the trap exterior and which would otherwise escape.

Screens must always be impregnated. The first trials with screens impregnated with the synthetic pyrethroid deltamethrin showed its powerful effect despite its tendency to knock down without killing at lower doses. In practice the knockdown effect is probably not disadvantageous because 85% of knocked-down flies are eaten by ants within 6 h, and virtually all have been eaten after 24 h.[74]

Synthetic pyrethroids also have an irritant effect (see Chapter 2), but this is not expressed until after tsetse have already picked up a lethal dose.[75]

Of the insecticides currently available, only pyrethroids are suitable for use on traps because of their effectiveness at low doses and short exposure times.[70,76] In contrast dieldrin acts slowly and requires high dosages.[62,77] When tested in the field on screens dieldrin produced a 26% reduction in a population of *G. tachinoides* in the time that it took deltamethrin to give a 96% reduction. Similar tests showed that endosulfan and cypermethrin are unsuitable for use with baiting methods.[70]

Pyrethroids are not toxic to humans at the doses used for impregnation, but it is necessary to warn people who will be handling them that they may cause sneezing and parasthesia (a tingling sensation). However, farmers are now accustomed to using much more dangerous agricultural insecticides.

The rate of degradation of insecticides is affected by solar radiation, rain, wind, and dust, and their persistence depends on the formulation: at equal doses and on the same type of sheeting a wettable powder of deltamethrin was more rapidly eliminated than an emulsifiable concentrate.[76] In contrast, on mosquito netting, the wettable powder persisted at least as well as the emulsifiable concentrate, but at the same dose the wettable powder had a lower insecticidal effect (Figure 6).

Persistence and effectiveness depend on the nature of the fabric. The short irregular fibers of cotton retain deltamethrin poorly whereas synthetic fibers retain it better. On blue, pure cotton sheeting the insecticidal effect of pyrethroids persists less well than on cotton polyester mixtures (Figure 7; see also Chapter 2). Bioassays confirm that even if the insecticide is still chemically present, its likelihood of being picked up by an insect depends as much on whether the weave is tight or loose as on the chemical nature of the fiber and the dye.

As might be expected, a given insecticidal effect lasts longer if the initial dose is higher (Figure 8). However, it has also been shown in controlled exposure of impregnated materials

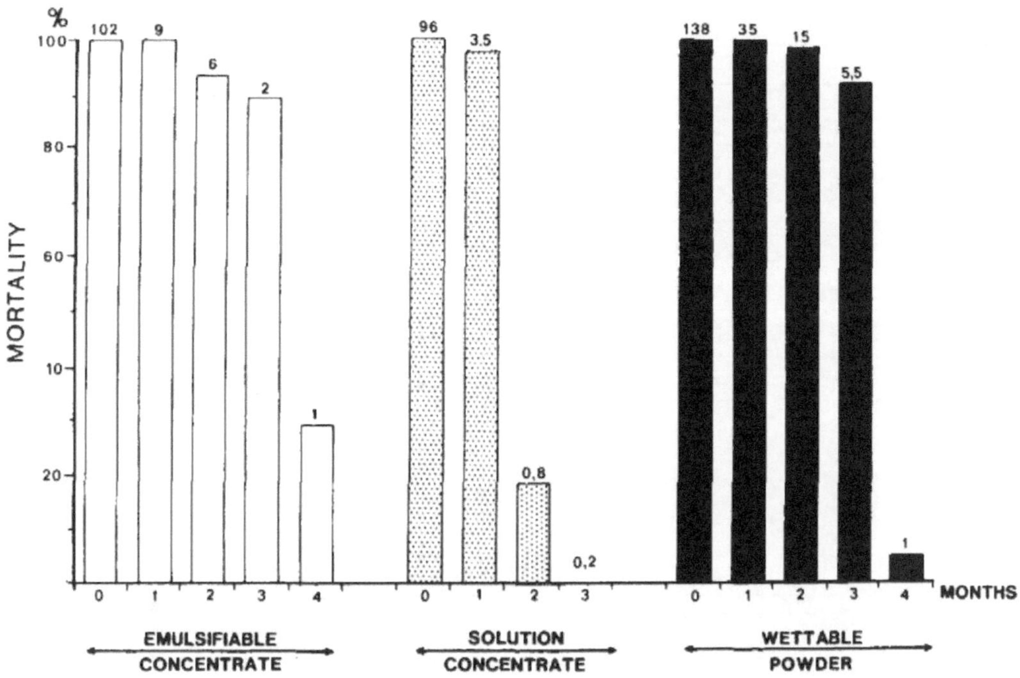

FIGURE 6. Comparison of the mortality of *G. palpalis* exposed for 10 s to three formulations of deltamethrin on polyester mosquito netting. The figures indicate the amount of active ingredient remaining.

that the persistence of chemically detectable active ingredient also varies with the initial dose. If samples of fabric are impregnated with 100 or 400 mg deltamethrin per square meter, the quantities of active ingredient retained after a month are not in the ratio 4:1 but 10:1, and this ratio rises to 32:1 and 64:1 after 2 and 3 months, respectively. From the same initial dosages of alpha cypermethrin the ratio had risen to 119:1 after 3 months.

In deciding on the most cost-effective dosage one must take account of not only the cost of the insecticide, but also the cost of the reimpregnation operation.[70]

I. DAMAGE TO, AND LOSS OF, TRAPS AND SCREENS

Apart from the loss of insecticide one also must consider loss or inactivation of screens due to fire or being overgrown. In the Sudan-Guinea zone it is best to collect the screens at the beginning of the heavy rains to avoid losing them (see Section IV.B). In doing so one must accept some degree of tsetse reinvasion, but this need not be serious. The tsetse population should have been reduced by 98 to 99% by the use of screens during the 6 to 7 months of the dry season. After effective medical inspections the reservoir of infection should be almost eliminated, and the risk of transmission negligible. Furthermore, the dispersion of riverine tsetse in the wet season occurs at a steady but low rate,[78] and there is no chance that in 4 months the population could return to its initial level. Monitoring of the consequences of several successive dry season campaigns against a tsetse population in a riverine gallery forest[79] showed that the density at the end of the wet seasons did not exceed 2% of the initial level. The recovery of the population is slower if the treated area is larger and the control system more efficient. In any case the replacement of the screens when the floods recede causes a renewed decline in the population to even lower levels.

J. INTERSPECIES SELECTION

The first trials of trapping showed that *G. tachinoides* was more affected by traps than *G. palpalis* (Figure 4). On the river Leraba in the north of Côte d'Ivoire the *tachinoides/*

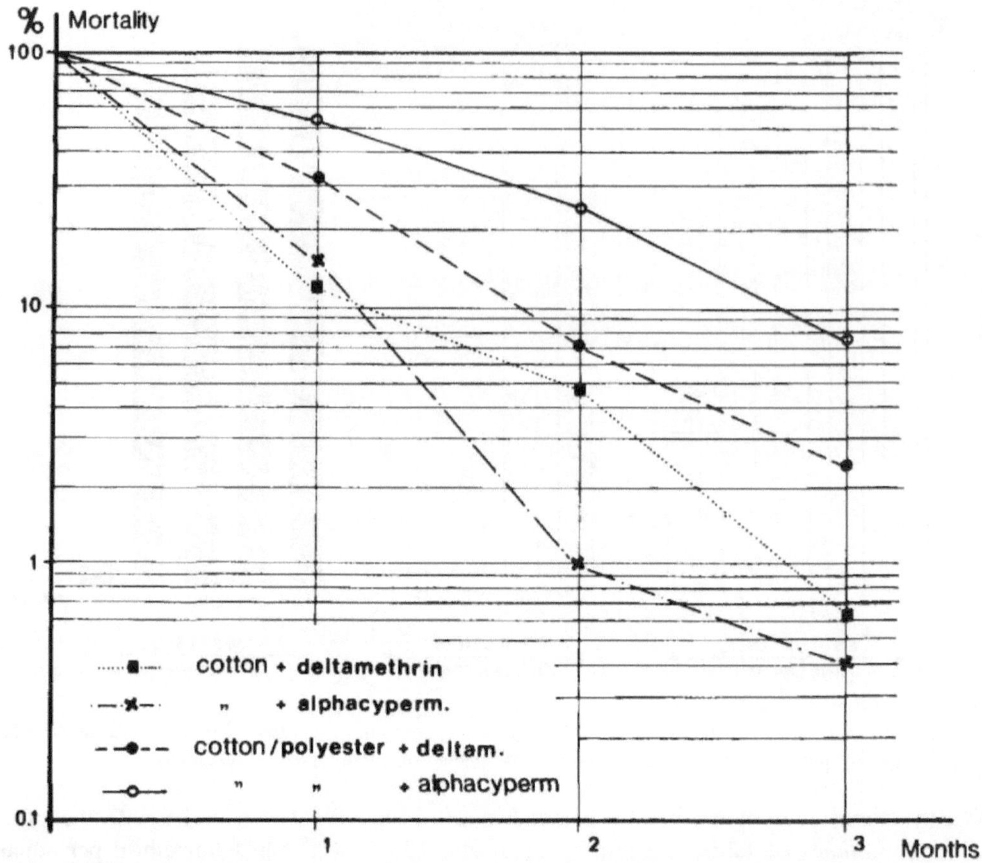

FIGURE 7. Mortality of *G. palpalis* exposed for 10 s to pyrethroids on different fabrics.

palpalis ratio was 24:1 at the beginning of the control operations but only 1.23:1 after 4 years of control and 1 year after its withdrawal.[79] Similar observations were made in the forest zone when screens were less effective against *G. pallicera* and *G. nigrofusca* than *G. palpalis*[66] with a consequent shift in the ratio in favor of the former species during 5 months of a control operation (Figure 9).

Such shifts in the species ratio would only matter if the species favored by selection also had a higher vectorial efficiency. Baiting methods have not been in use for long enough to allow an assessment of how much risk there is of failure due to selection for behaviorally resistant species or genetic varieties within species. In 1936 Zumpt[80] recognized the risk and recommended the simultaneous use of different trapping systems on the theory that the occurrence was unlikely of a species or variety simultaneously "resistant" to two or more different types of traps (see also discussion in Chapter 2 about insecticide resistance management with mixtures).

K. IMPROVEMENTS TO TRAPS

1. Color

As emphasized by Challier[81] the type of baiting methods currently used in West Africa depends on chance encounter between the fly and the bait. Improvements in the attractiveness or killing efficiency of traps may well be worthwhile (even if the cost per bait is somewhat increased) by greatly reducing the number of baits required per unit area.

Work in Côte d'Ivoire with the electrified grids designed by Vale[15] showed that the number of *G. palpalis* attracted by a blue screen was 13 times larger than the number landing

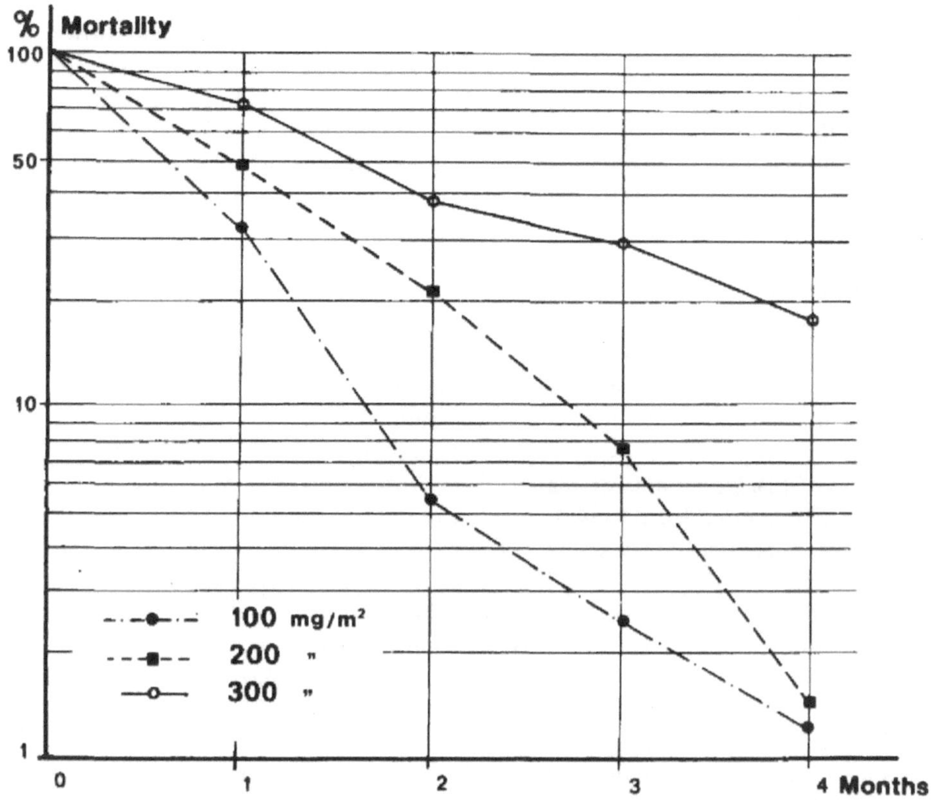

FIGURE 8. Effect of initial dose on the persistence of the insecticidal effect of deltamethrin in the dry season.

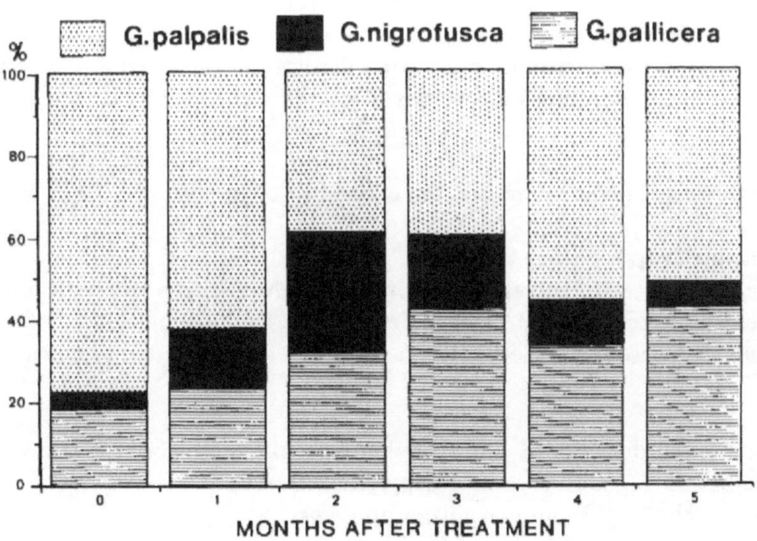

FIGURE 9. Proportions of three species of tsetse in a forest population after the introduction of impregnated screens.

on it. In tests of different colored screens the numbers attracted by blue appeared to be higher, but not significantly so, than those attracted by red, black, or white. However, the numbers landing on white cloth were greater than those landing on blue and were in general related to the ultraviolet reflectivity of the fabric.[82,83] The fly landing rate could be tripled by using a fabric with high reflectance in the near ultraviolet range. However, the color of fabrics may change to an important extent after a few months of use.[84] Long-term tests are needed, and one must reject fabrics which are not color fast, even if they are good substrates for insecticides.

2. Color Contrast

Contrast tests showed that blue and black was the best combination. This is surprising because black fabric does not reflect ultraviolet. Nevertheless, at least 75% of the flies were caught on the black segments of a blue-black screen. Tests with various arrangements of blue and black showed that the following were optimal:

1. A ratio of black:blue area of 1:2
2. The black part divided into two lateral strips
3. The black material translucent

Based on these studies the screen illustrated in Figure 10 was developed. The materials were chosen for their attractiveness to flies, their ability to retain insecticide, their durability, and lightness for transport by bicycle. The following features should be emphasized:

1. The blue section is made of a cotton-polyester mixture and is only moderately good as a substrate for insecticide and as a fly attractant; however, it is very durable and retains its color well.
2. The lateral black bands are of light nylon net which retains its color well and is excellent as a substrate for insecticide. A dose of 200 mg deltamethrin or 380 mg alpha cypermethrin per square meter remains effective for 5 or 6 months.
3. The support consists of a vertical piece of iron rod 1 cm in diameter, sharpened to drive into the earth, and soldered to it is a horizontal rod 0.8 cm in diameter.
4. To reduce the risk of tearing in the wind the screen is attached to the vertical support by tapes. Hanging screens from trees is not recommended as it increases the risk of tearing.

The cost in 1987 in Côte d'Ivoire was the equivalent of less than $2.70.

3. The Vavoua Trap

The biconical trap is relatively expensive and a cheaper trap has been produced based on the research on screens and insecticides and on the pyramidal trap of Gouteux and Lancien[85] which is described below. The Vavoua trap is illustrated in Figure 11 and consists of

1. A cone (basal diameter 81 cm) of nylon mosquito netting which is resistant to solar radiation and a good substrate for pyrethroids. The cone consists of three sections sewn to a circle of 3-mm galvanized iron wire.
2. Three screens sewn at angles of 120° to each other hanging 45 cm below the base of the cone. Each screen consists of a strip of blue cotton/polyester 28 cm wide and a strip of black nylon netting 12 cm wide. The black strips are located toward the center of the trap.

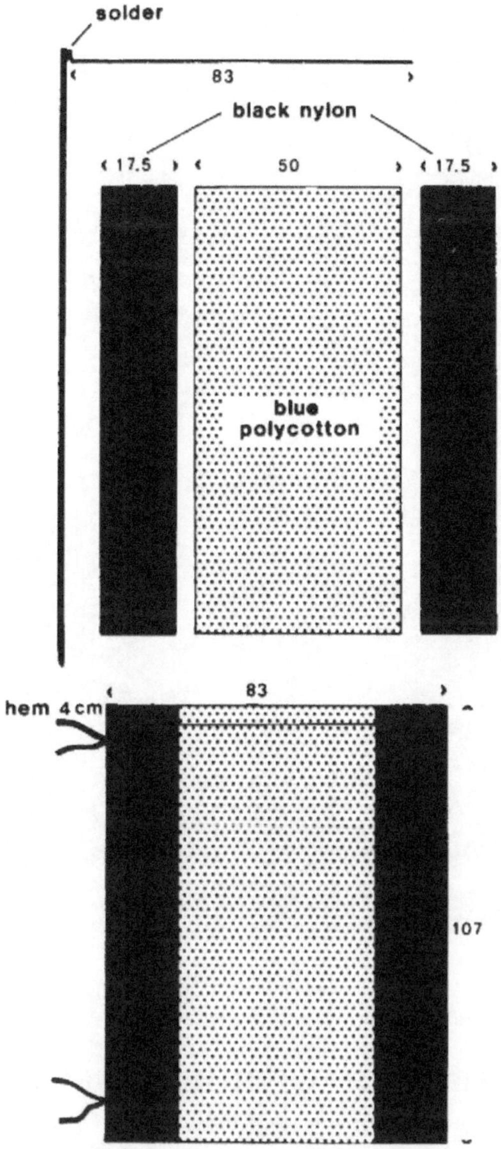

FIGURE 10. New type of black/blue/black screen used for control of *G. palpalis* in the forest zone of Côte d'Ivoire (dimensions in cm).

This arrangement, especially the size and positioning of the black strips, is optimal for catching *G. palpalis* which tend to rest on the black strips and then fly upward into the cone of netting.

The cost of a Vavoua trap in Côte d'Ivoire in 1987 was $3.90 (i.e. about half the price of a biconical trap and less than a pyramidal trap) excluding the cost of the metal stand and of the manpower.

The above-mentioned doses of insecticide on screens are doubled on traps, but for economy, only the black strips are treated, as this is where most of the flies rest.

FIGURE 11. A Vavoua trap (with longitudinal screens). (Photograph by C. Laveissière.)

VI. PYRAMIDAL TRAPS WITHOUT INSECTICIDE IN THE CONGO*

Figure 3B illustrates another variant of the biconical trap.[85] The lower cone is replaced by two intersecting blue and black screens, and for simplicity of manufacture, the upper cone is replaced by a pyramid. Biconical traps and their derivatives have previously been used impregnated with insecticide for population control purposes, though for sampling populations, permanent capture devices containing preserving fluid have been mounted on top of the trap so that all the flies killed by the trap can be seen and recorded.[46] In the pyramidal trap a plastic bottle containing a preserving fluid, based on fuel oil of low volatility, forms a permanent capture system. This has been moved inside the trap (Figure 3B) and provides a structural strengthening element.[85] Traps of this type can be suspended from tree branches, which simplifies their erection, by doing away with the need for metal supports, and gives them greater resistance to storms.

The fact that pyramidal traps do not require insecticidal impregnation and reimpregnation is an important advantage, especially in very humid areas such as the Congo, where re-

* J.-P. Gouteux

treatment of impregnated traps is required every 2 months.[86] The simplification of not having to reimpregnate makes it much more feasible to leave rural communities in sole charge of the traps.[87] The fact that the corpses of the captured flies can be taken out and counted gives a clear demonstration that the traps are effective. The absence of flies from traps in some sites shows that the traps are no longer needed there and can justifiably be moved somewhere else where a significant tsetse population may still exist.

Dagnogo and Gouteux[88] carried out comparative trials of various types of traps and found pyramidal traps to be two to five times more effective than biconicals.

Pyramidal traps have been selected for use by the Congolese National Program for Trypanosomiasis Control and are produced in a workshop in Brazzaville where a staff of four can turn out 800 traps per month. The workshop is financed by external aid. Each trap costs the equivalent of about $9.

There was a sleeping sickness focus where a main road crossed the Lefini river in the Congo. Insecticide-impregnated traps were used in the earlier phases of a control operation against the vector *G. fuscipes quanzensis,* but in the final phase unimpregnated pyramidal traps with a permanent capture system were introduced.[86] The results emphasized that traps at a few strategic points were capturing most of the remaining flies. Surveys using an immunofluorescence technique indicated interruption of transmission of sleeping sickness.

A larger operation with pyramidal traps was carried out in 1985 in 35 villages in the Niari sleeping sickness focus on the main route from Brazzaville to the sea.[89] The main vector was *G. palpalis palpalis,* and a total of almost 70,000 of these flies were trapped. In about half the villages the fly populations were reduced by more than 99%, and in the other villages the reduction was nowhere less than 86%.

The scale of the operation in the Congo was further increased in 1986 to 55 villages in the Bouenza sleeping sickness focus. Detailed management of such a large operation was only possible by handing over responsibility for looking after the traps to health committees which were set up in each village.[87,90] Each village is provided with 20 to 100 traps. The exact number is determined after discussion with the health committee and may be varied depending on the numbers of flies caught in the traps and recorded by the villagers. The health committees are provided with a kit of equipment to service the traps. They are visited every 2 or 3 months by staff of the national program to check on progress and to discuss changes in the number and placement of traps which may be necessary in the light of the catches of flies which are recorded. So far, 1263 traps have been placed in position, and there has been a total recorded catch of 240,514 *G. palpalis,* apparent elimination of tsetse in 9 villages, better than 90% suppression in 38 villages, and a welcome reduction of tsetse biting nuisance. Unfortunately, as the catches in the traps decline enthusiasm to continue servicing them also tends to decline.

Immunofluorescence tests on human subjects showed a decline in trypanosome positivity from between 2.4 and 4.0% in the years before the program began to 0.41% in 1986 and in January 1987. Checking the blood of "sentinel" animals kept in the villages where there has been apparent elimination of tsetse gave additional evidence for interruption of trypanosome transmission.

Although several sleeping sickness foci have been suppressed, so long as the human reservoir of trypanosomiasis exists, vector control will be needed, and continued support to the tsetse trapping operations based on the village committees is considered the most effective way of achieving this.

VII. ADVANTAGES OF BAIT METHODS*

Compared to the ground or aerial spraying of insecticide, the use of baits offers the following advantages:

1. Simplicity. The baits offer a simple mechanical technology which is appropriate to community participation in developing countries, as shown by the programs in Côte d'Ivoire and the Congo.
2. Local content. A relatively large proportion of the inputs can be met from local sources, avoiding the demand to import the large quantities of costly insecticide used in spraying campaigns.
3. Capital components. With spraying campaigns almost all of the inputs, such as cost of flying and insecticide, are nonrecoverable, but the cloth components of baits have an effective life of up to 2 years, and the metal components are serviceable for much longer.
4. Variable inputs. For spraying programs, virtually the same effort and expense is required to remove sparse tsetse populations as to remove very dense ones. Moreover, with aerial spraying, it is just as costly to spray a predominantly agricultural landscape, with only small pockets of tsetse habitat, as it is to spray a continuous habitat. Bait techniques allow inputs to be varied to suit the problem. With sparse infestations fewer baits will be needed, or the baits need to be maintained for short periods only. The baits need to be deployed only within the tsetse habitat.
5. Ecological risks. The widespread distribution of insecticide is likely to involve greater risks of ecological damage than the use of a sparse placement of baits that attract only tsetse in large numbers. Moreover, the study of the few nontarget organisms attracted to the baits can be more simple, economical, and precise than the study of the many creatures that can pick up the residues of widespread spraying.[91,92]
6. Seasonality. Whereas spraying campaigns are effective only in a few months of the year when the weather is suitable, in some parts of Africa bait techniques can be used throughout the year. Such bait operations can be started at any time that the resources become available (avoiding the tight deadlines for seasonal spraying programs), and the demand for equipment and transport is smoother and inherently more economical than the demands for the brief period of ground spraying.
7. Correction of errors. The gradual progress of a bait operation allows errors to be corrected as they emerge, whereas with spraying programs the operation may be complete before the error emerges, by which time it is too late for correction.
8. Barriers to invasion. With spraying campaigns, tsetse are free to invade the treated area as soon as the aerial spraying is complete or as soon as the residual insecticide applied by ground spraying has been washed away by rains.[93,94] The reinvasion of treated areas means that spraying programs in successive years must overlap by a considerable margin along the invasion front. To reduce the degree of overlap, there has been a trend to increase the area sprayed each year and emphasis has been given to exceedingly large operations to clear whole fly-belts and, hence, put an end to invasion problems. The opportunity to use a dense placement of baits to suppress or restrict tsetse invasions throughout the year offers the opportunity to clear and hold tsetse from relatively small areas.

Against the above advantages it must be recognized that bait methods of tsetse control have their own special problems. Thus, if artificial baits are employed, they may be stolen

* G. A. Vale and J.-P. Gouteux

if adequate measures are not taken to ensure the cooperation and involvement of the public in the program. Also, although bait operations require about as much access development as for ground spraying, continuous bait operations need wet-season maintenance of tracks whereas ground-spraying tracks need repair only in the dry season, and for aerial spraying only a few tracks are needed for ground-based teams to guide the aircraft.

On balance, there is much to recommend bait methods of tsetse control, but it must be recognized that these methods are relatively new in their present form and must be applied with special care to observe their effectiveness and practicability, and to assess means of improving their feasibility.

REFERENCES

1. **Vale, G. A. and Hargrove, J. W.,** unpublished data.
2. **Anon.,** Maneira practica de destruir as glossinas (moscas de Gabão), *Arch. Hyg. Pathol. Exoticas, (Lisbon),* 1, 291, 1906.
3. **Mendes, A. C., Mora, A. D.,Monteiro, A. S., and da Costa, D. B.,** La maladie du sommeil à l'île du Prince, *Arch. Hyg. Pathol. Exoticas, (Lisbon),* 2, 275, 1909; *Sleeping Sickness Bur. Bull.,* 2 (Abstr.), 26, 1910.
4. **Harris, R. H. T. P.,** The control and possible extermination of the tsetse by trapping, *Acta Conventus Tertii Trop. Morbis, (Amsterdam),* 1, 663, 1938.
5. **Henrard, C. L.,** Quelques essais de captures de *G. palpalis* au moyen de divers types de pièges Harris au Stanley Pool, *Ann. Soc. Belge Med. Trop.,* 14, 263, 1934.
6. **Lewillon, R.,** Le piégeage expérimental de *Glossina palpalis* à la mission médicale du Kwango, *Rec. Sci. Med. Congo Belge,* 4, 45, 1945.
7. **Challier, A.,** The ecology of tsetse (*Glossina* spp.) (Diptera, Glossinidae): a review (1970—1981), *Insect Sci. Appl.,* 3, 97, 1982.
8. **Swynnerton, C. F. M.,** Some traps for tsetse flies, *Bull. Entomol. Res.,* 24, 69, 1933.
9. **Rupp, H.,** Contribution à la lutte contre les tsétsés. Influences d'étoffes attractives, impregnées de DDT, sur *Glossina palpalis* spp. *martinii* Zpt., *Acta Trop.,* 9, 289, 1952.
10. **Hoof, L. van, Henrard, C., and Peel, E.,** Mekanische prophylaxie der slaapziekte. Strijd tegen de Glossinen in Belgisch Congo, *Acta Conventus Tertii Trop. Morbis, (Amsterdam),* 1, 641, 1938.
11. **Glover, P. E. and Langridge, W. P.,** An introductory note on modern methods of tsetse control, in Proc. 9th Meet. Int. Sci. Comm. for Tryponosomiasis Research, Conakry, 1962, 157.
12. **Morris, K. S. R.,** Trapping as a means of studying the game tsetse *Glossina pallidipes* Aust., *Bull. Entomol. Res.,* 51, 533, 1960.
13. **Smith, L. M. and Rennison, B. D.,** Studies of the sampling of *Glossina pallidipes* Aust. I. The number caught daily on cattle, in Morris traps and on a fly-round, *Bull. Entomol. Res.,* 52, 165, 1961.
14. **Gouteux, J. P. and Dagnogo, M.,** Ecologie des glossines en secteur pré-forestier de Côte d'Ivoire. XI. Comparaison des captures au piège biconique et au filet. Agressivité pour l'homme, *Cah. O.R.S.T.O.M. Ser. Entomol. Med. Parasitol.,* 24, 99, 1986.
15. **Vale, G. A.,** New field methods for studying the responses of tsetse flies (Diptera: Glossinidae) to hosts, *Bull. Entomol. Res.,* 64, 199, 1974.
16. **Vale, G. A.,** Field studies of the responses of tsetse flies (Diptera: Glossinidae) to odours of men, lactic acid and carbon dioxide, *Bull. Entomol. Res.,* 69, 459, 1979.
17. **Vale, G. A.,** Prospects for using stationary baits to control and study populations of tsetse flies in Zimbabwe, *Zimbabwe Sci. Newsl.,* 15, 181, 1981.
18. **Hargrove, J. W. and Vale, G. A.,** The effect of host odour concentration on catches of tsetse flies (Glossinidae) and other Diptera in the field, *Bull. Entomol. Res.,* 68, 607, 1978.
19. **Hargrove, J. W. and Vale, G. A.,** Aspects of the feasibility of employing odour-baited traps for controlling tsetse flies (Diptera: Glossinidae), *Bull. Entomol. Res.,* 69, 283, 1979.
20. **Vale, G. A.,** The trap-orientated behaviour of tsetse flies (Glossinidae) and other Diptera, *Bull. Entomol. Res.,* 72, 71, 1982.
21. **Vale, G. A.,** Field studies of the response of tsetse flies (Glossinidae) and other Diptera to carbon dioxide, acetone and chemicals, *Bull. Entomol. Res.,* 70, 563, 1980.
22. **Vale, G. A.,** The use of 1-octen-3-ol, acetone and carbon dioxide to improve baits for tsetse flies, *Glossina* spp. (Diptera: Glossinidae), *Bull. Entomol. Res.,* 75, 219, 1985.

23. **Owaga, M. L. A.,** Observations on the efficacy of buffalo urine as a potent olfactory attractant for *Glossina pallidipes* Austen, *Insect. Sci. Appl.*, 6, 561, 1985.
24. **Vale, G. A.,** The improvement of traps for tsetse flies (Diptera: Glossinidae), *Bull. Entomol. Res.*, 72, 95, 1982.
25. **Vale, G. A., Flint, S., and Hall, D. R.,** The field responses of tsetse flies, *Glossina* spp. (Diptera: Glossinidae), to odours of host residues, *Bull. Entomol. Res.*, 76, 685, 1986.
26. **Shereni, W.,** The use of cloth screens and acetone vapour as alternatives to a bait ox for sampling populations of tsetse flies (Diptera: Glossinidae), *Trans. Zimbabwe Sci. Assoc.*, 62, 22, 1984.
27. **Vale, G. A.,** unpublished data.
28. **Vale, G. A., Hargrove, J. W., Cockbill, G. F., and Phelps, R. J.,** Field trials of baits to control populations of *Glossina morsitans morsitans* Westwood and *G. pallidipes* Austen (Diptera: Glossinida), *Bull. Entomol. Res.*, 76, 179, 1986.
29. **Vale, G. A., Lovemore, D. F., Flint, S., and Cockbill, G. F.,** Odour baited targets to control tsetse flies, *Glossina* spp. (Diptera: Glossinidae), in Zimbabwe, *Bull. Entomol. Res.*, 78, 31, 1988.
30. **Willemse, L.,** unpublished.
31. **Slingenburg, J.,** unpublished.
32. **Politzar, H.,** unpublished.
33. **Dransfield, R., Brightwell, R., Golder, M. F., and Tarimo, S. A. R.,** The use of odour attractants for sampling *Glossina pallidipes* Austen (Diptera: Glossinidae) at Nguruman, Kenya, *Bull. Entomol. Res.*, 76, 607, 1986.
34. **Dolan, R.,** unpublished.
35. **Galley, J. B., Mérot, P., Mitteault, A., Filledier, J., and Politzar, H.,** Efficacité du dioxyde du carbone comme attractif pour *Glossina tachinoides* en savanne humide d'Afrique de l'Ouest, *Rev. Elev. Med. Vet. Pays Trop.*, 39, 351, 1986.
36. **Mérot, P., Galley, J. B., Politzar, H., Filledier, J., and Mitteault, A.,** Pouvoir attractif de l'odeur des hôtes nourriciers pour *Glossina tachinoides* en zone soudano-guinéene (Burkina Faso), *Rev. Elev. Med. Vet. Pays Trop.*, 39, 345, 1986.
37. **Bealby, K.,** unpublished.
38. **Takken, W.,** unpublished.
39. **Laveissière, C.,** unpublished.
40. **Thomson, M. C.,** The effect on tsetse flies (*Glossina* spp.) of deltamethrin applied to cattle either as a spray or incorporated into ear-tags, *Trop. Pest Manage.*, 33, 329, 1987.
41. **Wilson, A.,** unpublished.
42. **Brightwell, R., Dransfield, D., Kworku, C., Golder, T. K., Tarimo, S. A., and Mungai, D.,** A new trap for *Glossina pallidipes, Trop. Pest Manage.*, 33, 151, 1987.
43. **Challier, A. and Laveissière, C.,** Un nouveau piège pour la capture des glossines (*Glossina:* Diptera, Muscidae): description et essais sur le terrain, *Cah. O.R.S.T.O.M. Ser. Entomol. Med. Parasitol.*, 11, 251, 1973.
44. **Challier, A., Eyraud, M., Lafaye, A., and Laveissière, C.,** Amélioration du rendement du piège biconique pour glossines (Diptera: Glossinidae) par l'emploi d'un cône inférieur bleu, *Cah. O.R.S.T.O.M. Ser. Entomol. Med. Parasitol.*, 15, 283, 1977.
45. **Green, C. H. and Flint, S.,** An analysis of colour effects in the performance of F2 traps against *Glossina pallidipes* Austen and *G. morsitans morsitans* Westwood (Diptera; Glossinidae), *Bull. Entomol. Res.*, 76, 409, 1986.
46. **Gouteux, J. P., Challier, A., and Laveissière, C.,** Modifications et essais du piège à glossine (Diptera; Glossinidae) Challier-Laveissière, *Cah. O.R.S.T.O.M. Ser. Entomol. Med. Parasitol.*, 19, 87, 1981.
47. **Dransfield, R. D.,** The range of attraction of the biconical trap for *Glossina pallidipes* and *Glossina brevipalpis, Insect Sci. Appl.*, 5, 363, 1984.
48. **Owaga, M. L. A.,** Relative efficiency of some mechanical traps used in the study of the tsetse species, *Glossina pallidipes* Austen, *Insect Sci. Appl.*, 1, 197, 1980.
49. **Flint, S.,** A comparison of various traps for *Glossina* spp. (Glossinidae) and other Diptera, *Bull. Entomol. Res.*, 75, 529, 1985.
50. **Takken, W.,** Studies on the biconical trap as a sampling device for tsetse (Diptera; Glossinae) in Mozambique, *Insect Sci. Appl.*, 5, 357, 1984.
51. **Turner, D. A.,** Tsetse ecological studies in Niger and Mozambique. I. Population sampling, *Insect Sci. Appl.*, 1, 9, 1980.
52. **Ryan, L., Molyneux, D. H., and Kuzoe, F. A. S.,** Traps to control and estimate populations of *Glossina* species, *Trop. Med. Parasitol.*, 32, 145, 1981.
53. **Koch, K. and Spielberger, U.,** Comparison of handnets, biconical traps and an electric trap for sampling *Glossina palpalis palpalis* (Robineau-Desvoidy) and *G. tachinoides* Westwood (Diptera: Glossinidae) in Nigeria, *Bull. Entomol. Res.*, 69, 243, 1979.

54. **Küpper, W., Manno, A., Douati, S., and Koulibali, S.**, Impact des pièges biconiques imprégnés sur les populations de *Glossina palpalis gambiensis* et *Glossina tachinoides*, Résultat d'une campagne de lutte à grande échelle contre la trypanosomiase animale au nord de la Côte d'Ivoire, *Rev. Elev. Med. Vet. Pays Trop.*, 37, 176, 1984.
55. **Laveissière, C., Couret, D., and Kienou, J. P.**, Lutte contre les glossines riveraines à l'aide de pièges biconiques imprégnés d'insecticide, en zone de savane humide. IV. Expérimentation à grande échelle, *Cah. O.R.S.T.O.M. Ser. Entomol. Med. Parasitol.*, 19, 49, 1981.
56. **Laveissière, C., Couret, D., and Kienou, J. P.**, Lutte contre les glossines riveraines à l'aide des pièges biconiques imprégnés d'insecticide, en zone de savane humide. III. Résultats quantitatifs obtenus lors des premiers essais, *Cah. O.R.S.T.O.M. Ser. Entomol. Med. Parasitol.*, 18, 209, 1980.
57. **Tibayrenc, R. and Gruvel, J.**, La campagne de lutte contre les glossines dans le bassin du lac Tchad. II. Contrôle de l'assainissement glossinaire. Critique technique et financière de l'ensemble de la campagne. Conclusions générales, *Rev. Elev. Med. Vet. Pays Trop.*, 27, 81, 1977.
58. **Küpper, W.**, personal communication, 1987.
59. **Laveissière, C. and Couret, D.**, Essai de lutte contre les glossines riveraines à l'aide d'écrans imprégnés d'insecticide. *Cah. O.R.S.T.O.M. Ser. Entomol. Med. Parasitol.*, 14, 271, 1981.
60. **Laveissière, C. and Couret, D.**, Effet comparé des écrans et des pièges biconiques imprégnés d'insecticide sur les populations de *Glossina morsitans submorsitans* dans les galeries forestières, *Cah. O.R.S.T.O.M. Ser. Entomol. Med. Parasitol.*, 120, 63, 1982.
61. **Cuisance, D., Merot, P., Politzar, H., and Tamboura, I.**, Coût, de l'emploi d'écrans insecticides dans la lutte intégrée contre les glossines dans la zone pastorale de Sidéradougou, Burkina, *Rev. Elev. Med. Vet. Pays Trop.*, 37, 84, 1984.
62. **Cuisance, D. and Politzar, H.**, Etude sur l'efficacité contre *Glossina palpalis gambiensis* et *Glossina tachinoides* de barrières constituées d'écrans ou de pièges biconiques imprégnés de DDT, de deltaméthrine ou de dieldrine, *Rev. Elev. Med. Vet. Pays Trop.*, 36, 159, 1983.
63. **Politzar, H. and Cuisance, D.**, A trap-barrier to block re-invasion of a river system by riverine tsetse species, *Rev. Elev. Med. Vet. Pays Trop.*, 36, 364, 1983.
64. **Takken, W., Oladumade, M. A., Dengwat, L., Feldmann, H. U., Onah, J. A., Tenabe, S. O., and Hamann, H. J.**, The eradication of *Glossina palpalis* (Robineau-Desvoidy) (Diptera: Glossinidae) using traps, insecticide-impregnated targets and the sterile insect technique in Central Nigeria, *Bull. Entomol. Res.*, 76, 275, 1986.
65. **Sékétéli, A. Johannes, L., Van de Laar, M., and Kuzoe, F. A. S.**, Essais d'épandage au sol de la deltaméthrine poudre mouillable à différentes doses contre *Glossina palpalis (s.l.)* dans une zone pré-forestière de Côte d'Ivoire, *Insect Sci. Appl.*, 6, 187, 1985.
66. **Laveissière, C., Gouteux, J. P., and Couret, D.**, Essais de méthodes de lutte contre les glossines en zone préforestière de Côte d'Ivoire. III. Résultats qualitatifs obtenus sur les populations de *Glossina palpalis s.l.*, IV. Résultats quantitatifs obtenus sur les populations de *Glossina pallicera pallicera* et de *Glossina nigrofusca nigrofusca;* V. Note de synthèse, *Cah. O.R.S.T.O.M. Ser. Entomol. Med. Parasitol.*, 18, 307, 1980.
67. **Laveissière, C., Couret, D., and Eouzan, J. P.**, La campagne pilote de lutte contre la trypanosomiase humaine dans le foyer de Vavoua (Côte d'Ivoire). III. Résultats des évaluations entomologiques, *Cah. O.R.S.T.O.M. Ser. Entomol. Med. Parasitol.*, 24, 7, 1986.
68. **Hervouet, J. P.**, Systèmes agraires et trypanosomiase en Côte d'Ivoire forestière, in preparation, 1987.
69. **Laveissière, C., Hervouet, J. P., Couret, D., Eouzan, J. P., and Merouze, F.**, La campagne pilote de lutte contre la trypanosomiase humaine dans le foyer de Vavoua (Côte d'Ivoire). II. La mobilisation des communautés rurales et l'application du piégeage, *Cah. O.R.S.T.O.M. Ser. Entomol. Med. Parasitol.*, 23, 167, 1985.
70. **Laveissière, C., Couret, D., and Manno, A.**, Tests d'efficacité et de rémanence d'insecticides utilisés en imprégnation sur tissus pour la lutte par piégeage contre les glossines. III. Deuxième série: nouveaux composés, nouvelles formulations, additifs, effet de la dose, *Cah. O.R.S.T.O.M. Ser. Entomol. Med Parasitol.*, 24, 139, 1986.
71. **Gouteux, J. P. and Buckland, S. T.**, Ecologie des glossines en secteur pré-forestier de Côte d'Ivoire. VIII. Dynamique des populations, *Cah. O.R.S.T.O.M. Ser. Entomol. Med. Parasitol.*, 22, 19, 1984.
72. **Laveissière, C., Hervouet, J. P., Mérouze, F., and Cattand, P.**, La campagne pilote de lutte contre la trypanosomiase humaine dans le foyer de Vavoua (Côte d'Ivoire). IV. Bilan de la campagne: les prospections médicales et la participation de la population, *Cah. O.R.S.T.O.M. Ser. Entomol. Med. Parasitol.*, 24, 111, 1986.
73. **Laveissière, C. and Couret, D.**, La campagne pilote de lutte contre la trypanosomiase humaine dans le foyer de Vavoua (Côte d'Ivoire). V. Bilan financier, *Cah. O.R.S.T.O.M. Ser. Entomol. Med. Parasitol.*, 24, 149, 1986.
74. **Laveissière, C., Couret, D., and Traoré, T.**, Tests d'efficacité et de rémanence d'insecticides utilisés en imprégnation sur tissus pour la lutte par piégeage contre les glossines. I. Protocole experimental. L'effet "knock down" des pyréthrinoïdes, *Cah. O.R.S.T.O.M. Ser. Entomol. Med. Parasitol.*, 23, 61, 1985.

75. **Laveissière, C. and Couret, D.,** Observations sur l'effet irritant des pyréthrinoïdes de synthèse pour les glossines, *Cah. O.R.S.T.O.M. Ser. Entomol. Med. Parasitol.*, 23, 289, 1985.
76. **Laveissière, C., Couret, D., Manno, A., and Küpper, W.,** Tests d'efficacité et de rémanence d'insecticides utilisés en imprégnation sur tissus pour la lutte par piégeage contre les glossines. II. Première série de tests en saison humide et en saison sèche, *Cah. O.R.S.T.O.M. Ser. Entomol. Med. Parasitol.*, 23, 217, 1985.
77. **Laveissière, C. and Couret, D.,** Dieldrine et écrans pour la lutte contre les glossines riveraines, *Cah. O.R.S.T.O.M. Ser. Entomol. Med. Parasitol.*, 21, 57, 1983.
78. **Cuisance, D., Février, J., Dejardin, J., and Filledier, J.,** Dispersion linéaire de *Glossina palpalis gambiensis* et de *Glossina tachinoides* dans une galerie forestière en zone soudano-guinéenne (Burkina-Faso), *Rev. Elev. Med. Vet. Pays Trop.*, 38, 153, 1985.
79. **Laveissière, C. and Couret, D.,** Conséquences d'essais de lutte répétés sur les proportions de glossines riveraines, *Cah. O.R.S.T.O.M. Ser. Entomol. Med. Parasitol.*, Gustav Fischer, 21, 63, 1983.
80. **Zumpt, F.,** *Die Tsetsefliegen,* Fischer, G., Ed., Jena, 1936.
81. **Challier, A.,** Perspectives d'utilisation des systèmes attractifs toxiques dans la lutte contre les glossines (Diptera, Glossinidae), *Rev. Elev. Med. Vet. Pays Trop.*, 37, 31, 1984.
82. **Laveissière, C., Couret, D., and Grebaut, P.,** Recherche sur les écrans pour la lutte contre les glossines en région forestière de Côte d'Ivoire. Mise au point d'un nouvel écran, *Cah. O.R.S.T.O.M. Ser. Entomol. Med. Parasitol.*, 25, 145, 1987.
83. **Green, C. H.,** The effect of colour on trap and screen orientated responses of *Glossina palpalis palpalis* Robineau-Desvoidy (Diptera: Glossinidae) to traps and screens, *Bull. Entomol. Res.*, 78, 591, 1988.
84. **Laveissière, C., Couret, D., and Manno, A.,** Importance de la nature des tissus dans la lutte contre les glossines, *Cah. O.R.S.T.O.M. Ser. Entomol. Med. Parasitol.*, 25, 133, 1987.
85. **Gouteux, J. P. and Lancien, J.,** Le piège pyramidal à tsétsé (Diptera: Glossinidae) pour la capture et la lutte. Essais comparatifs et description de nouveaux systèmes de capture, *Trop. Med. Parasitol.*, 37, 61, 1986.
86. **Gouteux, J. P., Lancien, J., Noireau, F., and Sinda, D.,** Lutte antivectorielle par piégeage et impact sur la transmission de la maladie du sommeil dans une zone à forte densité de *Glossina fuscipes quanzensis*, *Trop. Med. Parasitol.*, 37, 101, 1986.
87. **Gouteux, J. P., Bansimba, P., Bissadidi, N., and Noireau, F.,** La prise en charge de la lutte contre les tsétsés par les communautés rurales: premiers essais dans cinq villages congolais, *Ann. Soc. Belge. Med. Trop.*, 67, 37, 1987.
88. **Dagnogo, M. and Gouteux, J. P.,** Comparaison de différents pièges à tsétsé (Diptera, Glossinidae) en Côte d'Ivoire et au Congo, *Rev. Elev. Med. Vet. Pays Trop.*, 38, 371, 1985.
89. **Gouteux, J. P., Noireau, F., Sinda, D., and Frézil, J. L.,** Essais du piège pyramidal contre *Glossina palpalis palpalis* (Rob.-Desv.) dans le foyer du Niari, *Cah. O.R.S.T.O.M. Ser. Entomol. Med. Parasitol.*, 241, 181, 1986.
90. **Duteurtre, P. and Gouteux, J. P.,** Les stratégies de lutte antisommeilleuse en République Populaire du Congo. Recherches et perspectives, *Med. Trop. (Marseille)*, 46, 375, 1986.
91. **Douthwaite, R. J.,** Effects of drift sprays of endosulfan, applied for tsetse-fly control on breeding little bee-eaters in Somalia, *Environ. Pollut.*, 41, 11, 1986.
92. **Phelps, R. J., Focardi, S., Fossi, C., Leonzio, C., and Renzoni, A.,** Chlorinated hydrocarbons and heavy metals in crocodile eggs from Zimbabwe, *Trans. Zimbabwe Sci. Assoc.*, 63, 8, 1986.
93. **Rogers, D. and Randolph, S. E.,** Population ecology of tsetse, *Annu. Rev. Entomol.*, 30, 197, 1985.
94. **Molyneux, D. H., Baldry, D. A. T., de Raadt, P., Lee, C. W., and Hamon, J.,** Helicopter application of insecticides for the control of riverine *Glossina* vectors of African human trypanosomiasis in the moist savannah zones, *Ann. Soc. Belge Med. Trop.*, 58, 185, 1978.

Chapter 4

NATURAL AND SYNTHETIC REPELLENTS

C. F. Curtis, J. D. Lines, Lu Baolin, and A. Renz

TABLE OF CONTENTS

I. NATURAL REPELLENTS

A. INTRODUCTION*

Since long before the advent of synthetic chemicals, people have been using plants and plant-derived substances to try to repel or kill mosquitoes and other domestic pest insects. The Ainu people[1] of Hokkaido, Japan, and the Micmac Indians[2] of Newfoundland both wore leggings of sedge, bark, or cloth to reduce insect biting nuisance which tends to be concentrated around the lower legs. The ancient Chinese classics contain many prescriptions for repellents against mosquitoes and other blood-sucking flies. According to Roark,[3] nearly 1200 plants have been listed in the literature as of potential insecticidal value. Most remain to be investigated either chemically or biologically. The work which has been done tends to fall into two categories: the analytic and the ethnobotanical.

As with other natural "medicines" the interest of chemists has been mainly analytic, centered on the hope of finding new repellent or insecticidal compounds which could be extracted or synthesized for commercial purposes. The best known examples of botanical insecticides are the plants *Derris, Tephrosia, Chrysanthemum,* and *Nicotiana,* which have given us the effective insecticidal compounds rotenone, pyrethrum, and nicotine. These compounds were once the basis for a major part of the pesticide industry. With the exception of pyrethrum, however, they never found wide application against insect vectors, largely because of their toxicity to nontarget organisms such as fish. They were largely superseded when synthetic insecticides arrived on the scene. The extensive literature on the biochemistry of botanical insecticides and repellents has been reviewed by Roark[3] and Jacobson and Crosby.[4]

Other workers have concentrated on ethnobotanical surveys, which document the use of plants by different cultures. A long list of traditional uses of plants against pests is given by Secoy and Smith,[5] and a data base on "plant species reportedly possessing pest-control properties" is also available.[6] Surprisingly, little attention has been given to the question of whether traditional preparations do, in fact, work when used in the traditional way, and hence to the possibility that they may show us how to use plants directly, without the need for petroleum products or industrial processes. By using home-grown repellents one could avoid the problems of shortage of foreign exchange and administrative failures which often bedevil conventional vector control programs.

B. PYRETHRUM*

Pyrethrum flowers contain a mixture of insecticidal substances called pyrethrins, which provided the model for synthetic pyrethroid insecticides. The biochemistry of the pyrethroids are discussed by Matsui and Yamamoto.[7] Unlike some other botanical insecticides, natural pyrethrins have a very low mammalian toxicity. They are highly photolabile and have very little residual effect. For this reason, pyrethrum has been mainly used against adult mosquitoes as an indoor fumigant in the form of smoke or sprays and against mosquito larvae as a nonresidual application with fuel oil.[8] Powder formulations were once common both in domestic and agricultural use, but synthetic chemicals have proved far more effective for such purposes.

Pyrethrins are normally extracted from the flowers with a solvent yielding a "pyrethrum concentrate" containing 20 to 25% of the active ingredient. Such extracts, dissolved in kerosene or a similar solvent (0.1 to 0.2% of pyrethrins), are the commonest ingredient of "flit" sprays, for the rapid knockdown of mosquitoes and other flying insects indoors.

* J. D. Lines

Sometimes a synergist, which has no insecticidal effect alone but makes the insecticide more effective, is added to the spray. The most common synergist is piperonyl butoxide, but sesame oil has been found to be a suitable substitute.[3]

Such sprays are useful for clearing screened rooms of mosquitoes in the evening, but the spraying has to be done every day. For this reason, this type of spraying is more suitable for personal or household protection than for controlling the mosquito population of an area, and good screening is essential if mosquitoes are not to reinfest the room as soon as the aerosol settles out of the atmosphere. Concerted daily spraying with pyrethrum in all the houses of a community, to kill mosquitoes resting indoors during the day, has been tried as a means of malaria control, but with mixed results. Whatever the circumstances, there seems little doubt that conventional residual insecticides, which can be sprayed once every few months instead of daily, would always work out cheaper and more effective for this purpose.[9-11]

For the mosquito entomologist, flit spraying remains a standard, quick, and easy method of sampling the indoor resting population, and in this case, the lack of residual effect can be an advantage. Details of the technique and its pros and cons can be found in Service.[12]

Like DDT and the synthetic pyrethroids, pyrethrum is an excito-repellent insecticide. This property has led to its combination with synthetic residual insecticides against bedbugs and cockroaches; the pyrethrum helps to flush the insects from their hiding places and, hence, to pick up a lethal dose of the synthetic insecticide.[8] Attempts to use pyrethrum in repellent preparations on the skin have been generally unsuccessful when it was formulated in an old-fashioned paraffin formulation or in a modern aerosol or mousse.[13] However, a formulation with gum tragacanth gave moderate repellency for up to 7 h.[14] Mosquitoes not repelled by pyrethrum on the skin tend to die some time after biting.[13]

Pyrethrum flowers form the traditional basis of mosquito coils. These probably originated as joss sticks in China, where smoking coils made of sawdust and sulfur are still employed as repellents.[15,16] People all over the world now use pyrethrum coils for protection against mosquito biting indoors. A major advantage is that the room does not necessarily have to be well screened, although it may help if it is. Coils tend to burn for 6 to 8 h, so if mosquito biting continues until the early morning, one can be left dangerously unprotected at this time. Use of a coil in the evening before going to bed, and then sleeping under a bednet, is much better.

Pyrethrum flowers have been grown commercially all over the world. Yugoslavia, Japan, the Middle East, Ecuador, and the plateaus of East Africa have all been major producers in the past. Mosquito coils are not usually manufactured as a cottage industry, but the equipment and processes required are not technically sophisticated. Basically, dried pyrethrum flowers, or pyrethrum extract, and jute or a similar filling are pressed together, sometimes including other agents to keep the rate of smoldering steady. That mosquito coils work is well attested by studies in Tanzania and Papua New Guinea, which have given measures of protection which ranged from approximately 40 to 80% reduction in biting.[17-19] In tests with *Anopheles gambiae* released into a 13-m^3 room, a coil containing 0.5% natural pyrethrins reduced the landing rate on human subjects by about 40%. Knockdown began 2 min after release of the mosquitoes, but a few mosquitoes managed to take some blood before this time and some recovered later.[20]

Better results with electrically heated vaporizing mats containing synthetic pyrethroids are described in Section III.

C. NATURAL PRODUCTS AGAINST ECTOPARASITES*

In the list compiled by Secoy and Smith[5] of 677 plants used against insect pests, 139 were applied against ectoparasites, as compared with only 50 as mosquito repellents.

* J. D. Lines

Traditional remedies for head lice exist in many cultures, and vestiges of these practices can still be seen. Pyrethrum powder, known as Persian powder, used to be imported for this purpose into Western Europe and may have been the origin of the practice of dusting hair and wigs with powdered chalk, to camouflage the insecticide. The use of hair oils, particularly neem *(Azidirachta indica)* and coconut oil (''macassar'') probably have a similar origin. These remedies may have some value, perhaps as repellents, but people wearing a lot of neem oil have been seen with heavy louse infestations.[21] The European plant fleabane *(Pulicaria dysenterica)* got its name for its use against lice and fleas, and derris has also been used against fleas. This must have been a dangerous practice as the active ingredient of derris, rotenone, is much more toxic to humans than most modern synthetic insecticides. This emphasizes the fallacy of the widespread belief that any natural product must naturally be safer than any synthetic.

D. VITAMIN B_1*

Some agricultural insecticides can be used systemically, i.e., they can be effective after absorption by plants. The only comparable claim with respect to biting insects concerns vitamin B_1, or thiamin, derived from yeast extract, in relation to mosquitoes and fleas.[22-25] Success in these cases was measured either by clinical improvement in lesions associated with bites or by the subjective impressions of the amount of biting. Whenever properly controlled comparisons have been made, counting bites as they occcurred, no effect of taking the vitamin has been demonstrated in either the field or the laboratory.[26-29] Neither was any effect noted on the amount of itching after being bitten.[13]

E. ESSENTIAL OILS**

The essential oils of a multitude of plants have been shown to have mosquito-repellent properties. Between 1935 and 1945, the *Review of Applied Entomology* noted many reports of this kind, in which repellency was ascribed to the oils of cassia, camphor, citronella, lemongrass, clove, thyme, geranium, bergamot, bay laurel, pine, wintergreen, pennyroyal, and eucalyptus. No doubt many others could be added to this list. These oils, and mixtures of them, were the basis of most commercial repellents, and many different varieties were produced and tested[30] until synthetic compounds such as diethyl toluamide (deet) took over most of the market.

The repellent properties of plants have usually been investigated by fractionation of the essential oil, followed by testing for repellency in the individual fractions. Entirely unrelated plants often share the same repellent constituents: eucalyptol, *d*-pinene, geraniol, citronellal, camphor, and linalool are just some examples of natural repellent substances which have been isolated from a wide range of plant species. Such substances are often active — as repellents or attractants — for nonbiting insects,[31] and many are insecticidal at higher concentrations. They are thought to be part of the plant's chemical defense system against herbivorous insects.

1. Citronella

Citronella oil is a popular repellent in India and is one of the few natural products which is still used in commercial repellents in Europe and North America. It has a more pleasant smell than most synthetics. A preparation made in Ghana and based on citronella has been found to be very effective against *Simulium damnosum s.l.*[22] Counts of the numbers of bites by anopheline mosquitoes with graded series of concentrations of repellents on human skin

* C. F. Curtis

** Lu Baolin and J. D. Lines

TABLE 1
Duration of Protection against Biting *Aedes aegypti* after Application to the Skin of the Principal Ingredients of Lemon Eucalyptus Oil[38]

Ingredient	Duration of protection (h)
Lemon eucalyptus oil	1
Citronellal	1
Geraniol	1
Isopulegel	0
d-Pinene	0
Sesquiterpene	0

showed that the dose for 50 or 90% repellency was much less for citronella than cedarwood oil.[33] By the same criterion citronella was about as effective, when freshly applied, as synthetics such as dimethyl phthalate (DMP), but it did not last as long. In an assessment of 12 repellent formulations on the European market, those with citronella as the active ingredient rated considerably worse than those with synthetic active ingredients.[34]

The main constituent of citronella oil is citronellal, a monoterpene aldehyde. A synthetic derivative of citronellal has been used as the active ingredient of a commercial repellent, a fresh application of which was found by Buescher et al.[35] and Rutledge et al.[36] to be about as effective as deet against sand flies and mosquitoes.

2. *Artemesia*

In China the burning of certain herbs, such as *Artemesia* and *Calamus* species, is still practiced in remote villages to banish mosquitoes and to protect cattle from blood-sucking insects. That this practice has some value is suggested by the results of Hwang et al.[37] Column chromatography of the essential oil of *Artemesia vulgaris* yielded several fractions active as repellents against *Aedes aegypti*. These included linalool, camphor, borneol, and terpinen-4-ol, of which the latter was the most active and was as effective as DMP. There does not seem to have been any investigation of whether burning *Artemesia* in the traditional manner is effective.

3. Lemon Eucalyptus

Since 1960, thousands of plants have been screened in China for repellency, and several materials have been reported to have good repellent properties against biting mosquitoes.

It has long been known that oil of the lemon eucalyptus plant *(Eucalyptus maculata citriodon)* has some repellent effect on mosquitoes. Chemical studies of this oil showed that its principal ingredients consisted of citronella, citronellol, geraniol, isopulegol, delta-pinene, and sesquiterpene. However, the results of laboratory assays on these ingredients for repellency to *Aedes aegypti* were disappointing as shown in Table 1.

However, an effective material was finally found in the waste distillate after the extraction of lemon eucalyptus oil, and it has been named *quwenling*, which means "effective repeller of mosquitoes". The chief ingredient of this material was found to be *p*-menthene-diol-3,8. It is a white or yellowish waxy material usually sold in 30 or 40% alcoholic solution, which is less oily than other conventional repellents and does not damage plastic materials, unlike deet. Its mammalian toxicity is lower than that of deet, with mouse oral and dermal LD_{50} values of 3200 mg/kg and 12,000 mg/kg, respectively.

In laboratory tests with mice *quwenling* was assessed for its "protection time", as defined by Schreck,[39] i.e., the time interval between application and the first confirmed mosquito bite. The protection time with *quwenling* was longer than with the same dose of

TABLE 2
Protection Time for Mice of *quwenling* and Deet against *Aedes aegypti* Biting[38]

Mosquito species	Protection time (h)	
	quwenling	Deet
Aedes aegypti	8.2	5.2
Ae. albopictus	12.5	8.3

Note: A 1.5 mg/cm² dosage of a 50% formulation of each repellent was used.

TABLE 3
Protection Time for Mice of Different Concentrations of *quwenling* against *Aedes aegypti*[38]

Conc. (%)	Dosage (mg/cm²)	No. tests	Average protection time (h)
15	0.26	17	4.4
20	0.35	18	5.3
30	0.53	18	9.4
50	0.88	12	13.0
95	1.77	12	22.0

TABLE 4
Field Tests with Human Subjects of *quwenling* against Biting by *Aedes* Species[38]

Mosquito species tested	Formulation	Dosage applied to face, neck, and hands (ml)	No. of tests	Average protection time (h)
Aedes vexans	30% in glycerol-alcohol	2—3	280	2.3
	30% in alcohol	2—3	254	2.7
Ae. albopictus	20% in alcohol	1—1.5	502	3.3
	30% in alcohol	1—1.5	357	5.5
	30% in glycerol-alcohol	1—1.5	686	5.7
	30% emulsion	1—1.5	706	5.9

deet (Table 2). As might be expected, the protection time was proportional to the concentration of the preparation (Table 3).

Field tests in northeast China against *Aedes vexans* and in south China, mainly against *Ae. albopictus*, showed protection times of 2 and 5.5 h, respectively (Table 4). *Quwenling* has a protective effect not only against mosquitoes, biting midges, and tabanids, but also against land leeches.

Cotton nets impregnated with *quwenling* are effective for 8.5 d on average (4- to 6- h usage daily with storage in a plastic bag between tests). This is slightly less than that of deet impregnation, which lasted 9.7 d on average.

Quwenling has been so successful that it has largely displaced the synthetic repellent DMP from the Chinese market. Recently, tests have been made with a formulation of *quwenling* containing a fixative which increases its persistence. This is not yet on the market.

4. Mint

An extract labeled 1247 has been made in China from wild mint *(Mentha haplocalyx)* and consists of *d*-8-aceto-oxycarbotanacetone.[40,41] In laboratory tests, its alcoholic solution was reported to have a protection time of 6 to 12 h against *Ae. albopictus*. Field tests in south China, mainly against *Ae. albopictus* and *Culex quinquefasciatus,* and in northeast China, mainly against *Ae. vexans,* showed protection times of 4 to 5 h and 1 h, respectively. Comparative laboratory tests with *Ae. aegypti* showed that it was slightly inferior to deet.[42] 1247 has not been produced on a commercial scale.

5. Oil of *Clausena*

A derivative designated 9525 has been made from the oil extracted from leaves and stems of the plant *Clausena kwangsiensis* which grows in Guangxi Province, China. Cage studies of a 50% solution of this material in alcohol, rubbed on the back of the hand at a dosage of approximately 0.001 mg/cm^2, showed a protection time of 10 h against *Ae. albopictus*. Field tests in Pingriang, Guangxi Province, where the dominant mosquito species was *Culex quinquefasciatus,* showed that it had a protection rate of 80.5% among 205 persons 10 h after application. This unusually high repellent effect was probably due to the fact that only a few mosquitoes were present during the tests. By contrast during the field trials in northeast China, where the dominant mosquito species was *Ae. vexans,* an 80% solution of this material gave a protection rate of only 88.7% among 559 persons 1.5 h after application.[43] The effective ingredient of 9525 is an isomer of *p*-menthene-diol-3,8. 9525 has not been produced on a commercial scale.

F. TRADITIONAL USE OF REPELLENTS*

1. Tests of Traditional Repellents

Use of traditional repellents is not only widespread among different cultures, but also may be very common within communities. Herbs of the basil family (Labiatae) have many traditional medicinal uses in Africa and Asia. In east and west Africa they are also used as mosquito repellents.[44,45] In a village in northern Tanzania, Fivawo[46] found that more than two thirds of women reported using some kind of natural insect repellent. Most used one or the other of these basil-like herbs (*Ocimum* spp. and *Hyptis soaveolens*), while a few used leaves of the neem tree *Azidirachta indica*.

White[47] showed that, when smeared on the legs, the juice of one species of *Ocimum* reduced biting by caged *Anopheles gambiae*. The repellency was later attributed to constituent eugenols,[48] which are major components of clove and other essential oils reputed to have repellent properties.

Smearing the juice on the skin is not the traditional way of using these plants. One such was is to bruise freshly picked sprigs and hang them in the house. When *H. soaveolens* (local name *hangazimu*) was hung in this way in an experimental hut, there was no significant reduction in biting.[49] *Hangazimu* is also sometimes burned on the evening fire, as occasionally are neem leaves.

On a series of evenings, counts were made of the numbers of *Cx. quinquefasciatus* biting two people inside a hut with a charcoal fire.[50] Three series of experiments indicated up to tenfold reduction in biting in the presence of *hangazimu* smoke, and lesser reductions when other plant material, such as grass, coconut husks, or neem was burned. However, a fourth series some months later showed no significant reduction due to the *hangazimu*. A possible explanation of this disappointing result is that this experiment was made at a season at which the plants had become senescent and lost much of their active ingredient, as happens with senescent *Artemesia*.[51]

Tests of *hangazimu* smoke with free flying mosquitoes in a laboratory room showed little reduction in biting,[13] but these exeperiments could only test biting inhibition in the

* J. D. Lines

vicinity of the host and could not test ability of the smoke to drive the mosquitoes away.

In The Gambia the woods and resins of aromatic trees are widely sold in the markets, under the collective term *churai,* to be burned as mosquito repellents. However, children in households using *churai* were shown to suffer from as much malaria as those in neighboring households which do not use it.[52] This does not prove that *churai* has no effect on mosquito biting — untreated mosquito nets can be assumed to cause some reduction in biting. However, as shown in Chapter 2, they were insufficient to have any appreciable impact on malaria in Gambian children in a trial conducted by the same methods as the *churai* trial. In trials with free flying *Anopheles gambiae* in a room, no reduction in biting due to the smoke of various types of *churai* could be detected,[13] but as pointed out above, these trials only test the mosquito's behavior when it is placed in the immediate vicinity of the host.

In the U.S.S.R., Rubtzov[53] tested a folk remedy against biting insects, consisting of thyme *(Thymus serpyllum)* sticks tied with thick cotton and dried in the air for a week. These were set smoldering and were reported to give 85 to 90% protection for 60 to 90 min in the open air.

That traditional practices and repellents can give real protection against biting, and even malaria, is indicated by the observation of Philip et al. in parts of India.[54] They associated lower spleen indices among women with the practice of smearing their bodies with turmeric and gingili or mustard oil before bathing without soap. Men, who do not do this, were observed being more frequently bitten by *An. fluviatilis* than were women sleeping nearby. A hand anointed in this way was not bitten by caged mosquitoes which fed avidly on an untreated hand.

2. The Need for Further Assessment of the Effectiveness of Traditional Repellents

Scientists working on plant repellents generally seem to have been searching for something which can be extracted (or synthesized) in a standard form, put in a bottle, and sold. As the examples above illustrate, active principles have been isolated from many plants traditionally used as repellents. However, relatively little seems to be known about the effectiveness of the traditional ways of using these plants.

Is there any need to know whether such practices work? We consider that there is. Such knowledge is desirable because these are the methods that people use without persuasion from scientific experts. A method which is freely available and of small benefit may be more useful than one which is more effective but unaffordable. Furthermore, an expert's advice about a new method will be far more convincing if he can say that his tests show that it is *x* times more effective than a traditional method with which his audience is familiar.

Where possible, standardized laboratory tests of traditional methods are desirable, but they may miss out important features of the field situation. In the field, on the other hand, there may be wide variations over both time and space in the biting densities of natural insect populations, seasonal changes in the plant, and unpredictable weather. However, these difficulties should not be insurmountable; the solution may often be no more — or less — than greatly increased replication. It is not altogether necessary to understand all the complex and interacting variables controlling the attraction of biting insects to hosts: an effective repellent should reduce biting for everyone in a range of conditions.

For instance, in many parts of Africa, just sitting under a neem tree or planting one outside the house is believed by some to give partial protection against biting. It is known that neem contains substances active against insects (see Chapter 8), but is it possible that these substances could act at a distance in this way? Testing this question is clearly not just a matter of comparing biting under a particular tree with that at another spot some distance away. There are far too many factors other than the tree which may be expected to influence the relative amount of biting at any two points in an area. However, it should be possible to test the question by replicating the paired comparison with different trees, each time measuring biting both under a tree and away from it. Proximity to people, other vegetation, breeding sites, and orientation to wind may all be expected to affect the relative numbers biting on any occasion, but over many occasions their influence should even out when the two positions, under or away from the tree, are compared. A weak trend due to the tree would be difficult to discern without many replicates, but if neem trees are as good as their reputation, a strong trend would be expected, and this should emerge reasonably quickly. Such studies will be necessary if informed comparisons are to be made between modern and traditional repellent methods.

II. SYNTHETIC REPELLENTS

A. COMPARISON OF DIFFERENT COMPOUNDS AND FORMULATIONS*

1. Liquid Formulations of Deet

Deet was introduced in the 1950s and shown to be a more effective repellent than DMP and ethyl hexanediol which were then the standard synthetic repellents.[55] These tests showed that the *meta* isomer of deet was more effective than the *ortho* isomer, and it is the *meta* isomer of deet which is the active ingredient of most commercially available repellents, though DMP and ethyl hexanediol are still available in some preparations.

Most tests of repellents have reported the "protection time" (as defined in Section I.E.3 above) against *Aedes aegypti*. This is a convenient measure for screening purposes but has been criticized[33,56] on the following grounds:

1. The results depend on the behavior of the few most tolerant individuals rather than the central tendency of the population.
2. It confounds differences in repellency per unit concentration applied with the rate of loss of repellent with time.
3. It does not allow for the fact that a mosquito responds more strongly to a given concentration of repellent if it has a choice of a lower or zero concentration than if it has no choice.[33]
4. Different species of blood-sucking insects have different responses to repellents.

Methods have, therefore, been developed for observing the response of insects with a graded series of dosages of repellent and for computing ED_{50} and ED_{90} values with confidence limits.[33,35] The procedure may be repeated at intervals after application of the repellent, and three-dimensional graphs of repellency, dose applied, and time after application may be constructed.[56]

Important vector species such as *Anopheles pulcherrimus*,[57] *An. albimanus*,[33,58] *An. gambiae*,[33] and *Glossina morsitans*[59] are less susceptible to deet than is *Aedes aegypti*. Nevertheless, deet is effective enough to be of practical use against all these vector species as well as against *Simulium* blackflies,[59,60] *Lutzomyia* sand flies,[35] and ticks (see Chapter 5). It is also useful against the following species which are not known to be vectors of human disease but are serious nuisances in some places: *Culicoides*,[61] bedbugs,[21] sweat bees (*Lasioglossum* spp.),[62] and leeches.[63]

Deet is available in various formulations, and the efficacies of some of these have recently been compared in China.[64] In laboratory tests against *Anopheles anthropophagus* (= *An. lesteri anthropophagus*) and *Aedes albopictus*, a protection time of up to 12 h was achieved, but the protection time varied for different formulations (Table 5).[64] However, in field tests with large numbers of people the duration of effective protection was less than in the laboratory tests. Protection times were particularly short against high densities of *Aedes* species in northeast China, but in these tests a dosage of only 1 to 1.5 ml per person was used (Table 6).

* C. F. Curtis and Lu Baolin

TABLE 5
Repellent Effect of Four Formulations of Deet in Laboratory Tests[64]

		Average protection time (h)[a]	
Formulations	No. of tests	*Aedes albopictus*	*Anopheles anthropophagus*
Undiluted	12	12.5	7.6
30% EC[b]	18	12.0	No data
30% Paste	12	9.3	6.5
30% Alc. solution	12	6.6	6.7

[a] Dosage of 0.8 mg/cm² was applied in each case.
[b] Emulsifiable concentrate.

TABLE 6
Results of Field Tests on Repellent Effect of Deet in Different Localities in China[64]

Locality	Dominant mosquitoes	Biting rate/min without repellent	Formulation	Dosage applied to face, neck and hands (ml)	No. of tests	Av. duration of effective protection (h)
Guangxi	*Culex quinquefasciatus* and *Cx. tritaeniorhynchus*	10—20	Undiluted	2—3	1996	5—6
			30% EC	2—3	3696	3—4
			30% Paste	2—3	3493	3.5—4.5
			30% Alc. solution	2—3	553	2.5—3.5
Yunnan	*Anopheles sinensis* and *Cx. quinquefasciatus*	8	Undiluted	2—3	336	4
			30% EC	2—3	693	4
			30% Paste	2—3	512	4
Northeast China	*Aedes vexans* and *Ae. dorsalis*	30	40% EC	1—1.5	609	1.1—2.1
			40% Paste	1—1.5	629	1.3—2.1
			40% Cream	1—1.5	802	1.5—2.8

2. Soap Formulation of Deet

A method of dissolving a high concentration of deet in soap has recently been developed,[65] and soap has been produced containing deet plus permethrin or permethrin alone. Spreading a film of the soap after wetting the skin is a convenient and familiar action, and field tests have shown the dried film to be an effective and persistent mosquito repellent.[66,67] The soap costs the equivalent of about (U.S.) $0.20/40-gm bar which is much cheaper than an equivalent amount of a normal commercially available repellent. The World Health Organization is planning to recommend a joint project with the Philippine Government for the production of mosquito-repellent soap in villages in a highly malarious part of the southern Philippines, using locally available coconut oil to make the soap.[67]

Users may think that if they use the soap for washing and then wash the soap off they will continue to be protected against mosquitoes. However, tests with *Anopheles gambiae* released into a room and given the opportunity to bite the bare legs of human subjects showed that the film of deet/permethrin and permethrin soaps both reduced biting considerably, but after rinsing with water all repellency was lost (Table 7).[20] The deet/permethrin soap gave significantly less repellency than the equivalent amount of deet in alcohol solution, and the latter had more wash resistance. Some people may view the easy removal of the

TABLE 7
Laboratory Test of Soap, Deet, and Permethrin Combinations with *Anopheles gambiae*[20]

	Treatment	Before washing			After washing		
		No. mosq.	% Fed	% Mort.	No. mosq.	% Fed	% Mort.
1.	None	200	68.5[a]	3.5[a]	—	—	—
2.	Soap	60	61.7[a]	6.7[a]	60	75.0[a]	3.5
3.	Soap with 20% deet and 0.5% permethrin	40	22.5[b]	5.0[a]	40	62.5[a]	0
4.	Soap with 1% permethrin	40	32.5[b]	27.5[b]	40	67.5[a]	2.5
5.	Liquid deet	60	6.7[c]	3.3[a]	60	20.0[b]	1.7

Note: The mosquitoes were flying freely in a room and human subjects with bare legs allowed the mosquitoes freedom to bite for 8 min and recorded mortality 24 h later. In tests 2, 3, and 4 about 1 g of soap was applied to each leg, thus in test 3 about 0.2 g of deet was applied to each leg; in test 5 about the same amount of deet was applied dissolved in alcohol. Different letters in the same column indicate significant differencs by chi-square tests.

deet in a soap film as an advantage because it allows one to easily avoid the sticky sensation on handling plastic material with liquid deet on the fingers.

In several experiments with permethrin soap, including that illustrated in Table 7, there was significant mosquito mortality. The mosquitoes which were killed were not those which took blood meals, but those which made fleeting contact with the treated skin. The deet vapor from the deet/permethrin soap prevented such contact, and there was therefore no mortality. From the community health point of view, mosquito mortality is highly desirable since the use of substances with a repellent action alone against the more exclusively man-biting species of mosquito would probably merely divert them from people who could afford to use repellents to less-privileged people who could not.

3. E-701

This substance is a derivative of tetrahydroquinoline. It has been formulated in China as the undiluted oil, 40% alcoholic solution, and 40% cream and tested for mosquito repellency.[68]

Tests were performed in a room into which artificially reared female *Aedes aegypti* were released at a density of 80 to 100/m^3. People with different formulations applied to exposed body surfaces stayed in the room and observed whether mosquitoes bit them. The effect of a formulation was expressed in terms of the percentage of "absolute protection", i.e., the percentage of subjects receiving no mosquito bites at specified times after the application of the repellent. Under these conditions the protection times of all the formulations were rather short (Table 8), in contrast to cage studies against the same species in which a protection time of several hours was reported.[42] In the room tests the cream formulation performed markedly worse than the alcoholic solution.

The results of field tests against the high *Aedes* densities of northeast China are shown in Table 9.[68] Once again short protection times were observed as in the tests with deet (Table 6). The cream formulation was less effective than the alcoholic solution. Protection times were shorter when the subjects carried out manual work, in agreement with studies on the effect of sweating on protection times with deet.[55]

TABLE 8
Percentage of "Absolute Protection" of Different Formulations of E-701 against *Aedes aegypti* Released into a Mosquito-Proof Room[68]

Formulation	No. of tests	Percentage of "absolute protection" at different times after application (h)			
		0.5	1.0	2.0	3.0
Undiluted	17	100%	100%	100%	35.3%
40% Alc. solution	18	100%	94.5%	61.2%	22.2%
40% Cream	14	100%	85.8%	0.0%	

Note: A dosage of 1 ml of solution or 1 g of cream was applied to the face and neck.

TABLE 9
Percentage of "Absolute Protection" of Different Formulations of E-701 in Field Tests in Northeast China[68]

Formulation	No. of tests	Percentage of "absolute protection" at different times after application (h)			
		0.5	1.0	1.5	2.0
Undiluted[b]	302	100%	100%	99.0%	95.7%
40% Alc. solution[a]	531	97.8%	70.4%	25.1%	3.8%
40% Alc. solution[b]	122	90.8%	18.9%	0.0%	—
40% Cream[a]	154	85.2%	35.1%	5.2%	0.0%

Note: 1 to 1.5 ml was applied to the face, neck, and hands in each test.

[a] Subjects sitting quietly.
[b] Subjects carrying out manual work.

B. IMPREGNATION OF CLOTHING AND OTHER FABRICS WITH REPELLENTS*

Deet and other volatile repellents last for no more than a few hours on the skin, many people do not like the sensation of repellents on their skin, and a few people react extremely badly to them.[69,70] However, deet or other volatile repellents can be absorbed into cotton fabric and can there form a reservoir which slowly evaporates, giving long-term repellency with minimal skin contact. Netting jackets with hoods impregnated at a rate of 0.25 g per gram of netting gave several weeks of protection from mosquitoes and other biting insects after one impregnation.[71] In these tests the garments were hung indoors in still air between successive field tests. If impregnated fabric is stored in a sealed polyethylene bag or, preferably, a tin with a tightly fitting lid between tests, little vapor is wasted and its "shelf life" is many months.[33] Deet-impregnated cotton strips, 10 cm wide, designed to be worn around the ankles ("anklets") were stored in a tin with a tight lid and only brought out for testing for 2 h each evening. They were tested in Tanzania by teams sitting on chairs. In this position, most biting by *Culex quinquefasciatus* and several other important mosquito species is on the feet and lower legs, i.e., the lowest part of the body. About 85% protection was obtained (Figure 1)[33] against *Cx.quinquefasciatus*, and it persisted at this level for many weeks. By targeting the repellent in its position on the body and time of day, the cost of deet, bought in bulk, need not exceed half a dollar a year per person protected. Similar

* C.F. Curtis

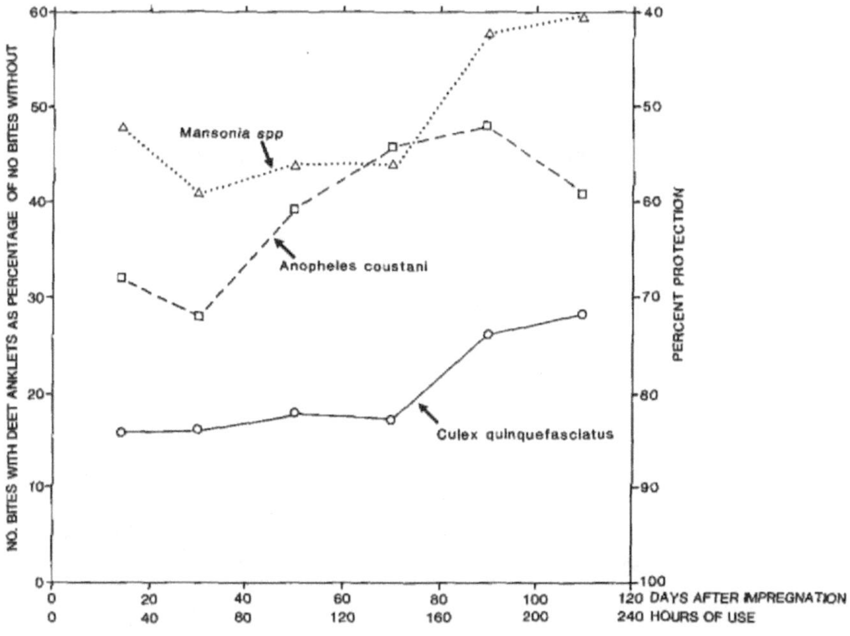

FIGURE 1. Results of tests with human subjects wearing deet-impregnated anklets.[33] The anklets were impregnated with 4 ml deet at the beginning of the study, were kept in a tin with a tight lid, and were brought out for testing for 2 h each evening. (From Curtis, C. F. et al., *Med. Vet. Entomol.*, 1, 116, 1987. With permission.)

good results were achieved against *Anopheles funestus,* but the protection against other species was not so good, presumably because they are not so rigidly programmed to bite on the lowest part of the body.

For protection from most malaria vectors there would only be any point in using repellents for protection in the evening if one was also well protected in bed. Deet has been used on broad mesh bed nets and has given many weeks of protection.[72] However, deet has now been superseded for this purpose by the impregnation of nets with synthetic pyrethroids, such as permethrin, which are almost completely nonvolatile and therefore last several months, and kill mosquitoes after brief contact, as described in Chapter 2.

Good results have also been reported with pyrethroids on clothing.[73] If the wearer remains within a limited area the mosquito attack rate declines over a few hours because an increasing proportion of the hungry mosquitoes have been incapacitated or killed by contact with the pyrethroid. In areas with intense mosquito biting a combination of permethrin-treated clothing plus deet on the exposed skin gave the best results.[74]

C. CLOTHING AND REPELLENTS AGAINST VECTOR POPULATIONS OF *SIMULIUM**

In the savanna regions of seven west African countries onchocerciasis-transmitting populations of *Simulium damnosum s.l.* are being successfully controlled by spraying the riverine breeding sites with larvicides from helicopters.[75] This method is unsuitable in other parts of the world where human population densities are very low or where the flies breed in small or overgrown streams. In such cases personal protection against biting may be a more affordable option. The effectiveness of a citronella-based repellent against *Simulium* has been mentioned in Section I.E.1. Figure 2 shows the results of tests in Cameroon of the

* A. Renz

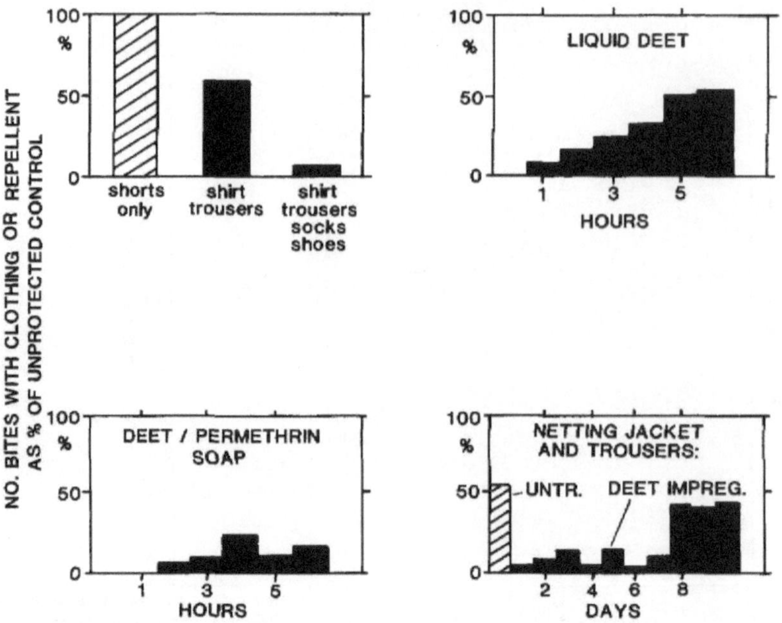

FIGURE 2. The effects of clothing, liquid deet, deet/permethrin soap, and deet-impregnated garments on biting by *Simulium damnosum* s.l. in Cameroon.[60,62,76] The amounts of repellent applied were 1 g liquid deet, soap containing 0.3 g deet and 0.04 g permethrin, netting jacket and trousers impregnated with 120 g deet.

effect on the biting rate of *S. damnosum* s.l. of applying liquid deet or deet and permethrin soap to the legs, or the wearing of unimpregnated clothing or netting jackets and trousers impregnated with deet.[60,62,76] These populations of *Simulium* bite predominantly on the lower legs and feet, and the most effective measure was the wearing of unimpregnated long trousers, socks, and shoes. Unlike mosquitoes *Simulium* cannot bite through cloth because of its short nonpenetrating mouth parts. The effectiveness of liquid deet was high at first, but declined markedly over a 6-h period; however, deet/permethrin soap remained effective throughout that period. Deet-impregnated garments remained effective for several days if stored in plastic bags when not in use.

Flies which were prevented from biting the lower parts of the body by any of these protective methods were not diverted to the hands or face. It appears that *Simulium* of a particular species is programmed to bite a specific area of the exposed human body, and it does not vary this behavior even to crawl under clothing.[60]

Until about 30 years ago the people in the area of Cameroon concerned were almost naked, but nowadays clothing covering much of the body including the lower legs has become common, especially among adults. Furthermore, visits to the riverside where *Simulium* densities are far higher than elsewhere[62] have become less frequent because of the digging of wells etc. The onchocerciasis-transmission potential of a fly population is conventionally estimated from biting catches by collectors wearing only shorts and stationed on riverbanks. On this basis the task of reducing transmission to tolerable levels seems almost impossible. However, one should allow for the actual extent to which people visit the riverbanks and wear clothes and for the fact that prolonged cumulative exposure to infection is necessary for the ocular lesions of onchocerciasis to appear. Taking these facts into consideration the task appears much less formidable. A real impact on disease transmission could be made by:

1. The increase, which will probably continue, in the amount of clothing worn.
2. Making deet available; it is appreciated as much for its repellency to the very annoying biting midges, *Chrysops,* tsetse, and sweat bees as to *Simulium.*
3. Making people aware of the danger of spending long periods near the rivers; young boys should be a particular target of such educational efforts as many of them spend many hours a day, almost naked, fishing and swimming in the rivers.

III. ELECTRICAL GADGETS FOR PROTECTION FROM MOSQUITOES*

Small electric heating plates for vaporizing mats (tablets) of volatile pyrethroids, such as bioallethrin, are now widely available. Tests[20] showed considerably better results than those from burning mosquito coils containing natural pyrethroids mentioned above. With *An. gambiae* released into a small room the vapor from the mats prevented all landings on human subjects, produced knockdown within half a minute, and none of the mosquitoes recovered. The vapor from the mats continued to be emitted for 10 h and was still effective when an extractor fan was switched on. Even after switching the heating plate off and running the extractor for many hours, the vapor had left a deposit on the walls of the room which was insecticidal. In suitable circumstances vaporizing mats could provide an alternative to the use of bed nets against night biting mosquitoes, as well as being useful indoors before people go to bed. For use where there is no mains electricity there is a 12-V battery-operated version and one which runs on methylated spirits without a flame, but their performance in drafty situations such as on verandas or in tents remains to be critically tested.

Traps designed to attract insects with ultraviolet light and then electrocute them are effective against houseflies and blowflies. However, in comparison with humans they have low attractiveness to mosquitoes, and running them in a backyard produced no detectable reduction in mosquito biting rate on human subjects.[77]

Buzzers have been widely sold as mosquito repellers with the claim that they simulate the sound of a male mosquito and that this sound is repellent to mated females. The sound of the wing beat of females is attractive to males (see Chapter 9), but females do not appear to respond much to sound. In any case, the buzzers are set at a much higher frequency than that of mosquito wing beats. At least 12 studies all agree that there is no difference in the mosquito biting rate with buzzers switched on or off (e.g., References 78 to 81). In England two companies have been fined for making unsubstantiated claims in advertising buzzers. It is desirable that similar legal proceedings are taken in other countries where this is possible since not only are people being cheated out of considerable sums of money, but some are being given a false sense of security that buzzers will protect them from malaria.

REFERENCES

1. **Batchelor, J.,** *Ainu Life and Lore: Echoes of a Departing Race,* Kyobunkwan, Tokyo, 1930.
2. **Chappell, E.,** *Voyage of His Majesty's Ship Rosamond to Newfoundland and the Southern Coast of Labrador,* J. Marvman, London, 1818.
3. **Roark, R. C.,** Some promising insecticidal plants, *Econ. Bot.,* 1, 437, 1947.
4. **Jacobson, M. and Crosby, D. G., Eds.,** *Naturally Occurring Insecticides,* Marcel Dekker, New York, 1971.
5. **Secoy, D. M. and Smith, A. E.,** Use of plants in control of agricultural and domestic pests, *Econ. Bot.,* 37, 28, 1983.

* C. F. Curtis

6. **Grainge, M., Ahmed, S., Mitchell, W. C., and Hylin, J. W.,** Plant Species Reportedly Possessing Pest-control Properties — an EWC/UH Database, Resource Systems Institute, EWC Honolulu, College of Tropical Agriculture and Human Resources, University of Hawaii, 1985.
7. **Matsui, M. and Yamamoto, I.,** Pyrethroids, in *Naturally Occurring Insecticides,* Jacobson, M. and Crosby, D. G., Eds., Marcel Dekker, New York, 1971, chap. 1.
8. Chemical Methods for the Control of Arthropod Vectors and Pests of Public Health Importance, World Health Organization, Geneva, 1984.
9. **Muirhead-Thomson, R. C.,** The effects of house spraying with pyrethrum and with DDT on *Anopheles gambiae* and *A. melas* in West Africa, *Bull. Entomol. Res.,* 38, 449, 1948.
10. **Russell, P. F. and Knipe, F. W.,** Malaria control by spray-killing adult mosquitoes. Third season's results, *J. Malar. Inst. India,* 4, 181, 1941.
11. **Senior-White, R.,** House-spraying with DDT and with pyrethrum extract compared: first results, *J. Malar. Soc. India,* 6, 83, 1945.
12. **Service, M. W.,** *Mosquito Ecology: Field Sampling Methods,* Applied Science, London, 1976.
13. **Curtis, C. F. and Hill, N.,** unpublished data, 1987.
14. **Johnson, C. G.,** Field tests on some pyrethrum anti-mosquito repellent creams against woodland species of *Aedes* in Britain, *Trop. Med. Hyg.,* 50, 32, 1947.
15. **Chadwick, P. R.,** Mosquito coils protect against bites, *Parasitol. Today,* 1, 90, 1985.
16. **Luh, P.-L., and Zhu, C.-P.,** Mosquito control in the People's Republic of China, in *Integrated Mosquito Control Methodologies,* Vol. 1, Laird, M. and Miles, J. W., Eds., Academic Press, London, 1983, 85.
17. **Hudson, J. and Esozed, S.,** The effect of smoke from mosquito coils on *Anopheles gambiae* and *Mansonia uniformis* in verandah trap huts at Magugu, Tanzania, *Bull. Entomol. Res.,* 61, 247, 1971.
18. **Smith, A., Hudson, J., and Esozed, S.,** Trials with pyrethrum mosquito coils against *Anopheles gambiae* Giles, *Mansonia uniformis* Theo. and *Culex pipiens fatigans* Wied. entering verandah trap huts, *Pyrethrum Post,* 11, 111, 1972.
19. **Charlwood, D. and Jolly, D.,** The coil works (against mosquitoes in Papua New Guinea), *Trans. R. Soc. Trop. Med. Hyg.,* 78, 678, 1984.
20. **Curtis, C. F. and Hill, N.,** Comparison of methods of repelling mosquitoes, *Entomol. Exp. Appl.,* 49, 175, 1988.
21. **Maunder, J.,** personal communcation, 1988.
22. **Shannon, W. R.,** Thiamin chloride: an aid in the solution of the mosquito problem, *Minn. Med.,* 26, 799, 1943.
23. **Eder, H. L.,** Flea bites: prevention and treatment with thiamin chloride, *Arch. Pediatr.,* 62, 300, 1945.
24. **Muting, D.,** Über die Verhütung von Mückenstichen durch Einnahme von Vitamin B_1, *Med. Klin. (Munich),* 23, 1023, 1958.
25. **Ruiz-Maldonado, R. and Tamayo, L.,** Treatment of 100 children with papular urticaria with thiamin chloride, *Int. J. Dermatol.,* 12, 258, 1973.
26. **Maasch, H. J.,** Freilanduntersuchungen zur mückenabweisenden Wirkung von Vitamin B_1, *Tropenmed. Parasitol.,* 4, 119, 1973.
27. **Khan, A. A., Maibach, H. I., Strauss, W. G., and Fenley, W. R.,** Vitamin B_1 is not a systemic mosquito repellent in man, *Trans. St. John's Hosp. Dermatol. Soc.,* 55, 99, 1969.
28. **Curtis, C. F.,** Fact and fiction in mosquito attraction and repulsion, *Parasitol. Today,* 2, 316, 1987.
29. **Bruun, J.,** Thiamin som middel mot mygg, *Nord. Med.,* 71, 650, 1964.
30. **Granett, P.,** Studies of mosquito repellents. II. relative performance of certain chemicals and commercially available mixtures as mosquito repellents, *J. Econ. Entomol.,* 33, 566, 1940.
31. **Dethier, V. G.,** *Chemical Insect Attractants and Repellents,* Blakiston, Philadelphia, 1947.
32. **Opoku, R. K., Raybould, J. N., and Kessie, D. K.,** Preliminary evaluation of the repellent "Simno" against the blackfly, *Simulium damnosum sensu lato,* a biting midge and mosquitoes, *Insect. Sci. Appl.,* 7, 31, 1986.
33. **Curtis, C. F., Lines, J. D., Ijumba, J., Callaghan, A., Hill, N., and Karimzad, M. A.,** The relative efficacy of repellents against mosquito vectors of disease, *Med. Vet. Entomol.,* 1, 109, 1987.
34. **Curtis, C. F. and Maxwell, C. A.,** Unpublished report to the Dutch and Belgian Consumers' Associations, 1987.
35. **Buescher, M. D., Rutledge, L. C., Wirtz, R. A., Blackin, K. B., and Moussa, M. A.,** Laboratory tests of repellents against *Lutzomyia longipalpis* (Diptera: Psychodidae), *J. Med. Entomol.,* 19, 176, 1982.
36. **Rutledge, L. C., Collister, D. M., Meixsell, V. E., and Eisenberg, G. H. G.,** Comparative sensitivity of representative mosquitoes (Diptera: Culicidae) to repellents, *J. Med. Entomol.,* 20, 506, 1983.
37. **Hwang, Y.-S., Wu, K.-H., Kumamoto, J., Axelrod, J., and Mulla, M. S.,** Isolation and identification of mosquito repellents in *Artemsia vulgaris, J. Chem. Ecol.,* 11, 1297, 1985.
38. **Li, Z., Yang, J., Zhuang, X., and Zhang, Z.,** [Studies on the repellent *quwenling*] (in Chinese), *Malar. Res.,* p. 6, 1974.

39. **Schreck, C. E.,** Techniques for evaluation of insect repellents: a critical review, *Annu. Rev. Entomol.*, 22, 101, 1977.
40. Yunnan Institute of Botany, [Preliminary studies on the ingredients of wild mentha] (in Chinese), *Malar. Res.*, p. 48, 1974.
41. Yunnan Institute of Botany, Kunming Institute of Medical Research, Yunnan Institute of Tropical Plants and Institute of the Chemical Industry of Shanghai, [Studies on wild mentha] (in Chinese), *Malar. Res.*, p. 33, 1974.
42. Shanghai Institute of Entomology, [Comparative studies on the repellent effect of seven repellents in the laboratory] (in Chinese), *Malar. Res.*, p. 142, 1974.
43. Guangxi Repellent Study Unit, [Studies on the repellent effect of "9525"] (in Chinese), *Malar. Res.*, p. 54, 1974.
44. **Dalziel, J. M.,** *The Useful Plants of West Tropical Africa,* Crown Agents, London, 1937.
45. **Kokwaro, J. O.,** *Medicinal Plants of East Africa,* East African Literature Bureau, Nairobi, 1976.
46. **Fivawo, M.,** Community Response to Malaria, Ph.D. thesis, University of Illinois, Urbana, 1985.
47. **White, G. B.,** The insect repellent value of *Ocimum* spp. (Labiatae): traditional anti-mosquito plants, *East Afr. Med. J.*, 50, 248, 1973.
48. **Chogo, J. B. and Crank, G.,** Chemical composition and biological activity of the Tanzanian plant *Ocimum soave, J. Nat. Prod.*, 44, 308, 1981.
49. **Lines, J. D. and Curtis, C. F.,** unpublished data, 1986.
50. **Nicholson, E. A. and Lines, J. D.,** unpublished data, 1987.
51. **Blust, M. H. and Hopkins, T. L.,** Gustatory responses of a specialist and a generalist grasshopper to terpenoids of *Artemesia ludoviciana, Entomol. Exp. Appl.*, 45, 37, 1987.
52. **Snow, R. W., Bradley, A. K., Hayes, R., Byass, R., and Greenwood, B. M.,** Does woodsmoke protect against malaria?, *Ann. Trop. Med. Parasitol.*, 81, 449, 1987.
53. **Rubtzov, I. A.,** [Remedies against black flies and mosquitoes] (in Russian), *Trav. Acad. Milit. Med. Kiroff Armee Rouge,* 19, 61, 1934; *Rev. Appl. Entomol. Ser. B,* 34 (Abstr.), 42, 1946.
54. **Philip, M. I., Ramakrishna, V., and Venkat Rao, V.,** Turmeric and vegetable oils as repellents against *Anopheles* mosquitoes, *Indian Med. Gaz.*, 80, 343, 1945; as cited in *Rev. Appl. Entomol. Ser. B,* 35, 33, 1947.
55. **Gilbert, I. H., Gouck, H. K., and Smith, C. N.,** New insect repellent, *Soap Chem. Spec.*, May, 115, 1957, and June, 95, 1957.
56. **Rutledge, L. C., Wirtz, R. A., Buescher, M. D., and Mehr, Z. A.,** Mathematical models of the effectiveness and persistence of mosquito repellents, *J. Am. Mosq. Control Assoc.*, 1, 56, 1985.
57. **Zhogolev, D. T.,** [The susceptibility to repellents of *Anopheles pulcherrimus* and mosquitoes of some other species] (in Russian), *Med. Parazitol. Parazit. Bolezni.*, 37, 493, 1968.
58. **Schreck, C. E.,** The status of deet (*N,N*-diethyl-*m*-toluamide) as a repellent for *Anopheles albimanus, J. Am. Mosq. Control Assoc.*, 1, 98, 1985.
59. **Schmidt, M. L.,** Relative effectiveness of repellents against *Simulium damnosum* and *Glossina morsitans* in Ethiopia, *J. Med. Entomol.*, 14, 276, 1977.
60. **Renz, A. and Enyong, P.,** Trials of garments impregnated with "Deet" repellent as an individual protection against *Simulium damnosum s.l.* the vector of onchocerciasis in the savanna and forest regions of Cameroon, *Z. Angew. Entomol.*, 95, 92, 1983.
61. **Harlan, H. J., Schreck, C. E., and Kline, D. L.,** Insect repellent jackets against biting midges in Panama, *Am. J. Trop. Med. Hyg.*, 32, 185, 1983.
62. **Renz, A. H. J.,** Adult *Simulium damnosum:* dispersal, migration, host-searching behaviour and vectorial capacity of flies in Cameroon, unpublished report to WHO/TDR No. 1D834456, 1987.
63. **Kochar, R. K, Dixit, R. S., and Somage, C. I.,** A critical analysis of deet as a repellent against arthropods of public health importance and water leeches, *Indian J. Med. Res.*, 62, 125, 1974.
64. Deet Study Unit, [Studies on the repellent effect of *N,N*-diethyl-*m*-toluamide] (in Chinese), *Malar. Res.*, p. 138, 1974.
65. **Simmons, T.,** Insect Repellent and Insecticidal Soap Composition, U.K. Patent 2/60216A, 1985.
66. **Yap, H. H.,** Effectiveness of soap formulations containing deet and permethrin as personal protection against outdoor mosquitoes in Malaysia, *J. Am. Mosq. Control Assoc.*, 2, 63, 1986.
67. New Weapon against Malaria, Health and Development, No. 56, World Health Organization, Western Pacific Region, Manila, 1988.
68. Shengyang Chemical Institute and Shengyang Institute of Medical Research, [Studies on the repellent effect of E-701] (in Chinese), *Malar. Res.*, p. 110, 1974.
69. **Heick, H. M., Shipman, R. T., Norman, M. G., and James, W.,** Reye-like syndrome associated with the use of insect repellent in a presumed heterozygote for ornithine carbamoyl transferase deficiency, *J. Pediatr.*, 97, 471, 1980.
70. **Miller, J. D.,** Anaphylaxis associated with insect repellent, *N. Engl. J. Med.*, 307, 1341, 1982.

71. **Grothaus, R. H., Haskins, J. R., Schreck, C. E., and Gouck, H.K.,** Insect repellent jacket: status, value and potential, *Mosq. News,* 36, 11, 1976.
72. **Gouck, H. and Moussa, M. A.,** Field tests with bed nets treated with repellents to prevent mosquito bites, *Mosq. News,* 29, 263, 1969.
73. **Schreck, C. E., Weidhaas, D. E., Smith, N., and Posey, K. H.,** Chemical treatment of wide-mesh net clothing for personal protection against blood-feeding arthropods, *Mosq. News,* 37, 455, 1977.
74. **Schreck, C. E., Haile, D. G., and Kline, D. L.,** The effectiveness of permethrin and deet alone or in combination for protection against *Aedes taeniorhynchus, Am. J. Trop. Med. Hyg.*, 33, 725, 1984.
75. Ten Years of Onchocerciasis Control in West Africa, WHO Doc. OCP/GVA/85.1B, World Health Organization, Geneva, 1985.
76. **Renz, A., Enyong, P., and Weyler, D.,** The efficacy of appropriate clothing and of DEET- *Simulium*-Repellent as an individual protection against the transmission of onchocerciasis, *Trop. Med. Parasitol.*, 38, 267, 1987.
77. **Nasci, R. S., Harris, C. W., and Porter, C. K.,** Failure of an insect electrocuting device to reduce mosquito biting, *Mosq. News,* 43, 180, 1983.
78. **Rasnitsyn, S. P., Alekseev, A. N., Gornostaeva, R. M., Kupriyanova, E. S., Potanov, A. A., and Razymova, O. V.,** [Negative results of tests of sound generator devices designed as mosquito repellents] (in Russian), *Med. Parazitol. Parazit. Bolezni,* 43, 706, 1974.
79. **Belton, P.,** An acoustic evaluation of electronic mosquito repellents, *Mosq. News.*, 41, 751, 1981.
80. **Curtis, C. F. and White, G. B.,** Once bitten, twice shy, *New Sci.*, 93, 328, 1982.
81. **Foster, W. W. and Lutes, K. I.,** Tests of ultrasonic emissions on mosquito attraction to hosts in a flight chamber, *J. Am. Mosq. Control Assoc.*, 1, 199, 1985.

Chapter 5

NEW STRATEGIES FOR CONTROLLING TICKS

Enala T. Mwase, Rupert G. Pegram, and Thomas N. Mather

TABLE OF CONTENTS

I. THE TICK PROBLEM

Ticks and the diseases they transmit are widely distributed throughout the world, particularly in tropical and subtropical countries. Ticks transmit numerous rickettsial, spirochetal, protozoan and viral pathogens to humans, many of which can produce fatal or debilitating infections. The importance of ticks in animal health and production lies in the enormous economic losses they cause by transmitting a wide variety of pathogens and by direct damage to their hosts. In addition, most livestock-parasitizing tick species are also capable of transmitting disease agents to man. In the developing world, the losses from ticks and tick-borne diseases are not merely economic. In many areas, malnourished people are deprived of animal protein and fat needed to increase their resistance to debilitating infectious diseases.[1]

Methods for the control of ticks have typically included the use of a variety of chemicals in dips, sprays, or dusts, and ecological manipulation of tick-bearing habitats. Chemical control has been most widely practiced and has proved to be the most successful method of controlling the tick-borne diseases of livestock. The combined use of cattle dipping in deltamethrin for tick and tsetse control is described in Chapter 3.

It is now being realized, however, that tick control by the standard use of acaricides alone will not provide adequate protection of human and livestock health. This is true in most African countries, where much of the livestock is owned by traditional farmers who face such problems as the cost of acaricides, acaricide resistance, poor dip-tank management, and social upheaval, such as was experienced in parts of Zimbabwe during the civil war of 1973—1978 when dipping was disrupted.[2] Control of the tick vectors of Lyme disease, babesiosis, and spotted fever rickettsias in temperate areas largely resist standard applications of pesticides through behavioral avoidance of the toxicants. Worldwide, the control of two- and three-host ticks has always proved difficult due to the range of alternative wild hosts available. Thus, new strategies and approaches to the control of ticks appear to be necessary. Besides the development of new acaricides and repellents to protect livestock and humans, new applications for existing methods of chemical treatment, when integrated with thoughtful ecological manipulations, may provide acceptable control of ticks and their related pathogens.

II. TICK REPELLENTS*

A. REPELLENTS AGAINST TICKS ON HUMANS

The search for repellents against ticks had for a long time been directed toward the protection of human beings. Investigation has largely been confined to impregnation of clothing with substances which will retain their effectiveness for long periods.[3] Hundreds of synthetic and plant-derived compounds have been assessed for their repellent effects against ticks.[4] Compounds or mixtures of compounds demonstrated to be 95 to 100% effective against *Amblyomma americanum* for up to several weeks are *N* butyl-acetanilide, 2-butyl-2-ethyl-1,3 propanediol, *N* butyl-4-cyclohexane-1,2 dicarboximide, undecylenic acid, 1-butyryl-1,2,3,4-tetrahydroquinoline, indalone, diethyl toluamide (deet), and dimethyl phthalate.[3] The repellent activities of deet, dimethyl phthalate, and benzyl benzoate against *Hyalomma excavatum* larvae and nymphs have also been assessed.[5] Bar-zeev and Bothilf (cited in Reference 6) reported that deet, pyrethrum, 2-benzyl-oxyethyl lactate ester with butyric acid, 2-dicyclo-hexyl-aminoethanol, alpha-tert-butyl pipenol, capric acid, ephedrine, dibutyl butanephosphonate, and *O*-vanilin were the most effective repellents against ticks. Kulkarini and Naik[7] also evaluated the repellent effects of dibutyl phthalate, dimethyl phthalate, deet plus dimethyl phthalate, dibutyl phthalate plus dimethyl phthalate, and pyrethrins against

* E. T. Mwase and R. G. Pegram

the larvae, nymphs, and adults of the ixodids *Haemaphysalis spinigera, H. turituris,* and *H. kinneari* and the argasids, *Argas persicus* and *A. arboreus,* and adults only of the ixodids *H. abesa, H. intermedia,* and *H. wellingtoni.* The overall relative repellency was in the order: (1) deet, (2) pyrethrins, (3) deet plus dimethyl phthalate, (4) dibutyl phthalate plus dimethyl phthalate, (5) dimethyl phthalate, (6) dibutyl phthalate.

For treatment or impregnation of clothing, however, the best repellents for general use have been considered to be deet and butopyronoxyl. More recent research using clothing impregnated with synthetic pyrethroids has indicated that long-lasting protection from tick bites can be achieved as a result of their toxicity.[8-10] For example, Schreck et al.[8] showed that permethrin treatment offered better personal protection than the more widely used repellents such as deet. Permethrin-treated clothing gave effective protection against ticks even after wearing it for 132 h (5.5 d) without washing, or after washing it three times in a machine.[9]

B. REPELLENTS AGAINST TICKS ON LIVESTOCK

No effective tick repellents are commercially available for use on livestock.[11] Some insecticides, however, have been reported to have repellent activities against ticks on domestic animals. Folz et al.[11] reported on the repellent activity of Amitraz against the brown dog tick, *Rhipicephalus sanguineus,* and the American dog tick, *Dermacentor variabilis.* A repellent activity of 92 to 95% was observed for 2 weeks posttreatment. The treatment was still moderately active (63%) during the third week, but at 4 weeks posttreatment the compound was inactive. Repellent activity (99%) was also detected against *D. variabilis,* but it was monitored for only 7 d. The liquid concentrate was also found to cause 100% tick detachment.[11] Some acaricides may not kill the vectors of disease before they have had time to attach and transmit infection. Thus, if a quick-acting but long-lasting tick repellent was available for use on livestock, it would be very valuable.

In the African traditional system where livestock are allowed to graze with wildlife during certain periods of the year, tick control by acaricides is temporary and expensive, and there is a serious risk that resistant strains of ticks may become established in the wild population. In this case repellents could be advantageous. Furthermore, they could be used at a time when animals are forced to travel long distances for grazing or watering through, and to, areas infested with tick-borne diseases. For example, Mulhearn (cited in Reference 3) noted the problem of marketing cattle from tick-free areas of Queensland in the principal consuming areas, which are tick infested.

C. SETBACKS IN THE USE OF REPELLENTS ON LIVESTOCK

The usefulness of synthetic chemicals or natural products as repellents against parasitic arthropods is limited by a number of factors. Several reports point out that the repellency of a particular compound varies among different tick species, as well as between developmental stages of the same species.[4] This has important implications in African conditions where cattle are infested with four or more species of ticks, the majority of which are the two- and three-host ticks. The effectiveness of standard repellents such as deet are known to vary with their chemical stability to environmental variables such as temperature, wind velocity, moistness of the skin, abrasion, and absorption into the skin. For human protection against disease-transmitting ticks or chiggers, successive applications at 2-h intervals may be necessary under humid tropical conditions. Such application frequencies are not feasible for the protection of livestock.[4] As with acaricides and insecticides, much of the loss of activity appears to be due to mechanical erosion, but loss of activity due to chemical breakdown or photooxidation, as in the case of pyrethrum, is also a problem. Hadani et al.[5] felt that the addition of nonpalatable deterrent agents to the repellents would slow the "decay" of the compounds on animals by minimizing licking. Some investigators have shown that various microencapsulation techniques can extend the period of protection offered by the natural pyrethrins. Such developments may lead to a more widespread use of repellents in the control of livestock pests, particularly those which are important as vectors of disease.[4]

D. TICK-REPELLENT PASTURES

In some parts of the world it has been known for many years that certain species of grasses are associated with low densities of ticks.[12] In Latin America, Menendez (cited in Reference 13) described the repellent properties of *Melinis minutiflora* Beauv. (molasses grass) against the tick *Margaropus annulatus-australis*. Later De Jesus (cited in Reference 13) also observed the inimical effects of molasses grass to *Boophilus australis*. Colombian cattlemen are well aware that cattle grazing on molasses grass carry fewer ticks than those on other improved or native pastures.[13] In Tanzania it was the custom to rub bruised leaves of molasses grass over the coats of animals as an arthropod repellent and to use the grass for nesting hens to control insect pests.[14] Thompson et al.[13] demonstrated the following associations of vegetation with tick populations. Of the six grass species analyzed, *Melinis minutiflora* showed the highest anti-tick properties while *Andropogon gayanus* Kunth was able to maintain field tick populations at low to moderate levels for long periods. *Brachiaria decumbens* Stapf, and *Hyparrhenia rufa* (Nees) Stapf were associated with very high initial infestations and appeared to maintain unacceptably high tick populations over long periods of time.

Obenchain[4] reported on investigations of the repellent activity of extracts of molasses grass. Leaf sheaths of this grass were found to be covered with fine hairs which exuded a sticky secretion with a characteristic odor. *Rhipicephalus appendiculatus* larvae appeared to be totally unable to climb the grass due to these physical properties, and this restriction of their movements was thought to be the cause of the reported absence of ticks from pastures planted with this species. Initial observations on extracts from this grass also indicated that such reductions in tick populations could be due to a repellent effect. Sutherst et al.[15] reported similar physical properites for *Stylosanthes* species as those noted for *Melinis minutiflora*. The genus *Stylosanthes* (Leguminosae) has been reported to have acaricidal properties against cattle ticks. Repellent properties against ticks have also been reported for the *Stylosanthes* species.[16] In Zambia the legume *Cassia absus* L. has also been found to possess the above-mentioned physical properties and to have some acaricidal action against larvae of *R. appendiculatus* ticks.[17] However, the repellent properties of this plant have not been investigated.

Following the discovery that herdsmen in northeastern Kenya used the gums of *Commiphora erythraea* (Ehrenb) Engl. and *C. molmol* Engl. in an attempt to control the ticks which attacked their livestock, investigations have been carried out on the acaricidal and repellent activity of the sesquiterpenoid compounds from these plants.[4] The gum resin of *C. africana* (A. Rich) Engl. has also been used as an insecticide, especially as a repellent of termites in west Africa.[14]

The use of repellent plants would seem to be of value only against those tick species, such as *R. appendiculatus*, that climb up vegetation to quest ("lie in wait") for their host. Most *Amblyomma* species do not quest in this manner, and so there would appear to be no prospect for their control by tick-repellent grasses.

The use of repellent plants to reduce numbers of ectoparasites on hosts, if combined with other tick control strategies such as pasture spelling (rotation), should yield good results in limiting tick populations. A form of pasture spelling (transhumance) is practiced in some parts of Africa where cattle are moved from the lowlands to the highlands during the wet season and vice versa in the dry season. For instance, in Namwala District of Zambia, there is an area that is traversed by the Kafue River Basin where the movement of cattle into the floodplain during the dry season and from the floodplain to higher ground along the periphery of the floodplain during the wet season has become a permanent feature of the seasonal grazing pattern.[17]

Presently, there are no really effective tick repellents for direct use on livestock or for incorporation into pastures. More research into tick-repellent agents and pastures could provide a valuable contribution to integrated biological/ecological control of livestock ticks. As more knowledge on the interaction of ticks with their environment becomes available, more unfavorable or tick-repellent conditions will certainly be identified.

III. APPLICATION OF ACARICIDES TO TICK HOSTS*

A. CONTROLLING THE VECTORS OF ZOONOSES

Many vectors of zoonotic diseases are the naturally occurring ectoparasites of a variety of small- and large-sized mammalian reservoirs. Unfortunately, through either a lack of host specificity or the existence of bridge vectors, certain zoonoses escape from their natural cycles to infect man or his domestic animals. Because of this, controlling ectoparasite infestations on such wild animals may be an efficient means of reducing vector abundance and, thus, protecting human or domestic animal populations from exposure to infected arthropods.

The methods available for applying pesticides directly to wild animal reservoirs are limited. Bait stations containing toxic insecticidal dusts have been widely used to control outbreaks of plague.[18,19] One recent modification of these bait stations used small tubes lined with cloth strips that had been soaked in peanut oil containing diazinon as bait.[20] Rodents that entered these stations were coated with the peanut oil/pesticide mixture as they brushed against the cloth. The use of this method proved effective in eliminating the immature stages of the dog tick, *Dermacentor variabilis,* from rodent hosts in small area tests.

B. USE OF TREATED NESTING MATERIAL

A method for attacking sand flies by distributing small pieces of cloth impregnated with insecticide near the burrows of desert rodents was mentioned in a discussion of strategies for the control of Leishmaniasis.[21] Presumably, it was intended that the rodents would take the treated cloth into their burrows, thus killing sand flies that were present. Depending on the nature of the insecticide formulation applied to the cloth, this method could have the potential to continue killing sand flies as they enter the burrow. As described, however, this method required precise knowledge of the location of the burrows.

More recently, a method was developed that eliminates tick infestations from nest-building rodents in an attempt to suppress transmission of Lyme disease, a tick-borne borreliosis.[22] The rationale for this intervention resulted from a thorough understanding of the epizootiology and transmission dynamics of the spirochetal pathogen. Wherever Lyme disease occurs, several species of wild animals may serve as hosts for the ixodid tick vectors of the disease,[23] but it appears that, at least in the northeastern U.S., one species, *Peromyscus leucopus,* acts as the principal reservoir of the spirochete *Borrelia burgdorferi.*[24-26] Thus, larval ticks ingest infective spirochetes while bloodfeeding on this mouse; the infection is then passed transstadially to the nymphal stage tick, which has proved an effective vector of the spirochete to mice, humans, and dogs.[27,28] Similar mouse reservoirs and tick vectors appear to be responsible for spirochete transmission to humans wherever Lyme disease is found.[29]

To expose these mice to a dose of pesticide which would be toxic to their ticks, cotton, for use as nesting material, was impregnated with permethrin and distributed throughout suitable mouse habitats.[30] The cotton had previously been soaked in an aqueous emulsion

* T. N. Mather

TABLE 1
Tick Infestations on Mice *(Peromyscus leucopus)* Captured in Locations Treated with Permethrin-Impregnated Nesting Material Compared with Infestations on Mice in Adjacent Untreated Sites

	Permethrin-treated nesting material			No treatment		
Location/date	No. mice examined	Percent infested	Av. no. ticks	No. mice examined	Percent infested	Av. no. ticks
Small plots						
1985[a]	125	11	1	150	99	17
1986[b]	45	7	under 1	53	100	47
Large plots						
Aug 1986[c]	40	2.5	under 1	34	97	20
May 1987[d]	15	0	0	8	75	4
June 1987[e]	17	0	0	15	100	41
Aug 1987[f]	20	0	0	22	100	13

[a] Summary of 5 trappings from June—September 1985; treated cotton was applied every 3 weeks.
[b] Summary of 7 trappings from May—September 1986; treated cotton was applied every 3 weeks.
[c] Treated cotton was applied in May 1986.
[d] The treated site had last been treated in August 1986.
[e] Treatment had been renewed in May 1987.
[f] Site was last treated in May 1987.

so that it contained 7.4% permethrin when dried. The treated nesting material was protected from the weather by placing it in dispensing tubes, 3.2 cm in diameter and 20 cm in length. The tubes were placed in a grid array with 10-m spacing between each tube. The selection of the material to be treated and the intervals at which to distribute the fiber were derived from studies on the nesting behavior of *P. leucopus*.[31] Presumably, modifications to this design, using different fibrous material or tube size, could be made to accommodate the behaviors of other rodent reservoirs.

Experiments conducted to demonstrate the effectiveness of this method for reducing *Ixodes dammini* infestations on *P. leucopus* have yielded dramatic results. Just three weeks after distributing treated cotton in 0.6-ha test plots, mice captured in those sites were infested with one tenth the number of ticks found infesting mice that inhabited nearby untreated plots.[22] Nearly 90% of mice, resident in the treated grids, were completely free of ectoparasites whereas virtually all mice in untreated grids were infested. The few tick-infested mice captured in the treated grids were not recognized (recognition was by ear-tagging) as residents and had probably immigrated into the small treated sites. To avoid confounding results from such immigrant mice, treated cotton was distributed throughout a 7.5-ha residential tract. Here, virtually all mice captured 2 months later were free of ticks.[32] Results were even better 1 year later with continued treatment. Table 1 summarizes the results of these trials. Clearly, distribution of permethrin-treated cotton nesting material was very effective at eliminating infestations of ticks on *P. leucopus*.

Eliminating larval ticks from reservoir mice also resulted in fewer infected nymphal ticks in the year following the first treatment. Because *I. dammini* has a 2-year life cycle, larval ticks that feed on infected mice during one season become infected host-seeking nymphs in the next season. Assessments of nymphal abundance were made by capturing host-seeking nymphs (by "flagging", i.e., dragging a piece of cloth over the vegetation) in the same locations that had previously been treated with impregnated cotton as well as in nearby untreated sites. The abundance of host-seeking nymphal ticks was 60 and 89% lower in the small and large treated sites, respectively, when compared with surrounding untreated locations.[32,33] Spirochete prevalence rates in those ticks, as determined using a

TABLE 2
Zoonotic Disease, The Rodent Reservoir or Critical Host for Vector Arthropods and the Vector Groups Responsible for Transmitting Pathogens

Zoonosis	Reservoir or critical host	Vectors
Lyme disease	Cricetid rodents	Ixodid ticks: *Ixodes ricinus* complex
Human babesiosis	Cricetid rodents	Ixodid ticks: *I. dammini*
Rocky Mountain spotted fever	Muroid rodents	Ixodid ticks: *Dermacentor variabilis, D. andersoni*
Colorado tick fever	Cricetid rodents	Ixodid ticks: *D. andersoni*
Scrub typhus	Muroid and cricetid rodents; certain insectivores	Trombiculid mites
Murine typhus	Muroid rodents	Fleas
Bubonic plague	Sciurid, muroid, and cricetid rodents	Fleas
Rickettsial pox	Muroid rodents	Dermanyssid mites: *Liponyssoides*
Relapsing fever	Muroid and cricetid rodents	Argasid ticks: *Ornithodoro* spp.
Tick-borne encephalitis	Muroid, cricetid, and sciurid rodents	Ixodid ticks: *I. ricinus, I. persulcatus*

direct fluorescent antibody assay,[34] was, likewise, lower in ticks collected from locations with treated nesting material. Tick infection rates were reduced by 60 and 72%, respectively, in the small and large treated plots.[32,33] Thus, following just one year of treatment, the risk of encountering a spirochete-infected nymphal *I. dammini* was reduced by 82 and 97% in locations treated with permethrin-impregnated nesting material.

The impregnated material is now available commercially to protect the users of public and private land from infection with Lyme disease.

C. POTENTIAL APPLICATIONS OF TREATED NEST MATERIAL TO OTHER VECTORS

Several important zoonoses that pose health threats to human populations are transmitted by the ectoparasites that infest nest-building rodent reservoirs. Therefore, it may be possible to limit the intensity of transmission of some of these zoonoses by using chemically treated nesting material. Table 2 provides a list of these zoonoses, the rodent reservoir or critical maintenance host of the vector (if known), and the arthropod group responsible for transmission. Such a directed intervention would, of course, exert a more powerful controlling effect on the intensity of transmission should the nest-building rodent serve as both arthropod host and pathogen reservoir than if the animal merely served as host for the arthropod. In addition to the rodent hosts listed, it may be possible to expose larger animals to pesticides using treated fiber should they either nest or consistently frequent a recognizable bedding area.

Thus, pesticide-treated nesting material can serve as a persistent control strategy where reservoir hosts are nest-building animals. One benefit of using chemically treated nesting material to control nidular ectoparasites is that the toxic dose can be reduced because the target arthropod is continually exposed to the toxicant and can accumulate a lethal dose over a period of time. In addition, because of the highly focused method of application, much less pesticide is used per unit area of land than would be needed for more conventional area-wide treatment. The nesting behavior exhibited by the host, as well as that species' degree of involvement in the transmission dynamics of the zoonosis, will ultimately determine the success of using treated nesting material to control a particular disease.

IV. PHEROMONES IN THE CONTROL OF TICKS*

Tick-derived pheromones offer new and innovative strategies for the control of ticks. Currently, pheromones are widely used in the control of certain other insect pests including gypsy moth and Japanese beetles. Exploitation of the arthropod's natural communication system to effect its own destruction theoretically presents a new dimension to pest control that would appear to be impervious to resistance by the target species. However, before pheromones can be expected to act effectively in a control program, knowledge of the biology and ecology of the target species is essential.

A. PHEROMONE-ASSISTED CONTROL OF TICKS

Pheromones are likely to be most effective in the control of ticks when used in combination with toxicants. In this way, the pheromone may act as a lure to attract ticks to the toxicant or to increase the exposure of the targeted pest with the acaricide. Gladney et al.[35] applied a mixture of Isobenzan® and the aggregation/attachment pheromone of *Amblyomma maculatum* to specific sites on cattle. Female *A. maculatum* were attracted to the pheromone-baited site, attached, and died. Specific body sites of cattle treated with extracts of fed male *A. hebraeum* and Toxaphene® attracted and killed both adults and nymphs of this species for up to 4 d in field and laboratory experiments.[36] The tick sex pheromone, 2,6-dichlorophenol, has been combined with propoxur to suppress populations of the American dog tick, *Dermacentor variabilis*.[37] The pheromone, when encapsulated in gelatin microcapsules[38] has been shown to persist in the fur of treated dogs and to disrupt mating for at least 18 d. Dogs treated with the pesticide/pheromone mixture produced significantly fewer fed ticks than did those treated with the pesticide alone. Furthermore, extreme reductions in the mating success of those ticks surviving the acaricidal treatment were also observed. The increased effectiveness of combining sex pheromone with acaricide appeared to be due almost entirely to the greater kill of male ticks.

B. APPLICATIONS WITH OTHER PHEROMONES

It is clear that several other compounds, in addition to the sex pheromone, serve in communication between ticks,[39] and these may also be useful tools in a variety of control strategies. Tick assembly pheromone serve to arrest movement of ticks, thus resulting in concentrations of specific species and stages of ticks on surfaces containing the pheromone. When mixed with acaricides, assembly pheromones may prove useful for control of soft ticks or certain ixodid species found in confined environments, such as in caves, burrows, nests, or similar habitations of wild or domestic animals. These pheromones possess broad interspecific activity and, thus, would prove effective to lure ticks to toxicants. However, a major factor limiting commercial development of the technique is the identity of the pheromone(s), which remain(s) unknown.

In certain specialized situations, pheromones may also function as repellents. Excessive concentrations of 2,6-dichlorophenol effectively deters searching behavior by male *Hyalomma dromedarii* and *H. anatolicum excavatum*.[40] Extension of this observation to treatment of livestock with excessive concentrations of sex pheromone to repel homo- and heterospecific males may well serve to disrupt mating and tick reproduction without the use of toxic chemicals.

* T. N. Mather

REFERENCES

1. Ticks and Tick-borne Disease Control, a Practical Field Manual, Food and Agricultural Organization, Rome, 1984.
2. **Norval, R. A. I.,** Tick infestations and tick-borne diseases in Zimbabwe Rhodesia, *J. S. Afr. Vet. Assoc.*, 50, 289, 1979.
3. **Barnett, S. F.,** Control of Ticks on Livestock, Food and Agriculture Organization, Rome, 1961, 24.
4. **Obenchain, F. D.,** Non-acaricidal chemicals for the management of Acari of medical and veterinary importance, *Recent Adv. Acarol.*, 1, 35, 1979.
5. **Hadani, A., Cwilich, R., and Rechav, Y.,** Laboratory studies of tick repellents and acaricides, in *Proc. 2nd Int. Congr. Acarology, 1967*, 1969, 523.
6. **Kulkarini, S. M.,** Laboratory evaluation of some repellents against larval trombiculid mites, *J. Med. Entomol.*, 14, 64, 1977.
7. **Kulkarini, S. M. and Naik, V. M.,** Laboratory evaluation of six repellents against some Indian ticks, *Indian J. Med. Res.*, 82, 14, 1985.
8. **Schreck, C. E., Snoddy, E. L., and Mount, G. A.,** Permethrin and repellents as clothing impregnants for protection from the lone star tick, *J. Econ. Entomol.*, 73, 436, 1980.
9. **Schreck, C. E., Mount, G. A., and Carlson, D. A.,** Wear and wash persistence of permethrin used as a clothing treatment for personal protection against the lone star tick (Acari: Ixodidae), *J. Med. Entomol.*, 19, 143, 1982.
10. Chemical Methods for the Control of Arthropod Vectors and Pests of Public Health Importance, World Health Organization, Geneva, 1984, 72.
11. **Folz, S. D., Ash, K. A., Conder, G. A., and Rector, D. L.,** Amitraz: a tick and flea repellent and tick detachment drug, *J. Vet. Pharmacol. Ther.*, 9, 150, 1986.
12. **Beesley, W. N.,** The ecological basis of parasite control: ticks and flies, *Vet. Parasitol.*, 11, 99, 1982.
13. **Thompson, K. C., Roa, E. J., and Romero, N. T.,** Anti-tick grasses as the basis for developing a practical tropical tick control package, *Trop. Anim. Health Prod.*, 10, 179, 1978.
14. **Watt, J. M. and Breyer-Brandwijk, M. G.,** *The Medicinal and Poisonous Plants of Southern and Eastern Africa*, 2nd ed., E & S. Livingstone, Edinburgh, 1962, 152 and 477.
15. **Sutherst, R. W., Jones, R. J., and Schinitzerling, H. J.,** Tropical legumes of the genus *Stylosanthes* immobilize and kill cattle ticks, *Nature (London)*, 295, 320, 1982.
16. **Norval, R. A. I., Ntando Tebele, Short, N. J., and Clatworthy, J. N.,** A laboratory study on the control of economically important tick species with legumes of the genus *Stylosanthes*, *Zimbabwe Vet. J.*, 14, 26, 1983.
17. **Mwase, E. T. and Pegram, R. G.,** unpublished data.
18. **Kartman, L.,** An insecticide bait-box method for the control of sylvatic plague vectors, *J. Hyg. Camb.*, 56, 455, 1958.
19. **Kartman, L.,** Further observations on an insecticide-bait-box method for the control of sylvatic plague vectors; effect of prolonged field exposure to DDT powder, *J. Hyg. Camb.*, 58, 119, 1960.
20. **Sonenshine, D. E. and Haines, G.,** A convenient method for controlling populations of the American dog tick, *Dermacentor variabilis* (Acari: Ixodidae in the natural environment, *J. Med. Entomol.*, 22, 577, 1985.
21. **Marinkelle, C. J.,** The control of leishmaniases, *Bull. W.H.O.*, 58, 807, 1980.
22. **Mather, T. N., Ribeiro, J. M. C., and Spielman, A.,** Lyme disease and babesiosis: acaracide focused on potentially infected ticks, *Am. J. Trop. Med. Hyg.*, 36, 609, 1987.
23. **Anderson, J. F. and Magnarelli, L. A.,** Avian and mammalian hosts for spirochete-infected ticks and insects in a Lyme disease focus in Connecticut, *Yale J. Biol. Med.*, 57, 627, 1984.
24. **Levine, J. F., Wilson, M. L., and Spielman, A.,** Mice as reservoirs of the Lyme disease spirochete, *Am. J. Trop. Med. Hyg.*, 34, 355, 1985.
25. **Donahue, J. G., Piesman, J., and Spielman, A.,** Reservoir competence of white-footed mice for Lyme disease spirochetes, *Am. J. Trop. Med. Hyg.*, 36, 92, 1987.
26. **Anderson, J. F., Johnson, R. C., Magnarelli, L. A., Hyde, F. W., and Myers, J. E.,** Prevalence of *Borrelia burgdorferi* and *Babesia microti* in mice on islands inhabited by white-tailed deer, *Appl. Environ. Microbiol.*, 53, 892, 1987.
27. **Steere, A. C., Broderick, T. F., and Malawista, S. E.,** Erythema chronicum migrans and Lyme arthritis: epidemiologic evidence for a tick vector, *Am. J. Epidemiol.*, 108, 312, 1978.
28. **Piesman, J., Mather, T. N., Dammin, G. J., Telford, S. R., III, Lastavica, C. C., and Spielman, A.,** Seasonal variation of transmission risk of Lyme disease and human babesiosis, *Am. J. Epidemiol.*, 126, 1187, 1987.
29. **Spielman, A., Wilson, M. L., Levine, J. F., and Piesman, J.,** Ecology of *Ixodes dammini*-borne human babesiosis and Lyme disease, *Annu. Rev. Entomol.*, 30, 439, 1985.
30. **Mather, T. N., Ribeiro, J. M. C., and Spielman, A.,** U.S. Patent 4,662,104, 1987.

31. **Nicholson, A. J.,** The homes and social habits of the woodmouse *Peromyscus leucopus noveboracensis* in southern Michigan, *Am. Midl. Nat.*, 25, 196, 1941.
32. **Mather, T. N., Ribeiro, J. M. C., Moore, S. I., and Spielman, A.,** Reducing transmission of Lyme disease spirochetes in a suburban setting, *Proc. N. Y. Acad. Sci.*, 539, 402, 1988.
33. **Mather, T. N., Moore, S. I., Piesman, J., and Spielman, A.,** unpublished data, 1987.
34. **Piesman, J., Mather, T. N., Donahue, J. G., Levine, J., Campbell, J. D., Karakshian, S. J., and Spielman, A.,** Comparative prevalence of *Babesia microti* and *Borrelia burgdorferi* in four populations of *Ixodes dammini* in eastern Massachusetts, *Acta Trop.*, 43, 263, 1986.
35. **Gladney, W. J., Ernst, S. E., and Wehler, D. D.,** The Gulf Coast tick: evidence of a pheromone produced by males, *J. Med. Entomol.*, 11, 303, 1974.
36. **Rechav, Y. and Whitehead, G. B.,** Field trials with pheromone-acaricide mixtures for control of *Amblyomma hebraeum*, *J. Econ. Entomol.*, 71, 149, 1978.
37. **Sonenshine, D. E., Taylor, D., and Corrigan, G.,** Studies to evaluate the effectiveness of sex pheromone-impregnated formulations for control of populations of the American dog tick, *Dermacentor variabilis* (Say) (Acari: Ixodidae), *Exp. Appl. Acarol.*, 1, 23, 1985.
38. **Ziv, M., Sonenshine, D. E., Silverstein, R. M., West, J. R., and Gingher, K. H.,** Use of sex pheromone, 2,6-dichlorophenol, to disrupt mating by American dig tick, *Dermacentor variabilis* (Say), *J. Chem. Ecol.*, 7, 829, 1981.
39. **Sonenshine, D. E.,** Tick pheromones, in *Current Topics in Vector Research*, Vol. 2, Harris, K. F., Ed., Praeger, New York, 1984, 225.
40. **Khalil, G. M., Sonenshine, D. E., Sallam, O. M., and Homsher, P. J.,** Mating regulation and reproductive isolation in the camel ticks, *Hyalomma dromedarii* and *H. anatolicum excavatum* (Ixodoidea: Ixodidae), *J. Med. Entomol.*, 20, 136, 1983.

Chapter 6

CONTROL OF *AEDES* MOSQUITOES BY THE COMMUNITY

Kai-Lok Chan, Moh-Seng Chang, Marshall Laird, and Boonluan Phanthumachinda

TABLE OF CONTENTS

I. *AEDES* MOSQUITOES: VIRUS TRANSMISSION AND CONTROL METHODS

Mosquitoes of the genus *Aedes* bite mainly in the daytime and include *Aedes aegypti*, "the yellow fever mosquito", and the related virus vector *Ae. albopictus*. These two species are the subjects of this chapter, but the genus *Aedes* also includes other virus vectors such as *Ae. simpsoni* and *Ae. triseriatus*, vectors of filariasis such as the *Ae. scutellaris* complex in the Pacific area, and the main nuisance mosquitoes of salt marshes and tundras.

Ae. aegypti is responsible for transmission of yellow fever virus when it reaches urban areas of Africa and the Americas, but other mosquitoes transmit it continually between monkeys and sometimes from monkeys to man in forest and rural areas.[1] Nowadays *Ae. aegypti* is more important as the vector (together with *Ae. albopictus*) of dengue virus which causes dengue fever (DF) and in the last three decades has also caused frequent outbreaks of the much more serious dengue hemorrhagic fever (DHF) in Southeast Asia[2] and, more recently, in the Caribbean area.[3] DF is an acute and painful disease which is often not lethal, but DHF may lead to dengue shock syndrome (DSS), i.e., circulatory failure which is often lethal within 12 to 24 h. Dengue virus is transmitted almost wholly from man to mosquito to man, but there is now evidence from Malaysia[4] for a monkey reservoir. Unlike yellow fever there is no effective vaccine against the four dengue viruses,[5] and DHF is one of the ten leading causes of hospitalization in eight Asian countries.[6]

Ae. aegypti and *Ae. albopictus* breed mainly, though not entirely, in man-made habitats, such as water-storage vessels and flower vases, and water accumulations in discarded tins, tires (Figure 1), vegetable waste, and blocked gutters. In principle, breeding in these sites can be prevented by screening and/or regular scrubbing of water storage pots and vases, unblocking gutters and disposing of rubbish in any way that prevents it ever retaining freshwater again (bearing in mind that *Ae. aegypti* eggs can survive dry and hatch when water returns). The organophosphate insecticide temephos (trade name Abate) is also widely available. It is safe for use in drinking water[7] and persists for several months if applied to water-storage vessels in granular formulations.[8,9] *Ae. aegypti* remains susceptible to it for the most part, though some resistance has been detected in the Caribbean area, Malaysia, and New Caldedonia.[10]

Alternative means of *Aedes* larval control exist in the form of the insect growth regulator methoprene, an analogue of juvenile hormone,[11] and the toxin of the bacterium *Bacillus thuringiensis* ssp. *israelensis* (= serotype H-14, often called Bti).[11] (Here and throughout this book this bacterium is named in accordance with the International Code of Nomenclature of Bacteria as advocated by Laird et al.)[12] The toxin of *B. sphaericus* is effective against *Culex* mosquitoes (see Chapter 10) but not *Aedes*.[13] Use of predatory fish and dragonfly larvae against *Aedes* is described in Chapter 8, and the use of predatory larvae of the mosquitoes of the genus *Toxorhynchites* is also briefly mentioned there. Another possibility is luring *Aedes* females to lay their eggs in highly attractive black-colored ovitraps in which the hatching larvae are prevented from developing.[14,15] Where an epidemic of an *Aedes*-borne virus is already in progress, one needs to kill the already-infected adult mosquitoes, and for this, effective methods of insecticidal fogging have been developed.[9,16,17]

Thus, methods for dealing with domestic *Aedes* exist. The problem lies in organizing their application on a sufficiently intensive and sustained basis to eliminate the risk of virus transmission. This chapter describes attempts at such organization in a number of different countries.

FIGURE 1. Piles of used tires in Bouaké, Côte d'Ivoire. Such tires are traded within and between many countries for reuse on vehicles or making into shoes and are a major *Aedes aegypti* breeding place. (Photograph by Marshall Laird.)

II. SARAWAK, MALAYSIA*

A. THE METHODS USED FOR *AEDES AEGYPTI* AND *AE. ALBOPICTUS* CONTROL

Ae. aegypti and *Ae. albopictus* are both vectors of DHF in Sarawak, the latter species probably being the more important.[18] *Ae. aegypti* is highly domesticated and breeds principally in man-made containers in or near houses. Most of the people are still dependent on rainwater kept in storage tanks or jars, and these provide permanent mosquito-breeding places. These are supplemented by artificial receptacles such as flower vases and ant traps, and some of them are small and difficult to locate.

Ae. albopictus is dominant in the inland hilly and rural areas of Sarawak[19] and is less domesticated, breeding mainly out of doors and in a wider variety of breeding places than *Ae. aegypti*, including discarded containers of artificial origin, and also split cocoa pods, ornamental plants with enclosed leaf axils, large fallen leaves, tree holes, etc. Thus, in towns it is found breeding mainly in parks, wasteland, graveyards, illegal dumping grounds and large compounds with abundant vegetation. In rural areas it breeds in cocoa and rubber plantations, secondary jungle, and forest fringes.

The long-term solution to the *Ae. aegypti* problem should come from the replacement of rainwater storage by piped water systems. Interim measures for *Ae. aegypti* control are the use of temephos (Abate) and installation of mosquito-proof rainwater tanks. Temephos is supplied in the form of sand granules impregnated with 1% of the insecticide (Abate 1% SG). It is made up in batches of 20 g (2 teaspoonfuls) which is sufficient for 40 l of water for 3 months. Batches of 10 and 5 g are also made up for smaller sources. The batches of granules are packed in cloth bags so that they can be easily removed from a tank when it is cleaned and replaced afterward.

The mosquito-proof rainwater storage tanks are illustrated in Figure 2. They have a capacity of 150 l and are made of high-density polyethylene. The lid is made mosquito-

* M. S. Chang

FIGURE 2. A mosquito-proof water-storage tank used in Sarawak, showing the fiberglass screen which admits rainwater, but not mosquitoes. (Photograph by M. S. Chang.)

proof while still admitting rainwater collected off roofs etc. by incorporating fiberglass mosquito screening into the lid. Under normal conditions this remains effective for 3 to 4 years.

Ae. albopictus larvae are controlled by the following main methods:

1. Cleaning of water-storage containers (where they have not been made mosquito-proof) and refilling them with clean water
2. Removal and burial or burning of discarded tins, bottles, large fallen leaves, etc. (Figure 3)
3. Splitting of cocoa pods into four pieces to prevent water accumulating in them
4. Cutting bamboo fencing down to the nodes to prevent collection of water in it

In the event of DF or DHF epidemics larval control is supplemented by fogging using 4% malathion in diesel oil.

B. ORGANIZATIONAL METHODS

In the urban and suburban areas of Sarawak, including the capital and 101 small towns, *Aedes* control is centrally organized and is based on enforcement of laws which enjoin householders not to allow mosquito breeding on their premises. The laws are based on those of Singapore (see Section VI).

There are 4764 villages in Sarawak, and in 2200, a piped gravity-fed water supply has been provided. The source of water is generally a dammed mountain stream, and such supplies are only feasible in the inland mountainous areas. Mosquito-proof water tanks have been provided in 659 villages where rainwater is stored. The members of the community are expected to participate in water-supply installation under the supervision and technical guidance of the Health Department staff. They are also expected to provide a monetary

FIGURE 3. Community campaign in Sarawak to burn or bury rubbish which could be used as *Aedes* breeding places. (Photograph by M. S. Chang.)

contribution for purchasing essential items such as fittings and faucets and for maintenance of the system. However, prior to the construction of a water-supply system, the community is advised and directed to construct sanitary pour-flush latrines (see Chapter 10), house drains, fencing for animal enclosures, and generally to clean up the village.

For day-to-day *Aedes* control in villages, especially those not accessible by road, a centrally organized program would not be appropriate. Since 1986-1987 community-based programs have been instituted in 371 villages, and these programs are expected to be fully operational by 1989-1990.

The Health Department acts as a motivator and provider of technical assistance to these village programs, but the community is expected to participate actively. The programs are under the general supervision of the District Anti-Dengue Committee which is advised by the District Medical Officer of Health, and each program is based on the nearest *Klinik Desa* or Rural Health Center which serves a population of 1500 to 3000 within a radius of about 10 km and provides basic curative and preventive services. *Aedes* control is thus incorporated as an integral part of the general health program. Such integration is considered crucial to the success of *Aedes* control programs, and wherever possible links are also made with rural water-supply and sanitation projects.

The selection of the villages in which community based *Aedes* control programs have so far been instituted has depended upon (1) prevalence of DF and DHF, (2) density of *Aedes* populations, (3) existence of related health projects, and (4) proximity to a *Klinik Desa* so that continued monitoring and evaluation of the program is reasonably likely.

Within each village selected for an *Aedes* control program an *Aedes* control subcommittee is appended to the existing Health Committee, or if no such committee exists steps are taken to form one. The Health Committee is briefed on the objectives and strategy of the *Aedes* control program, and they then organize a discussion with the whole community. During this session a description is given of the biology and life cycle of *Aedes* and the available control methods. The signs, symptoms, and dangers of DF and DHF are stressed. At the end of the session a demonstration is given of larval survey and control techniques.

Larger villages are divided into units of 10 to 12 houses, each in charge of a leader appointed by the community. The unit leader is responsible to the village *Aedes* Control

Committee who are in turn responsible to the Government Health Department. Periodic discussions within the community and visits from staff of the *Klinik Desa* and Health Department are arranged to evaluate progress and check on the availability of temephos. This insecticide is at present supplied free by the Health Department to the village *Aedes* Control Committees for distribution to the community. The total cost in 1987 was the equivalent of U.S. $2400 for a total of 737 kg of Abate 1% SG, or about $0.17 per treated house. In the long term the policy of the Health Department is to have the insecticide sold in village shops and cease the free supply. In one district this system already operates.

A public health assistant from the *Klinik Desa* or, where one exists, a village health representative appointed by the villagers, is responsible for intruction of villagers, including school children, in the dosage to apply to different sizes of water containers and for ensuring that retreatment is carried our every 3 months. In the event of DF or DHF epidemics, responsibility for larviciding and fogging is taken over by the Health Department.

Especially in inland areas where *Ae. albopictus* is the main problem, the principal community-based control method is the destruction of water-holding receptacles and split cocoa pods in regular clean-up campaigns named *Gotong Royong* (Figure 3). Reminders about the destruction of cocoa pods are directed at cocoa farmers through the farmers' bulletins issued by the Agriculture Department.

The Health Department has set a target for the community-based *Aedes* control program of a reduction in the Breteau Index (number of infested containers per hundred premises) of 15% per year. Assessment teams from the State Headquarters or the Divisional Committees will check whether this target is being reached. It is too early to give any detailed data from these assessments, but provisional indications are encouraging.

III. THAILAND*

Ae. aegypti mosquitoes have been recognized as the vectors of DHF in Thailand for over 20 years.[2] The first outbreak of the disease took place in 1957, and extensive studies on the ecology of the mosquito and control measures were carried out by the W.H.O./*Aedes* Research Unit between 1966 and 1972.[8,9,16,17,20-22] From 1972 to the present time the Division of Medical Entomology of the Thai Department of Medical Sciences has continued the studies,[23-28] especially on integrated control measures and community participation in control of *Ae. aegypti* in line with World Health Organization recommendations.[29,30]

From 1983 to 1985 a study[31] was conducted in the Phanus Nikhom district of Chonburi province which is 90 km from Bangkok on the east coast of Thailand. The study was in the Phanus Nikhom municipality and in two rural subdistricts, Wat Boat and Wat Luang. There is a piped water supply in the urban area. Nevertheless, an average of 8.2 water storage jars are kept by each householder. In the rural areas rain, canal water, and well water are used, and there are almost twice as many water storage jars (15.9) per house. These are a favored site of *Ae. aegypti* breeding.

Meetings, radio broadcasts, and street parades (Figure 4) were used to inform the community about control of the DHF vector and to motivate their involvement in the proposed control campaigns. For the first campaign adult volunteers were recruited to visit houses, to eliminate unneeded water containers, and to apply temephos (in the form of Abate 1% SG) or tablets containing methoprene to water-storage pots found to contain *Aedes* larvae. In subsequent campaigns the enthusiasm of adult volunteers to take part waned rapidly (Table 1), and volunteer school children were therefore recruited instead. Volunteers and householders were generally more enthusiastic and cooperative in the rural areas than in the town. In the campaigns in 1984, instead of the volunteers applying the insecticide, they counted the number of infested containers and gave an appropriately marked form to the householder

* B. Phanthumachinda

FIGURE 4. Street parade in Thailand to advertise the start of *Aedes* control campaigns. (Photograph by Boonluan Phantumachinda.)

TABLE 1
Summary of the Numbers of Volunteers Involved at Each Campaign in Chonburi Province, Thailand, the Houses Visited, and the Abate 1% SG Issued and Applied

Urban/rural	Campaign no. (year)	No. volunteers: Adults	No. volunteers: Children	Houses visited (%)	Abate SG issued (kg)	Abate applied (%)
Urban (2221 premises)	1 ('83)	112	0	86	126	100
	2 ('83)	34	80	70	119	100
	3 ('83)	26	100	72	153[a]	100
	4 ('84)	0	329	74	165	34
	5 ('84)	0	196	67	124	44
	6 ('85)	Abate application and fogging by health officers				
Rural (1224 premises)	1 ('83)	115	0	81	75	100
	2 ('83)	100	99	81	125	100
	3 ('83)	35	40	87	103[b]	100
	4 ('84)	10	68	66	71	100
	5 ('84)	Abate application by volunteers organized by health officers				
	6 ('85)					

[a] Plus 44 kg of methoprene tablets.
[b] Plus 25 kg of methoprene tablets.

who could then take it to a nearby office where the necessary amount of temephos was provided, together with instructions on how to apply it. Although about the same amount of temephos was issued as in the previous year, it was discovered in a postcampaign survey by staff of the Division of Medical Entomology that less than half of this was applied to infested containers (Table 1). In the rural areas in 1985 there was a change of personnel, and the *Aedes* control campaign was neglected.

The *Aedes* population was monitored by staff of the Division of Medical Entomology and Figure 5 shows the results in terms of the Breteau index (infested containers per 100 houses) in the rural part of the experimental area. The urban results were generally similar. In most cases the control campaigns were followed by reductions in the level of *Aedes* infestation. However, at best, the average was not reduced below 100 i.e., one infested

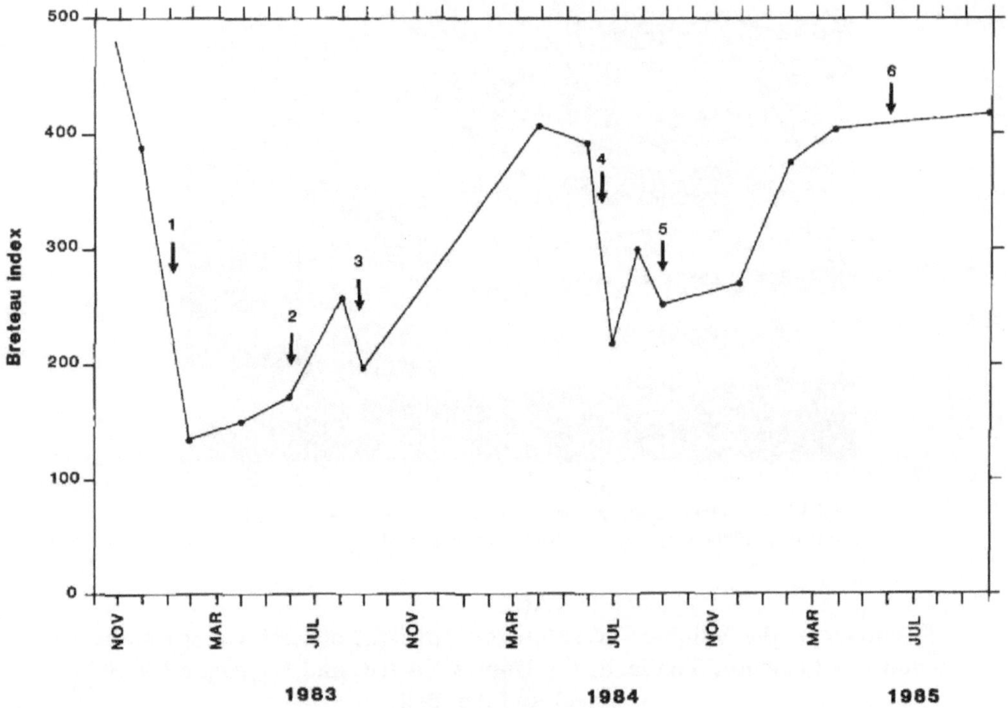

FIGURE 5. *Aedes* larval infestation in the rural part of the experimental area in Thailand in relation to successive control campaigns (indicated by arrows).

container per house. In an area where people have become accustomed to mosquito nuisance, it is difficult to persuade them that *Ae. aegypti* is such a serious health hazard that they should maintain a high level of vigilance against it.

Figure 6 shows the number of DHF cases over a 9-year period in the experimental area, in the surrounding district, in the province, and in the whole of Thailand. No effect of the *Aedes* control campaigns is apparent.

IV. TUVALU*

The nine coral atolls of the state of Tuvalu (the former Ellice Islands) form a chain about 578 km in length between 5 and 11°S and 176 and 180°W. Their total land area is only about 26 km^2, and the highest point is only 4.6 meters above sea level. The population of resident Tuvaluans is about 7100 with over 2000 living on the main islet (Fongafale) of Funafuti, the atoll which is the administrative center of the state. The people are of Polynesian race, and their level of literacy is high. In 1979 about 95% of children aged 6 to 13 years were attending school, and almost all adults, even in the oldest age groups had received at least some years of primary education.

Cases of DHF have occurred in Tuvalu, and between 1981 and 1983 a project was carried out there by a joint team from Canada and Polynesia to test *Ae. aegypti* control by source reduction and application of various agents which had then been recently developed. These agents included the insect growth regulator methoprene (in the form of Altosid briquettes and tablets), which was evaluated in Funafuti, *Bacillus thuringiensis* ssp. *israelensis* (= serotype H-14), in the form of Teknar® (which was tested on the island of Vaitupu) and Vectobac® which was used with success against the nocturnal pest *Ae. vexans nocturnus* on

* M. Laird

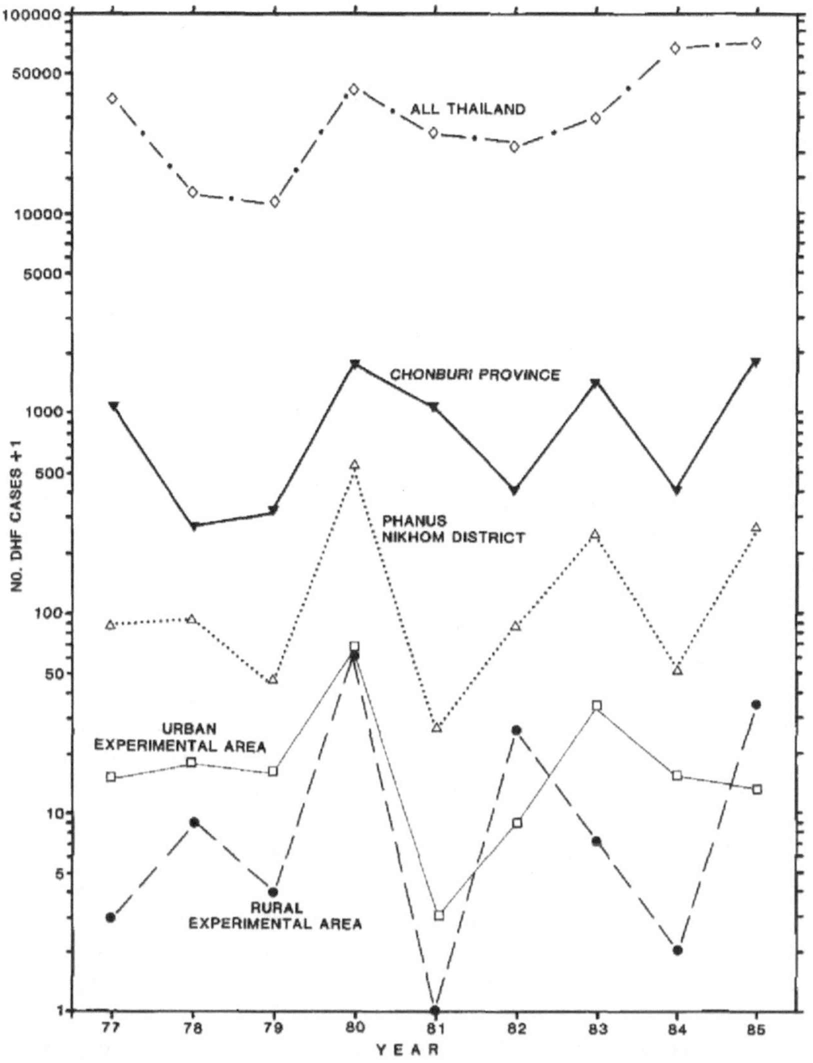

FIGURE 6. Cases of DHF in the urban and rural experimental areas in Thailand, in the whole of Phanus Nikhom district, in Chonburi Province, and in the country as a whole.

Funafuti, and the mermithid nematode *Romanomermis culicivorax*, which was tested on the atoll of Nukulaelae. In the event *R. culicivorax* proved disappointing and following preliminary work in all three experimental sites in 1981 and 1982, the actual control operation was carried out on Funafuti in 1983. In this operation a combination of source reduction and application of Altosid and Teknar was used against larvae and pupae of *Ae. aegypti*, together with the residual adulticide carbamate bendiocarb (Ficam W) which was sprayed in all buildings. The project results have already been summarized,[11] and the remainder of this section concerns only the community participation aspects.

During the first phase of the work in 1981 which aimed at public health education it was found that in addition to large cisterns and water tanks requiring larviciding, there were numerous discarded 200-l fuel drums, plastic containers, coconut shells, and beer cans (some dating back to World War II) scattered in undergrowth which together comprised a major source of *Ae. aegypti*. In Tuvalu, however, discarded automobile tires were still relatively rare. It was also found that there was considerable public disquiet about possible adverse effects of the novel control agents being added to drinking water. These worries resulted from reports on radio and in newspapers and magazines about strong antipesticide views being expressed elsewhere. On Nukulaelae, when *R. culicivorax* was being evaluated, penetrating questions were asked by people well aware of the dangers of *Wuchereria bancrofti* and intestinal worms, about the possibility of human infection.

Therefore, in time for the second phase of the project in 1982, a brochure was prepared in the Tuvaluan dialect of the Polynesian language which stressed source reduction for mosquito control and provided a basis for discussion about safety of the control agents. The message about source reduction was presented in the form of a puzzle picture, the title of which translates as "How many mosquito-breeding places are in this picture?". The illustration included objects familiar to all Tuvaluans: coconut shells, cans, bottles, plastic salt-beef containers, 200-l drums (one receiving its water via a traditional coconut leaf conduit from a palm trunk), a concrete water tank with a metal header tank, and a canoe of traditional local design with its stern terminating in a carving representing a kingfish mouth.

The section on safety in the brochure and the ensuing local discussions started from authoritative data from the World Health Organization etc. but were reinforced by general observation by the islanders that the Teknar and Altosid already used had no offensive or noticeably harmful effects. Furthermore, the specificity of these agents to mosquitoes had been noticed by islanders, who found larval chironomids, dragonfly larvae, etc. in treated containers.

Excellent relations having been established with a very interested public, the next step was a massive source reduction effort with full community participation, including Girl Guides, members of the Junior Red Cross, church groups, and the Women's Committee (a powerful means of exerting pressure on feckless members of the male population, as in other Polynesian islands). Collected rubbish was dumped in a saltwater site where further *Ae. aegypti* breeding was impossible.

The decision to use residual house spraying with bendiocarb in 1983 arose from the observation that not all *Ae. aegypti* breeding was in man-made containers — the species was found to readily utilize the rainwater-filled trunk cavity of pollarded or insect-bored papaya trees. The felling of such trees, which are an important source of fruit, was out of the question, and it was therefore decided that attacking the adult mosquitoes was the only solution.

The domestic use of a powerful carbamate insecticide required the preparation of a further safety leaflet, based on those used in other countries, but translated and adapted to local circumstances. Quite contrary to the unpopularity of house spraying in some countries where it has long been used (see Chapters 1 and 2), the use of house spraying in Tuvalu proved very popular because it helped to achieve an obvious reduction in the populations not only of day-biting *Ae. aegypti*, but also of the night-biting pest mosquito *Ae. vexans nocturnus*. The sight of accumulations of dead ants and cockroaches on the floors of sprayed houses was a source of much satisfaction.

Apart from the entomologists' measures of reductions in mosquito populations,[11] public perception of the success of the campaign was reflected in a decline in the sales of pyrethrum-based mosquito coils (see Chapter 4) from 80 before the start of the project to 6 per week 3 months after it ended.

Despite all the favorable factors in Tuvalu (high levels of literacy and public concern, advanced social organization, well planned preparations, etc.), the project failed to achieve its target of a fully participatory source reduction effort daily for a month prior to the arrival of the overseas members for the 1983 control operation of Tuvalu. The team was left with a pessimistic estimate of the chances of achieving the necessary minimum level of public cooperation for vector control by purely voluntary means in less healthy, literate, and socially conscious communities.

V. THE AMERICAS

The most spectacular successes and failures in *Aedes* control have been recorded in the Americas. *Ae. aegypti* was probably introduced there by the slave ships from Africa, and from the 17th to 19th centuries it caused disastrous yellow fever epidemics centered in the Caribbean, but reaching as far as Boston, Rio de Janeiro, and European ports trading with the Americas.[1,32-34] Yellow fever, together with malaria, defeated the attempt by de Lesseps to build a Panama Canal in the 1880s,[32] but the proof early in this century of transmission of yellow fever by *Ae. aegypti* was soon followed by suppression of this mosquito (and, hence, of yellow fever) in its main Caribbean focus, Havana, and in Panama, thus allowing completion of the canal.[33] Both in Havana and Panama the antimosquito operations were under military control.

Subsequently, the work of highly disciplined teams under the command of F. L. Soper[34,35] led to the eradication of *Ae. aegypti* from the whole of Brazil. This extraordinary feat was achieved in its early stages "the hard way",[35] by searching for and destroying breeding places. An "especially disagreeable mixture of fuel oil and kerosene"[35] was applied to all water containers with larvae, which forced householders to scrub the containers, thus dislodging the eggs. Such a procedure could not be expected to endear the control teams to the public. In the later stages spraying of residual adulticides was also used. By the early 1960s *Ae. aegypti* eradication was reported from most of the other countries of Latin America, but not from the Caribbean islands or from the U.S., where its eradication had "no priority" status among many officials.[36] In 1964 an attempt at eradication in the U.S. was belatedly taken up[36] at the request of the Pan American Health Organization, only to be abandoned in the early 1970s.

Already by 1965 El Salvador was reinfested, supposedly via a shipment of used tires containing dry *Ae. aegypti* eggs,[37] and Brazil, Colombia,[38] and most of the rest of Latin America have recently seen the return of *Ae aegypti*, with consequent outbreaks of DF and DHF.[3,38] The reappearance of *Ae. aegypti* was presumably mainly due to international transport, but there seems a possibility that residual populations had remained undetected in nondomestic breeding sites (papaya tree holes etc. as in Tuvalu, see Section IV). However, Soper[37] considered that in the New World *Ae. aegypti* had never adapted to life in the forest. On the other hand, the flower vases in the cemetery in Caracas, Venezuela have been estimated to contain up to 50 million mosquito larvae, including *Ae. aegypti*.[39]

In the Americas in the 1980s the trend in *Ae. aegypti* populations has been downward only in Cuba. Following the DF and DHF epidemic of 1981[3] a major effort reduced the mean Breteau index there to 0.09, compared with indices ranging from 2 to 65 in other Caribbean countries.[40] The methods used in the "attack phase" in Cuba by 15,000 specially mobilized Civil Defense staff etc. included treatment of all water containers with temephos sand granules, residual spraying with fenthion, and weekly insecticidal fogging with malathion in houses and streets.[3] In the "consolidation phase" methods have included puncturing of unneeded water containers, use of artificial flowers in houses and cemeteries to avoid vases with water, compulsory use of mosquito-proof plastic containers for water storage,[41] and legal sanctions against householders with *Ae.aegypti* breeding on their premises.[3]

After the abandonment of *Ae. aegypti* eradication attempts in the U.S. worse was to follow in the 1980s, with the introduction of *Ae. albopictus* from Asia, apparently in shipments of used tires containing dry eggs or water with larvae.[42,44] The first major American infestation by this mosquito was found in Texas,[42] but it has rapidly spread as far as Chicago. Though much concern has been expressed, little appears to have been done, especially with regard to enforcement of *Aedes* control on those who store used tires (Figure 1). It would appear that there has, therefore, been a permanent addition to the potential dengue vector population in the U.S., and it can be expected to spread unchecked to the tropical countries to the south.[45,46]

VI. SINGAPORE*

Singapore, the "pearl of the orient", is reputed to be the cleanest, greenest, and most economically robust metropolis in Southeast Asia. It is also reputed to be the most orderly and disciplined society in the world, and it has one of the best public health programs. The *Aedes* house index (the percentage of premises with one or more infested containers) is 1 to 2% compared with 30% or more in other Southeast Asian urban areas. (The house index is numerically virtually the same as the Breteau index, as defined above, in situations such as Singapore where infestation rates are so low that the premises with two or more infested containers are very rare.) In Singapore the DF and DHF incidence is the lowest in the region despite its high population density (4500/km^2) which might be expected to favor dengue virus transmission.

The basis of the success of Singapore in *Aedes* control is source reduction accomplished by community participation that has health education and law enforcement as its twin pillars. Chemical control, the mainstay of the vector control programs in other Southeast Asian countries, plays a minor role in Singapore.

The methodology of the Singapore program has been explained in detail by Chan.[47,48] Here the means of fostering community participation are emphasized.

Jackson[49] and Aitken and Leigh[50] have pointed out the direct relationship of poverty to DF and DHF in the major cities of Southeast Asia. In these cities the proliferation of slums and squatter settlements lacking piped water provides ideal breeding sites for *Ae. aegypti* in the jars in which people have to store washing and drinking water. In Singapore, slum areas have been demolished since the 1960s and replaced by high-rise apartment blocks.[51] In 1960 10% of the population lived in such blocks, and by 1980 this proportion had risen steadily to 75%. These apartments all have piped water. Thus, water-storage pots are not required, and 95% of *Aedes* breeding is in other types of artificial domestic containers in and around apartments and commercial premises,[52] including flower vases and ant traps inside apartments and water tanks, blocked gutters, and discarded items outside them.

DHF appeared in Singapore in 1960, and subsequent epidemics occurred almost every year. The epidemics of 1966 to 1968, for example, claimed, respectively, 24, 21, and 18 lives. With the formation of the Vector Control Unit in 1966 a pilot project was launched in a representative area to determine whether source reduction and health education, without the use of insecticides, could bring about a significant reduction in the populations of *Ae. aegypti* and *Ae. albopictus*. The results showed that these measures could reduce the house index from 16 to 2% after three rounds of house inspections at monthly intervals.[53] However, this level of control was not maintained when the intensive rounds of house inspections ended.[54] It was decided that in order to maintain adequate *Aedes* control it was necessary to stimulate active and continuous community participation through public health education supported by law enforcement.[52,54]

The legislation concerned is the Destruction of Disease-Bearing Insects Act (D.D.B.I.A.) which was passed in September 1968. The law empowers vector control officers to enter and examine premises, to take appropriate vector control measures, and to issue summonses to people refusing or neglecting to take action to prevent mosquito breeding on their premises. The procedure is that if an officer finds breeding he will serve an order on the property owner or householder instructing him or her to eliminate the breeding within 14 d. There is the right of appeal to the Minister of the Environment against such orders. In the absence of such an appeal and if mosquito breeding is still found on a second inspection 14 d after the first, a summons to court will be served. If the case is proved in court, the magistrate may impose a fine for a first offense of the equivalent of up to U.S. $500 or 3 months in

* K. L. Chan

jail. Generally, building contractors are fined more heavily than private householders. Between 1973 and 1981 54,297 orders and 7047 summonses were served, resulting in the imposition of fines amounting to about U.S. $800,000; this money was channeled into public health education and source reduction operations.

There are obvious risks in such a system that householders or property owners will attempt to avoid a substantial fine by offering a smaller sum as a bribe to the vector control officer or, conversely, that officers will attempt to extort such bribes by making unfounded accusations. Great care is taken to prevent such corruption, starting by insisting that officers work in pairs, wear uniforms, carry identification cards, explain the reasons for their inspection, and make their inspections in the company of the householder so that everything is done "aboveboard".[55]

Unlike the eradication program in Brazil (Section V), the work of eliminating *Ae. aegypti* breeding is not primarily in the hands of the vector control officers, but in those of the community. The role of the officers is mainly as health educators and, in the last resort, enforcers of the law.

Effective health education requires not merely that the facts are imparted, but that they are clearly understood and motivate changes of behavior by members of the community. This requires that written material is provided in all the main languages of Singapore — Chinese, English, Malay, and Tamil — and that the messages are presented in a way that the recipient sees the full implications of what is being taught in the context of his or her own experience.

Particular emphasis is given to the teaching of children, who are more receptive than older people. The main points which are presented are

1. DHF remains a threat, especially to children, even in present-day Singapore.
2. There is no drug treatment or vaccine for this disease, hospitalizing patients with DSS may be too late, and preventive vector control is the only available protection.
3. *Ae. aegypti* bites in the daytime and frequently breeds inside apartments, so bednets and window screening provide little or no protection.
4. *Ae. aegypti* adults come from aquatic larvae and pupae by processes of metamorphosis; these processes are demonstrated in schools and in public film and television shows.
5. Flower vases should be prevented from producing adult mosquitoes by frequent water changes and scrubbing to dislodge *Aedes* eggs.
6. Ant traps (pots of water in which the legs of food cupboards stand) may be kept free of larvae by adding salt.
7. Unwanted or unused containers should be disposed of in plastic bags via the public refuse disposal system, or at least, stored out of the rain.
8. Gutters should be inspected and blockages cleared to prevent them becoming breeding places.
9. Market stallholders should dispose of vegetable waste etc. in such a way as not to clog drains and create mosquito-breeding places.
10. Building contractors should prevent accumulation of stagnant water on construction sites or, at least, treat it with a suitable insecticide such as temephos (Abate SG).
11. *Aedes* adults fly from one house to another (even before completion of one bloodmeal if they are disturbed); thus, individual protection is not enough, but everyone has a duty to prevent mosquito breeding for the good of all.

In making contact with individual householders vector control officers need to take account of the diversity of the Singapore community (76% Chinese, 15% Malays, 6.5% Indians, 2.5% others) and the customs of each group, if goodwill is to be maintained. For example, the Chinese New Year is a busy period of housecleaning, and no Chinese family would welcome a visit from vector control officers at this time, nor would such a visit serve any useful purpose because the Chinese population will certainly destroy all *Aedes* breeding places as they clean up their homes for their New Year. Similarly, inspections of Hindu and Muslim homes during the times of the Hindu New Year or Ramadan would be neither welcome nor worthwhile.

It is important that highly organized public services such as those of Singapore are not seen as being manipulative, but rather that the public can participate in decision making after taking into account the advice of experts. Therefore, four types of "grassroots organization" have been set up to comment on a wide range of government policies, including those on public health. These organizations are

1. Citizens' Consultative Committees in each parliamentary constituency
2. Residents' Committees in each Housing and Development Board (publicly owned) estate
3. The People's Association which runs 128 community centers, each with its own management committee
4. The Feedback Unit which receives suggestions and complaints from the public and transmits them to the appropriate government department

Apart from organizing and funding the vector control officers who visit houses, the government has several other roles in the *Aedes* control system. These include:

1. Provision of up-to-date information to the mass media on the location and numbers of DF and DHF cases and cases where fines have been imposed under the D.D.B.I.A. to instill a sense of urgency
2. Mosquito-proof screening of water tanks and checking of stopcock pits and gully traps on and around publicly owned housing
3. Provision of a highly efficient refuse disposal system (collections from all premises in Singapore *every day*) so that householders and commercial and industrial organizations can readily dispose of unneeded containers
4. In the event of outbreaks of DF/DHF to rapidly organize the insecticidal fogging of the houses of victims of the disease and their neighbors, using hand-portable cold aerosol generators (e.g., the Fontan) or thermal foggers (e.g., the Swingfog SN 11) as these can be carried up to apartments in high-rise blocks whereas the aerosol emitted by vehicle-mounted machines would not reach them

As in other countries, the breeding places of *Ae. albopictus* overlap with those of *Ae. aegypti* in Singapore, but also include more natural sites such as tree holes, leaf axils, large fallen leaves, etc.[52,56] Such sites are much more difficult to control than domestic ones and even if *Ae. aegypti* were eradicated from the city, *Ae. albopictus* in suburban areas might be able to continue a low level of transmission of dengue viruses.

The present system of *Aedes* control in Singapore was introduced in the late 1960s at a time of serious DHF outbreaks. The system is based on source reduction rather than chemical control, with strong leadership, good management, and generous funding provided by the government (which pays 1500 vector control staff) and with health education and law enforcement to foster community participation. All these factors have been essential to the successful control of *Aedes* vectors and the diseases which they carry (Figure 7). Since the outbreak of 1973 the diseases and the vectors have been maintained at consistently low levels in Singapore in contrast to the situation in many other countries.

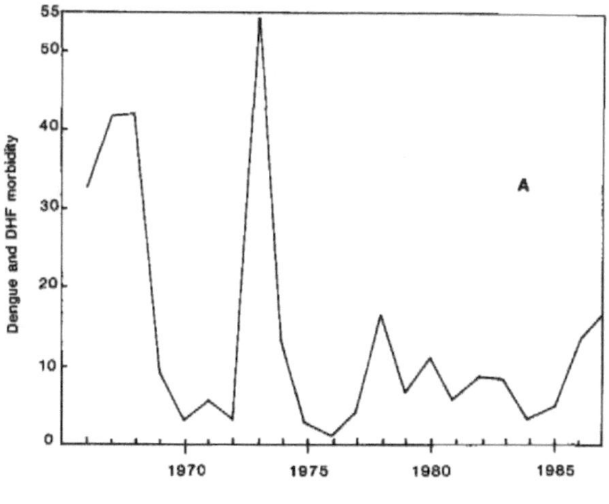

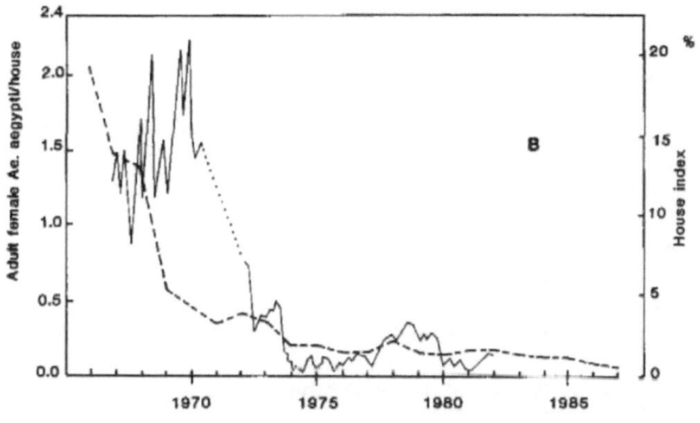

FIGURE 7. (A) Singapore DF/DHF morbidity per 100,000 population; (B) solid line — 3-months moving average of total numbers of adult female *Aedes aegypti* per house (collected by two men using torches, aspirators, and nets) in Singapore. Broken line — yearly average of house index in Singapore (% of houses with one or more containers infested with *Ae. aegypti* and/or *Ae. albopictus*).

REFERENCES

1. **Strode, G. K., Ed.,** *Yellow Fever,* McGraw-Hill, New York, 1951.
2. **Halstead, S. B.,** Mosquito-borne haemorrhagic fevers of South and South-East Asia, *Bull. W.H.O.,* 35, 3, 1966.
3. **Tonn, R. J., Figueredo, R., and Uribe, L. J.,** *Aedes aegypti,* yellow fever and dengue in the Americas, *Mosq. News,* 42, 497, 1982.
4. **Rudnick, A.,** Dengue virus ecology in Malaysia, in *Dengue Fever Studies in Malaysia,* Rudnick, A. and Lim, T. W., Eds., Bull. No. 23, Institute Medical Research, Kuala Lumpur, Malaysia, 1986, 51.
5. **Brandt, W. E.,** Current approaches to the development of dengue vaccines and related aspects of the molecular biology of Flaviviruses, *J. Infect. Dis.,* 157, 1105, 1988.
6. Guide for Diagnosis, Treatment and Control of Dengue Haemorrhagic Fever, 2nd ed., World Health Organization, Geneva, 1980.
7. Chemical Methods for the Control of Arthropod Vectors and Pests of Public Health Importance, World Health Organization, Geneva, 1984.

8. **Bang, Y. H. and Pant, C. P.,** A field trial of Abate larvicide for the control of *Aedes aegypti* in Bangkok, Thailand, *Bull. W.H.O.*, 46, 416, 1972.
9. **Bang, Y. H., Gratz, N., and Pant, C. P.,** Suppression of a field population of *Aedes aegypti* by malathion thermal fogs and abate larvicide, *Bull. W.H.O.*, 46, 554, 1972.
10. Resistance of Vectors and Reservoirs of Disease to Pesticides, *W.H.O. Tech. Rep. Ser.*, No. 737, World Health Organization, Geneva, 1986.
11. **Laird, M., Mokry, J., Semese, A., and Uili, R.,** Integrated control operations against *Aedes aegypti* in Tuvalu, Polynesia, in *Integrated Mosquito-Control Methodologies*, Vol. 2, Laird, M. and Miles, J. W., Eds., Academic Press, London, 1985, 395.
12. **Laird, M., Lacey, L. A., and Davidson, E. W.,** Preface, in *Safety of Microbial Insecticides*, Laird, M., Lacey, L. A., and Davidson, E. W., Eds., CRC Press, Boca Raton, FL, in press.
13. **Davidson, E. W.,** Microbiology, pathology and genetics of *Bacillus sphaericus:* biological aspects which are important to field use, *Mosq. News*, 44, 147, 1984.
14. **Chan, K. L.,** The eradication of *Aedes aegypti* at the Singapore Paya Lebar International Airport, in Vector Control in Southeast Asia, Proc. 1st SEAMED-TROPMED Workshop, Singapore, 1972, 85.
15. **Cheng, M. L., Ho, B. C., Bartnett, P. E., and Goodwin, N.,** Role of a modified ovitrap in the control of *Aedes aegypti* in Houston, Texas, U. S. A., *Bull. W.H.O.*, 60, 291, 1982.
16. **Pant, C. P. and Mathis, H. L.,** Residual effectiveness of ULV aerosols against *Aedes aegypti* in Bangkok: a study of sumithion and malathion applied by a portable ULV machine,, WHO mimeographed doc. WHO/VBC/72.340, World Health Organization, Geneva, 1972.
17. **Pant, C. P., Mathis, H. L., Nelson, M. J., and Phanthumachinda, B.,** A large-scale field trial of ultra-low-volume fenitrothion applied by a portable mist blower for the control of *Aedes aegypti*, *Bull. W.H.O.*, 51, 409, 1974.
18. **Chang, M. S. and Jute, N.,** Dengue and dengue haemorrhagic fever outbreak in Lawas District, Sarawak, East Malaysia, *Med. J. Malays.*, 4, 310, 1986.
19. **Chang, M. S. and Jute, N.,** Distribution and density of *Aedes aegypti* (L.) and *Aedes albopictus* (Skuse) in Sarawak, *Med. J. Malays.*, 3, 205, 1982.
20. **Sheppard, P. M., Macdonald, W. W., Tonn, R. J., and Grab, B.,** The dynamics of an adult population of *Aedes aegypti* in relation to dengue haemorrhagic fever in Bangkok, *J. Anim. Ecol.*, 38, 661, 1969.
21. **Tonn, R. J., Sheppard, P. M., Macdonald, W. W., and Bang, Y. H.,** Replicate surveys of larval habitats of *Aedes aegypti* in relation to dengue haemorrhagic fever in Bangkok, Thailand, *Bull. W.H.O.*, 40, 819, 1960.
22. **Yasuno, M. and Tonn, R. J.,** A study of the biting habits of *Aedes aegypti* in Bangkok, *Bull. W.H.O.*, 43, 319, 1979.
23. **Phan-Urai, P. and Phanthumachinda, B.,** A survey of breeding habitats and biting rate of *Aedes aegypti* (L.) in Bangkog-Thonburi, 1973—1974, *Bull. Dept. Med. Sci., Thailand*, 18(3), 1976.
24. **Phanthumachinda, B., Phanurai, P., Samutropong, W., Charoensook, O., and Rielrangboonya, P.,** Surveillance and control of the vectors of dengue and chikungunya in Thailand 1973—1978, *Bull. Dept. Med. Sci., Thailand*, 22(3), 1980.
25. **Samutrapongse, W. and Pant, C. P.,** Sequential applications of ULV Sumithion for sustained control of *Aedes aegypti* Linn: use of the Fontan back pack portable mist blower, WHO mimeographed doc. WHO/VBC/73.432, World Health Organization, Geneva, 1973.
26. **Phanthumachinda, B., Samutrapong, W., Phanurai, P., Chattrarabhuti, R., Ruengsri, A., and Rielrangboonya, P.,** Ultra low volume malathion application for the emergency control of *Aedes aegypti* during a DHF epidemic at Chantaburi, WHO mimeographed doc. WHO/VBC/74.477, World Health Organization, Geneva, 1974.
27. **Samutrapong, W., Boonyabuncha, S., Matcum, B., Chenchittikul, M., Rielrangboonya, P., and Phanthumachinda, B.,** Tests of insecticides applied by ULV ground and thermal fogging equipment for the control of *Aedes aegypti* in Thailand, WHO mimeographed doc. WHO/VBC/82.869, World Health Organization, Geneva, 1982.
28. **Phanthumachinda, B., Samutrapong, W., Phanurai, P., and Boonyabuncha, S.,** A New Approach for *Aedes aegypti* Control to Combat DHF in Four Provinces on the Eastern Part of Thailand, Res. Rep. Medical Entomology Division, 1981.
29. Vector Control in Primary Health Care, Rep. of a WHO Scientific Group, *W.H.O. Tech. Rep. Ser.*, No. 755, World Health Organization, Geneva, 1987.
30. Integrated Vector Control, 7th Rep., WHO Expert Committee of Vector Biology and Control, *W.H.O. Tech. Rep. Ser.*, No. 688, World Health Organization, Geneva, 1983.
31. **Phanthumachinda, B., Phanurai, P., Samutrapongse, W., and Charoensook, O.,** Studies on community participation in *Aedes aegypti* control at Phanus Nikhom district, Chonburi province, Thailand, *Mosq. Borne Dis. Bull.*, 2, 1, 1985.
32. **Cloudsley-Thompson, J. L.,** *Insects and History*, Weidenfeld and Nicolson, London, 1976.
33. **Smith, H. N. and Downs, W. G.,** Historical perspectives on yellow fever vector research, in *Current Topics in Vector Research*, Vol. 1, Harris, K. F., Ed., Praeger, New York, 1983, 1.

34. **Soper, F. L.,** *Aedes aegypti* and yellow fever, *Bull. W.H.O.*, 36, 521, 1967.
35. **Soper, F. L.,** The prospects for *Aedes aegypti* eradication in Asia in the light of its eradication in Brazil, *Bull. W. H. O.*, 36, 645, 1967.
36. **Schliessmann, D. J.,** Initiation of the *Aedes aegypti* eradication programme of the U.S.A., *Bull. W.H.O.*, 36, 604, 1967.
37. **Soper, F. L.,** Dynamics of *Aedes aegypti* distribution and density, *Bull. W.H.O.*, 36, 536, 1967.
38. **Groot, H.,** The reinvasion of Colombia by *Aedes aegypti*: aspects to remember, *Am. J. Trop. Med. Hyg.*, 29, 330, 1980.
39. **Barrera-Rodríguez, R., Machado-Allison, C. E., and Bulla, L. A.,** Criaderos, densidad larval y segregación de nicho en tres Culicidae urbanos (*Culex fatigans*, Wied., *C. corniger*, Theo. y *Aedes aegypti* L.) en el cemeterio de Caracas, *Acta Cient. Venez.*, 30, 418, 1979.
40. **Knudsen, A. B.,** *Aedes aegypti* and dengue in the Caribbean, *Mosq. News*, 43, 269, 1983.
41. **Varma, M. J. R.,** personal communication, 1987.
42. **Reiter, P. and Darsie, R. F.,** *Aedes albopictus* in Memphis, Tennessee (U.S.A.): an achievement of modern transportation?, *Mosq. News*, 44, 396, 1984.
43. **Sprenger, D. and Wuithiranyagool, T.,** The discovery and distribution of *Aedes albopictus* in Harris County, Texas, *J. Am. Mosq. Control Assoc.*, 2, 217, 1986.
44. **Craven, R. B., Eliason, D. A., Francy, D. B., Reiter, P., Campos, E. G., Jakob, W. L., Smith, G. C., Bozzi, C. J., Moore, C. G., Maupin, G. O., and Monath, T. P.,** Importation of *Aedes albopictus* and other exotic species into the United States in used tires from Asia, *J. Am. Mosq. Control Assoc.*, 4, 138, 1988.
45. **Knudsen, A. B.,** The significance of the introduction of *Aedes albopictus* into the southeastern United States with implications for the Caribbean, and perspectives of the Pan American Health Organization, *J. Am. Mosq. Control. Assoc.*, 2, 420, 1986.
46. **Miller, B. R. and Ballinger, M. E.,** *Aedes albopictus* mosquitoes introduced into Brazil; vector competence for yellow fever an dengue viruses, *Trans. R. Soc. Trop. Med. Hyg.*, 82, 476, 1988.
47. **Chan, K. L.,** *Singapore's Dengue Haemorrhagic Fever Control Programme: a Case Study on the Successful Control of Aedes aegypti and Aedes albopictus Using Mainly Environmental Measures as a Part of Integrated Vector Control*, Publ. No. 45, Southeast Asian Medical Information Center, Tokyo, 1985.
48. **Chan, K. L. and Bos, R.,** Control of dengue vectors: Singapore's success story, *World Health Forum*, 8, 101, 1987.
49. **Jackson, J. C.,** Urban squatters in Southeast Asia, *Geography*, 59, 24, 1975.
50. **Aitken, S. R. and Leigh, C. H.,** Dengue haemorrhagic fever in Southeast Asia, *Trans. Inst. Br. Geographers*, 3, 476, 1978.
51. **Chan, K. L. and Counsilman, J. J.,** Effects of slum clearance, urban development and vector control on densities of *Aedes* in Singapore, *Trop. Biomed.*, 2, 139, 1971.
52. **Chan, K. L., Chan, Y. C., and Ho, B. C.,** *Aedes aegypti* (L.) and *Aedes albopictus* (Skuse) in Singapore: II. Larval habitats, *Bull. W.H.O.*, 44, 629, 1971.
53. **Chan, K. L.,** A report on the pilot control scheme in mosquito eradication at the Geylang-Tanjong Rhu area in Singapore, *Singapore Public Health Bull.* 1, 52, 1967.
54. **Chan, K. L.,** Control of vectors towards a better Singapore, in Towards a Better Singapore, Proc. 1st Conv. Singapore Professional Centre, October 9—13, 1975, Tan It Koon, Ed., 1975, 107.
55. **Chan, K. L.,** Ecological approach to integrated control of *Aedes aegypti* and *Aedes albopictus* in Singapore, *J. Singapore Natl. Acad. Sci.*, 3, 116, 1973.
56. **Chan, K. L.,** Control of dengue vectors, *Asian J. Infect. Dis.*, 2, 139, 1978.
57. **Chan, K. L., Ng, S. K., and Chew, L. M.,** The 1973 dengue haemorrhagic fever outbreak in Singapore and its control, *Singapore Med. J.*, 18, 81, 1977.

Chapter 7

ENVIRONMENTAL AND WATER MANAGEMENT FOR MOSQUITO CONTROL

P. K. Rajagopalan, P. K. Das, K. N. Panicker, R. Reuben, D. Raghunatha Rao, L. S. Self, and J. D. Lines

TABLE OF CONTENTS

I. INTRODUCTION

Mosquitoes of different species require water with particular qualities for their larval and pupal stages. These qualities include:

1. Persistence of the water body for a week or more so that development from egg to adult can be completed
2. A low or zero flow rate
3. Organic pollution level, high levels being favored by *Culex quinquefasciatus*, low levels by other vector species
4. Access to the air surface for breathing (all vector genera except *Mansonia*)
5. Suitable vegetation, required by *Mansonia* for breathing and as a refuge from predators by many species (shade is either required or detrimental to various mosquito species)
6. Salinity, fatal to most vectors except coastal species such as *Anopheles melas* and *An. sundaicus*

Early in this century it was realized that, in principle, all that is required to eradicate mosquito-borne disease is to manage the environment so that no water of the quality required by the local vector species is available. This is more easily said than done, even though high densities of vector species are not natural but result from human alteration or mismanagement of the environment.

For most poor people, for most of the time, control of vector-borne disease does not have a high priority compared with earning a living. If poor communities are to sustain control for any length of time it will probably be essential for the environmental changes to be part of a package with no net drain on economic resources and which includes income generating features. This chapter includes some case studies of such packages; none is universally applicable, and each has to be designed with close attention to the local ecological, socioeconomic, political, and cultural setting.

II. INTEGRATED CONTROL OF *CULEX QUINQUEFASCIATUS* IN THE CITY OF PONDICHERRY, INDIA*

A. ASSESSING THE MOSQUITO PROBLEM

Pondicherry, on the southeast coast of India has a population of about 275,000. A survey showed that 2.3% of females and 13.7% of males had symptoms of Bancroftian filariasis. These included elephantiasis of the legs (Figure 1), but the organ most commonly affected was the scrotum. This explains the difference in prevalence of symptoms between the sexes. In 1981 it was found that the prevalence of microfilaremia averaged over all age groups was 8.4% in the Pondicherry urban area. This is a much higher rate than in surrounding rural areas because in the town the vector mosquito, *Cx. quinquefasciatus*, bred in polluted water which stagnated in innumerable places throughout the town because of poor engineering and maintenance of the wastewater disposal system. In this respect Pondicherry is typical of many expanding tropical urban areas, especially where the community has demanded and obtained a reasonably abundant water supply but the means of disposing of the water after use are inadequate. In 1981, the Vector Control Research Centre (V.C.R.C.) began a 5-year project designed to demonstrate that a combination of appropriately chosen larval control methods could greatly reduce the vector population at an affordable cost. The following account is based on a full report on the project.[1] The project was carried out in four of the six zones of the city, the other two being left under the routine operation of the Indian National Filaria Control Programme.

* P. K. Rajagopalan, P. K. Das, and K. N. Panicker

FIGURE 1. A severe case of leg elephantiasis. (Photograph by Vector Control Research Center.)

Only about 1000 households in Pondicherry are connected to an underground piped sewerage system. Excreta disposal is mostly into septic tanks, with vent pipes which are frequently unscreened against mosquito entry (see Chapter 10). The overflow from the septic tanks generally discharges into the surface-water drainage, as does sullage (wastewater from washing, cooking, etc.)

In many areas housing has been built without any provision for drainage. This is contrary to bylaws, but these are seldom enforced because of corruption of officials by contractors and property owners. In such areas each household provides itself with a soakage pit to dispose of sullage water.

When there are surface drains, they may be unlined *(kutcha)* ditches (former natural water courses), or they may be L-shaped (i.e., gutters formed in the angle between the curb and cambered roadways) or U-shaped cement-lined ditches. Some of the drains have been left incomplete with no outlet, leading to inevitable formation of ponds of polluted water, and many of the drains have too little slope to ensure that water flows and clears obstructions. Solid waste is often dumped in drains. U-shaped drains, though apparently more "modern", were found to suffer the worst blockages and stagnation problems and were the most difficult to clear. This was especially true when they had been roofed over with concrete slabs (i.e., converted to box drains) and then built over without leaving any convenient access for cleaning. In these cases the bed of the drain had often been obstructed by pillars to support the buildings above, thus increasing the likelihood of blockages which could not be dug out.

The drains were intended to discharge into 2- to 10-m-wide canals which had originally been designed for storm water, but are now heavily polluted. When the final stage of the canal system reached 150 m from the shore, it was found that the canal was 2 m below the level of the sea. Further construction was abandoned at that point, and the polluted water was left to spread out over a low-lying area to form a 70-ha "lagoon" or swamp. In this area about 500 pits had been dug for the soaking of cocunut husks to make coir.

In the higher income areas plots are frequently left vacant for many years between developed land. Contrary to bylaws, these are frequently below the level of the roadway so that polluted water finds its way onto the plots and forms large stagnant areas of flooded land.

With the provision of a piped water supply, old wells have been left open but disused and have become breeding places, not only for *Cx. quinquefasciatus*, but also for anopheline mosquitoes.

The unregulated keeping of animals in the urban area aggravates mosquito problems because animal excrement is a major source of water pollution and because wandering pigs and cows destroy the banks of unlined drains.

B. THE CONTROL ACTIVITIES

In planning control operations it was clear that the 70-ha "lagoon" or swamp required priority treatment. It was proposed that this should be dealt with by preventing the water reaching the area below sea level near the seawall, and constructing a lined canal 2.6 km down the natural gradient to a point where a permanent outfall to the sea could be created. This proposal was accepted by the local government but never implemented. Therefore, the V.C.R.C. arranged that in the monsoon season, when the "lagoon" was full, an opening to the sea should be dug out manually through a sandbar which, however, was regularly reformed by wave action at spring tides. In the dry season the area became a swamp, and pumps were installed to lift the water over the seawall and into the sea. The coir pits were filled with sand, and part of the area was dredged to make it deep enough for the rearing of fish. All these activities reclaimed a large land area, the value of which, if sold for housing, would be about twice as great as all the costs of the project for the 5 years of its life.

To assess priorities for control operations in other areas of the town, estimates were made of the output of *Cx. quinquefasciatus* adults from each type of breeding site.[2,3] These were based on estimates of the percentage of each type of site which was mosquito infested (in the absence of control measures) and the mean number of pupae per unit area (as measured by taking large numbers of samples with a dipper) when a site was infested. The results varied with season,[2] and Table 1 shows an example of such estimates with their application to one of the districts of the city for which data had been collected on the total water surface of each type of site. It is noteworthy that though soakage pits covered only a small total area, they were so productive when infested that they made the largest single contribution of mosquitoes. The other important categories of site were flooded plots and U-shaped drains, which contributed more than the supposedly more primitive L-shaped or unlined drains.

Once the problem of the "lagoon" had been dealt with, the next priority was the soakage pits, and these were treated weekly with the organophosphate insecticides fenthion or phenthoate, or malariol (oil with a spreading agent). Monitoring susceptibility to the two organophosphates showed no marked trend in increasing tolerance over the 5 years of the project. The LC_{50} values tended to be 1.5- to 4-fold higher in the sectors of the city where the project was located than in those under routine control, where insecticide usage was much less intensive. However, no increase in dosage was required to obtain effective control in the field.

Insecticidal treatment of flooded plots would not have been cost effective. Initially, some of these were pumped dry, but a more satisfactory permanent solution was filling the hollows and raising the land level above the level of the drains. This proved highly cost

TABLE 1
Infestation of Various Types of Site with Pupae of *Culex quinquefasciatus*, the Total Water Surface of Each Type in One District of Pondicherry, and, Hence, the Absolute and Relative Contribution of Each to the Adult Female Mosquito Production[2,3]

Type of site	Samples with mosquito breeding (%)	Pupae/m² in infested samples	Area of water surface in one district (m²)	Daily output[a] of female adults: In thousands	Daily output[a] of female adults: Percent of whole
Soakage pits	5.12	2949	525	19.82	32.0
Flooded plots	1.58	633	5217	13.04	21.1
U-drains	3.28	1007	1829	15.10	24.4
L-drains	0.84	524	5823	6.41	10.3
Unlined drains	0.84	1360	1485	4.24	6.8
Canals	1.69	925	740	2.89	4.7
Wells	0.83	278	756	0.44	0.7
				61.94	

[a] A pupal period of 2 d and a 1:1 sex ratio is assumed, so the number of pupae is divided by 4 in calculating the output of female adults.

TABLE 2
Approximate Water Surface Subject to, and Cost over 5 Years of, Each Type of Mosquito Control in Pondicherry

Activity	Area (ha)	Cost ($)[a]
Draining "lagoon"/swamp	70	6,700
Land filling	4	4,600
Channelization	2	900
Desilting drains	3	52,000
Insecticidal treatment	5	144,600[b]
Introduction of fish	under 1	6,900

[a] Calculated assuming 10 Indian rupees = $1.
[b] About 65% of cost was for insecticides and the rest for labor.

effective (Table 2). In other cases the larvivorous fish, *Gambusia affinis*, was introduced into the flood water. The fish were mass reared in a functionless arm of the drainage canal system which was walled off from the rest of the system and was thus turned into a fish pond. The fish were also used to stock disused wells in which there was mosquito breeding.[4]

A regular program of unblocking, desilting, and deweeding drains was introduced, and the debris was carted away to use in the landfill program (see above). This prevented it being left beside the drains from where it would eventually fall back and reblock the drains. In the case of box drains, flushing with impounded drain water or, preferably, seawater was used to clean the accumulated blockages. Seawater had the additional advantage of killing larvae by raising the salinity. Even after desilting, the dry season flow rate in many drains was insufficient (less than 30 cm/s) to prevent mosquito breeding. Once the major blockages were cleared the routine weekly desilting was rather expensive (Table 2) and is now considered not to have been cost effective. A major program of drain reconstruction would be necessary to solve this problem permanently. The gradients would have to be increased,

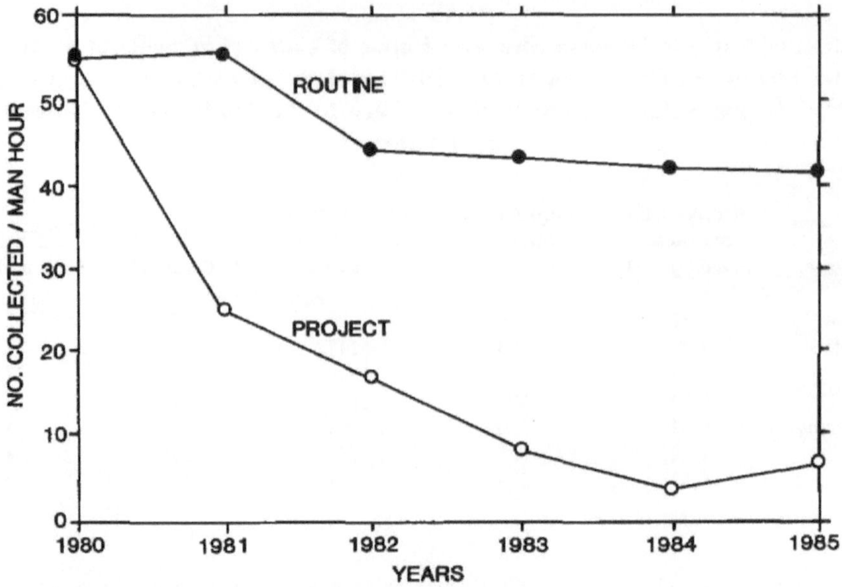

FIGURE 2. Mean numbers of female *Culex quinquefasciatus* caught resting in houses per collector hour in the project and routine control sectors of Pondicherry.[1]

and the cross sections of the drains made V-shaped so that the dry season flows could clear blockages, but the drains would also have the capacity to deal with storm water. In the absence of such a program, insecticidal treatment was extended to the drains. In some cases teams with knapsack sprayers, who dealt with the soakage pits, also sprayed the drains. However, it was found more cost effective to use drip applicators made from polyethylene bags with jute wicks. These required refilling every 10 d.

Unlined drains were channelized (narrowed and deepened) to increase the flow rate. The work needed frequent repair because of the activities of roaming animals.

Table 2 gives an approximate idea of the water surfaces treated by the various means and the direct costs of these activities over the 5-year project.[5]

C. ASSESSING THE RESULTS

Efforts were made to assess the results of the control project on vector density and filariasis prevalence and incidence. Comparisons were made between the sectors of the city where the project was located and the sectors under routine operations, though it was recognized that the movement of mosquitoes and people between these sectors would to some extent obscure the effects of the project. Figures 2 and 3 show the numbers of female *Cx. quinquefasciatus* collected by standard procedures resting in houses and biting, respectively. There appears to be a very marked effect of the project on vector densities, though unfortunately the 1980 (precontrol) data cannot be separated into those from the sectors which were later subjected to the project and the routine control operations.

Since filarial worms have a life of several years, the prevalence of microfilaremia and infected mosquitoes would be expected to respond only slowly to vector control in the absence (as in Pondicherry) of intensive antifilarial drug treatment of the human population (see Chapter 10 for an account of a project integrating antivector and antifilarial operations). In Pondicherry the greatest effect would have been expected in children born during the course of the 5-year project. Table 3 shows that the prevalence of microfilaremia declined between 1981 and 1986 in both sectors of the city. The decline appeared much greater in the project sector, but the difference was not quite statistically significant.[6] Figure 4 shows

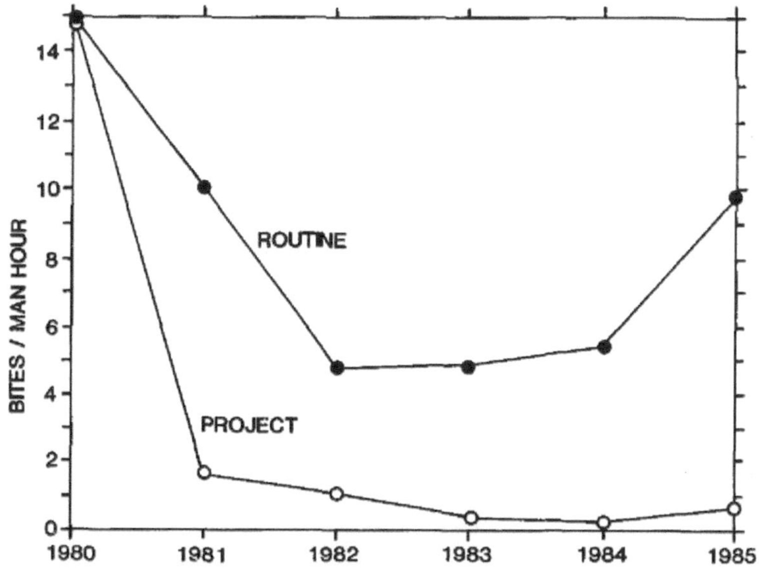

FIGURE 3. Mean numbers of female *Culex quinquefasciatus* caught biting at night per collector hour in the project and routine control sectors of Pondicherry.[1]

TABLE 3
Numbers of Children in Age Groups 0 to 5 Positive and Negative for Microfilaremia at the Beginning (1981) and End (1986) of the Pondicherry Project in the Sectors under the Project and Routine Control

	Project			Routine		
	+ve	−ve	% +ve	+ve	−ve	% +ve
1981	23	940	2.39%	8	426	1.84%
1986	3	1402	0.21%	4	614	0.65%

Note: The significance of the greater apparent change over the 5-year period in the project sector was tested[6] as follows. For the project sector: $\log_e$ of odds ratio = $\log_e(23/940 \times 1402/3) = 2.4367$, with variance $(1/23 + 1/940 + 1/1402 + 1/3) = 0.3786$. Corresponding parameters for the routine sector are 1.0587 and 0.3790. Difference between the sectors gives standard normal deviate: $\frac{2.4367 - 1.0587}{\sqrt{(0.3786 + 0.3790)}} = 1.58$, not significant.

data on the arithmetic mean microfilarial count in all age groups in 1981 and 1986. In calculating these means the total number of people sampled are included in the denominator, i.e., they reflect both the intensity of infection in infected people and the percentage of infected people. The mean counts were generally higher in 1981 in the project sector (mean over all age groups = 0.98) than the routine sector (mean = 0.80). However, by 1986 they had become about equal in the two sectors (mean = 0.71 in each). This presumably indicates that with a greatly reduced input of new infection in the project sector the death of many worms in previously infected people caused a lowering in the mean level of infection.

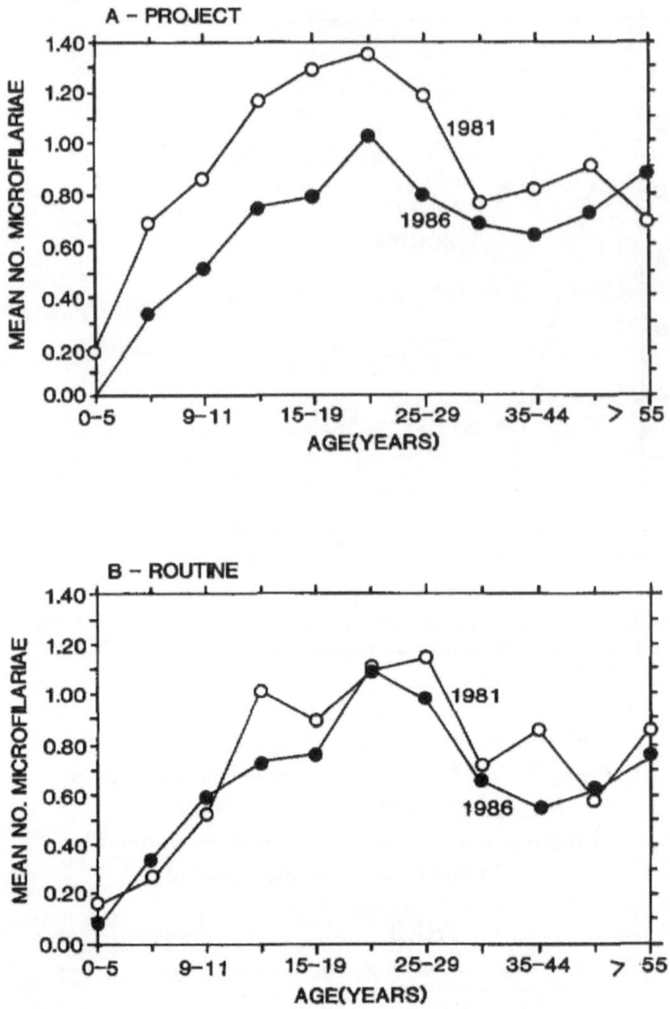

FIGURE 4. Mean microfilarial counts by age group in 1981 and 1986 in the project and routine control sectors of Pondicherry.[1]

III. INCOME GENERATION AND *ANOPHELES* CONTROL IN SOME SOUTH INDIAN VILLAGES*

Although malaria control by conventional house spraying with residual insecticides such as DDT has generally been successful in south India, malaria has persisted along the Coromandel (southeast) coast.[7] *Anopheles subpictus* has been thought of as an abundant but innocuous species, but in this area it has been shown to be a malaria vector[8] and to be a distinct sibling species from inland populations of *An. subpictus*.[9]

In the fishing communities in villages in Pondicherry Union Territory malaria is frequently reintroduced by fishermen returning from malaria endemic ports, and it is locally transmitted by *An. subpictus*. This mosquito breeds in many small bodies of water around the villages, but the principal breeding site was in shallow coastal lagoons ("backwaters") in which algal growth was abundant except during the monsoon season when the lagoons deepened and became confluent with the sea. The algae sheltered the larvae from fish and other natural predators.

* P. K. Rajagopalan, P. K. Das, and K. N. Panicker

The staff of the V.C.R.C. had considerable difficulty in gaining the confidence of the people in the villages who are divided into factions and suspicious of supposedly helpful schemes imposed by government officials without consultation with the villagers. Drinking water, electricity, and roads were given higher priority than control of vector-borne diseases, except during occasional malaria epidemics. The V.C.R.C. assisted by making representations to the relevant government departments to secure action on water and electricity supply etc., and this gained sufficient local confidence to allow the initiation of a number of small-scale community-based vector control activities; for example, the digging of soakage pits to eliminate standing water around wells and communal stand pipes. Predatory larvae of the mosquito *Toxorhynchites splendens* were also used to consume the larvae of the vector mosquitoes which live in earthenware pots in which the villagers ret coconut husks to make coir.[10]

However, there was opposition to placing the larvivorus fish *G. affinis* in the water-filled pits in which young casuarina trees are grown, and it was not felt that the other antivector activities had "caught on" sufficiently to be sustained by the community without outside intervention. At this point two ways were discovered of dealing with the major problem of mosquito breeding in the backwaters, and both simultaneously generated income.[11,12] The first of these was the economic exploitation of the algae on which the larvae depend for shelter. It was found that the alga (*Enteromorpha compressa* var. *pinqulata*) could be used to make a mottled art paper and cardboard. The algae were manually cleared in the premonsoon period when they flourish, and sold by a cooperative society which now makes the equivalent of about $500/year by sale of the algae to a paper factory.

In the dry season, part of the backwaters becomes a chain of disconnected pools in which mosquitoes breed profusely. After lengthy discussions with the village council *(panchayat)*, it was decided to deepen part of the area to create a 1-ha pond for breeding prawns and to dig a channel to drain the remainder of the marshy area into the pond. Each family contributed one man to the labor force, and in 20 d 28,000 m^3 of mud were removed from the pond area and used for filling other low-lying areas. The sale of the prawns yields an income for the village of the equivalent of $1000 to $2000 per year.

With the mutual confidence engendered by these successful operations other problems are now soluble. For example, science clubs in the village schools have taken on the task of stocking the hundreds of casuarina pits with larvivorous fish.[13]

IV. COMMUNITY PARTICIPATION IN THE CONTROL OF VECTORS OF *BRUGIA MALAYI* IN SHERTALLAI, KERALA, INDIA*

Recently the V.C.R.C. has attempted to transfer the principle of vector control by activities which generate income for the community from the Pondicherry area to the very different biological and socioeconomic situation in Kerala State in southwest India. Shertallai, in Kerala, is probably the only remaining pocket of *Brugia malayi* infection in India. The vector mosquitoes, *Mansonia* species, breed in innumerable ponds infested with the aquatic weeds *Pistia, Eichhornia,* and *Salvinia.* The mosquito larvae breathe by attaching to the roots of these plants, and once the weeds are removed, the vector species cannot breed.

The problem was recognized as early as 1855, when Waring, the Durbar physician of what was then Travancore State, recorded that 1 in 23 of the population was affected by chronic filarial disease in Shertallai. The first state-wide survey was carried out in 1931 by M. O. T. Iyengar, who reported an endemicity rate of 48% for Shertallai region. Because of the high prevalence of chronic manifestations of this disease in this area, they became

* P. K. Rajagopalan, P. K. Das, and K. N. Panicker

known as "Shertallai leg" (Figure 1). During the last three decades, a Filaria Control Organization under the State Government and a *B. malayi* research unit of the National Institute of Communicable Diseases have been functioning in the area, but there has been no impact on the disease. An extensive survey is now being carried out by the V.C.R.C. in Shertallai. The preliminary data show that, based on a sample of over 25,000 people, the microfilaria rate ranged from 1.9 to 6.1% and the disease rate ranged from 9.8 to 15.2%. It has been claimed that Brugian filariasis is on the decline, but the annual incidence of microfilaraemia for the year 1986, calculated by the method of Kimura et al.[14] ranges from 2.04 to 7.59 per 1000 population in the age group 1 to 7, which indicates that transmission is still going on in the area.

The total elimination of the foci of infection from this relatively small area is beset with difficulties due to certain peculiarities of this region. Shertallai taluk in Alleppey district lies in the low-lying region between Vembanad Lake in the east and the Arabian Sea in the west. It covers an area of 304 km^2 with a population of 400,000. The area is waterlogged with numerous canals and over 75,000 small ponds in which *Pistia* and *Eichhornia*, the water hyacinth, proliferate. *Salvinia* is a relatively recent introduction. Three species of *Mansonia* — *M. annulifera, M. uniformis*, and *M. indiana* — breed in these ponds. Because of sandy soil the only crops grown are coconut and colocasia. The ponds are the primary source of water for the inhabitants, and the aquatic vegetation is used for green manuring the coconut plantations. Therefore, there was strong opposition to the use of chemical larvicides or herbicides in these ponds, especially as the people believe that the weeds keep the water clean.

To prevent vector breeding it was first necessary to convince the people that compost is better than green manure for fertilizing the coconuts, and the fast-growing terrestrial legume *Glyricedia* can be grown for composting, thus allowing the ponds to be cleared of weeds without economic loss. It is not possible for any governmental organization to sustain a program of regular removal of the weeds from the ponds. Therefore, the V.C.R.C. has introduced fish culture into the area as an incentive for the people to remove the weeds (Figure 5) which is necessary if the fish are to grow. Fingerlings (young fish) of five species of fast-growing edible species (*Catla catla, Cyprinus carpio, Labeo rohita, Labeo fimbriatus*, and *Cirrhina mrigala*) were purchased from the State Fisheries Department and distributed free to those people who agreed to deweed their ponds. People were provided with seedlings of *Glyricedia* and taught to prepare compost from the mature plants. The obvious economic gains motivated the community to undertake fish culture in their ponds, thereby making them free from vector breeding. Up to October 1987, about a million fingerlings had been stocked in over 20,000 ponds at the rate of 0.5 fish per square meter. For a 100-m^2 pond, with an average expenditure of the equivalent of $1.15 (cost of fingerlings, transportation, and overhead expenses), the net return was calculated to be $25.76/year. When this program covers the entire area, the net return for the people is likely to be over $1.5 million in 5 years.

The strategy of the V.C.R.C. was, through a team of qualified sociologists, to pass on the technology to the people and create sufficient interest in the program so that the entire area could be rid of *Mansonia* breeding. Instead of creating a new infrastructure to achieve this, the various channels of communication already available in the area were mobilized. First, it was found that there was a filariasis patients' association, with a membership of over 125,000, all of whom had chronic manifestations of the disease. Its president, Mr. Benjamin, a retired school headmaster, had contested a seat in parliament and had written a parody based on Milton's *Paradise Lost* to highlight the plight of filariasis patients. His services were utilized for a major health education campaign in schools. The campaign was directed primarily toward girls because of a common belief that people will not choose brides from Shertallai due to filariasis. The movement has therefore caught the imagination

FIGURE 5. Community action for the clearing of aquatic weeds in Shertallai, Kerala. (Photograph by Vector Control Research Center.)

of women in the area. Since Kerala has the highest literacy rate in India, particularly among women, it was not difficult to change some of their cultural practices.

The Urban Basic Services (U.B.S.) of the municipality (under a UNICEF program) has established in the area a system of neighborhood committees to which one representative is elected for every 50 families. Vector control through weed removal and fish culture was propagated through these committees, with the active cooperation of the U.B.S. The National Service Scheme (N.S.S.) Units of all the educational institutions were utilized for general environmental improvement in the area through voluntary labor *(shramdan)*. A filaria-control movement was started by some members of the public and has been registered as a society, and many other organizations have become affiliated to it. This movement, which presses for measures to be taken to prevent filariasis-vector breeding, is slowly gaining momentum. The interns of Alleppey Medical College Department of Community Medicine help with monthly health camps in selected localities and run weekly night clinics. In this process, apart from training the interns in detection and treatment of microfilaria carriers, there was a sense of participation for the Medical College, and the people who come to the health camps and clinics get a feeling that something is being done for them. The Departments of Fisheries and Agriculture are also assisting the program through their extension officers, by popularizing fish culture and compost making, respectively. Hoardings depicting the plight of people with elephantiasis have been erected in many places.

It is hoped that as a result of all these activities, by the end of 1990, at which time the first phase of activities will come to an end, the program will have become self-sustaining in the community.

The disease-vector control strategy followed by the V.C.R.C. in Shertallai is unconventional because the methods prescribed by the national programs cannot be adopted here. If the simple technology can be implemented by the people themselves and the program is sustained on a continuing basis, there is no reason why this strategy will not succeed here and in other areas with similar problems. The key to community participation in vector control is to ensure that there are economic benefits accruing to the people. Unlike malaria, the epidemiology of filariasis is such that disease reduction cannot be demonstrated in a season or in a year. The government programs, carried out according to a time-honored routine, have not made any impact. As a result people were despondent when the present program started. At the first meeting, the Director of Health Services of Kerala was very critical of the word "research" since a research unit of a national institute had been functioning in the area for many years, but the disease had remained unabated. What the V.C.R.C. is doing is also research, but it is socio-economic research within a time frame — to make vector control operations an attractive proposition where the people derive direct benefit in the form of high yield of fish and better manure for their coconut plantations. In the course of time it is hoped to eliminate the stigma associated with the town, Shertallai leg, by long-term suppression of the vectors of the disease.

V. ENVIRONMENTAL AND CULTURAL METHODS FOR VECTOR CONTROL IN PADDY FIELDS*

A. THE PATTERN OF MOSQUITO BREEDING IN PADDY FIELDS

Mosquito larval control in paddy (irrigated rice) fields is a problem of a completely different magnitude from that in other habitats because of the enormous breeding surface involved. In China the malaria vectors *Anopheles sinensis* and *An. anthropophagus* breed principally in paddy fields (see Chapter 2), and in Indonesia *An. aconitus* does the same. In Africa increases in the cultivation of irrigated rice are providing perennial breeding sites for *An. gambiae* (especially, in West Africa, one particular chromosomal form)[15] and *An. arabiensis*.[16,17] A shift in the mosquito fauna in favor of vector species was associated with rice irrigation in Kenya.[18] The natural history of *An. gambiae s.l.* breeding in paddy fields has been studied in The Gambia[17] and in Kenya[16,19,20] where the weeks following transplantation were regularly associated with peaks of vector population.

In India a number of malaria vectors of both major and minor importance have been reported to breed in paddy fields. *An. culicifacies* in Pattukottai Taluka, Thanjavur District, Tamil Nadu;[21] *An. fluviatilis* in North Kanara[22] and the Nilgiris districts of Tamil Nadu;[23] *An. philippinensis* and *An. annularis* in Bengal.[24] However, it was only in Pattukottai Taluka that malaria-vector breeding in paddy fields was extensive enough to be considered a major problem. Fortunately, recent studies have shown that *An. culicifacies* no longer breeds prolifically in this habitat, although it does so in small numbers,[23-27] and the main anopheline whose density has gone up as a result of the increase in the area under rice cultivation in recent years is *An. subpictus*, which is not a recognized vector in inland areas.[28] The main vector control problem in India now associated with paddy fields is that of the culicine vectors of Japanese encephalitis virus (JEV), *Culex tritaeniorhynchus, Cx. pseudovishnui*, and *Cx. vishnui*, all of which breed abundantly wherever rice is cultivated.

Although effective methods of chemical larviciding have been developed against vectors of JEV, it is generally accepted that these would be uneconomical for routine control and should be considered only if limited high-risk areas can be identified. This is never so in India, where the disease is diffused over extended rural areas. However, it has been suggested that agricultural use of insecticides should be coordinated with vector control efforts for maximum effect.[29] Both in Japan and in Korea the serial application of agricultural insecticides, either from a helicopter or from the ground using knapsack sprayers, led to a substantial reduction in vector density as well as JEV infection.[30,31] A variety of organophosphate and carbamate insecticides was used, and four to ten applications were carried out against rice pests per growing season. However, the vector *Cx. tritaeniorhynchus* developed resistance after about 10 years. As a result there was an outbreak of JE in Korea in 1982 with 2975 cases and 280 deaths. Quite apart from the danger of development of

* R. Reuben, D. Raghunatha Rao, L. S. Self, and J. D. Lines

resistance by vectors to agricultural pesticides, this type of intensive spraying of crops is now not possible in India. There is already serious concern about the resurgence, attributed to indiscriminate use of pesticides, of sucking pests of rice, among them the green plant hopper *(Nephotettix virescens)*, the vector of rice tungro virus. The present recommendations of the Indian Agricultural Department are to use insecticides only if signs of pest damage exceed economic threshold levels, which have been defined for most of the key pest species. Thus, some fields may not be sprayed at all, and others may receive one or two treatments within each growing season. The use of organophosphate and carbamate compounds in the paddy fields led to a sharp, but transient, reduction of the mosquito larval population density, because, being intended as either systemic insecticides against borers or leaf-dusts against sucking pests, most of the pesticides remained in the soil or on the leaf blades, respectively, and relatively little reached the water in the field.[32]

Death of many natural predators of mosquitoes has been observed in fields of Kenya sprayed with Dimecron, with strong indications that within 2 weeks of spraying the *An. arabiensis* larval population was larger than before.[20,33] Reiter[34] reported that 8 d after spraying, the anopheline larval population was five times greater than in unsprayed fields.

Thus, use of synthetic chemical insecticides is not an appropriate response to the problems of vector control in paddy fields, and solutions have to be sought elsewhere. Whatever the technology, however, in order to be effective, it has to be acceptable to the agriculturalist. For this purpose it must be developed in collaboration with agricultural scientists as part of a combined field-management strategy which will result in economic benefit to the farmer, who ultimately must apply it.

B. INTERMITTENT IRRIGATION

Intermittent irrigation of paddy fields for control of malaria vectors was suggested as early as 1922, and there is a statement in a League of Nations report[35] about the success of this method in Indo-China. A careful study in Portugal[36] from 1935 to 1939 proved the value of intermittent irrigation for slow-growing varieties of rice in that country. In those days the rice-growing parts of Portugal (unlike other parts of the country) were hyperendemic for malaria transmitted by *An. labranchiae*. In a series of comparisons of intermittent vs. continuous irrigation, yield was generally higher in the former, the quality of the grain was at least as good, and there was a saving in the amount of water used. It is reported that the aquatic stages of anophelines lasted for at least 18 d, and an intermittent system was adopted in which water was intended to be present for no more than 16 or 17 d at a time. When small plots dried out completely, all the larvae were killed, and no later instars were found in the water 4 d after the fields were reflooded. However, in large fields there were usually small depressions which did not dry out completely, and 80% larval control was the level generally obtained. It was confirmed that larvae could not survive when the surface-water film dried off the soil. It was also noted that, unlike fish, the larvae did not have much capacity to locate and congregate in persisting accumulations of water. The disadvantage of converting to intermittent irrigation was that the irrigation and drainage ditches had to be rearranged, but once this was done, the upkeep was no greater than with continuous irrigation. As a result of these studies the Portuguese Government passed a law in 1938 governing rice cultivation which included a clause making intermittent irrigation obligatory when so ordered by the Malaria Service. At the time of writing[36] the law had not been invoked since it was felt that persuasion was better than compulsion.

In China and Japan intermittent irrigation is a well-known technique,[37-39] mainly because of its benefits to the rice plant rather than for mosquito control. In both countries, for 3 to 4 weeks after transplantation and for a short period during ear sprouting, continuous irrigation is considered necessary, but midseason drying ("land baking" or *nakaboshi*) is customary. This is considered beneficial to the crop because it releases toxins such as H_2S and oxygenates

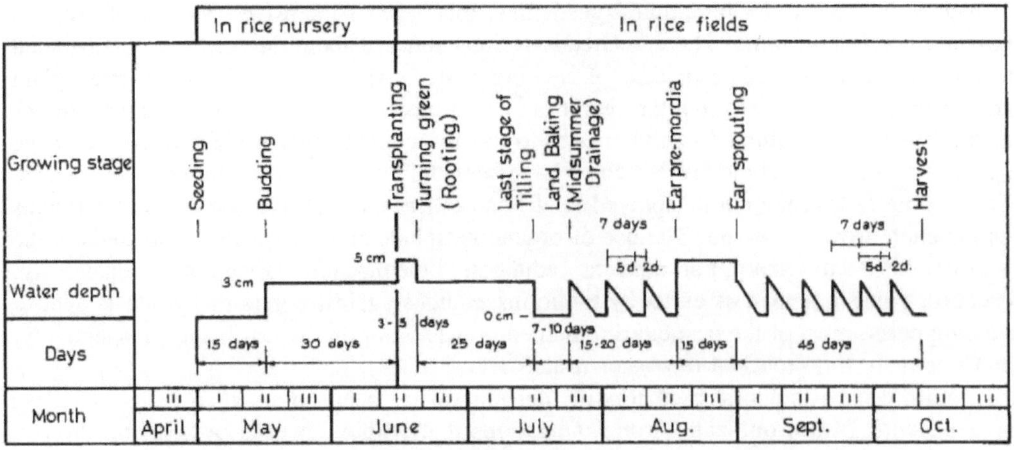

FIGURE 6. Schedule for intermittent irrigation of rice in use in the Jining area of China.[37]

the soil. Since 1977 in the Jining area of China a system has been adopted of admitting water only once a week in the 3 weeks after "land baking" and the 5 weeks before harvest (Figure 6).[41] The fields dry by evapotranspiration and percolation resulting in near dryness for 2 d out of each 7. However, the fields do not become dry enough for long enough to kill mosquito larvae, and it has been suggested that the amount of water admitted should be reduced by 15 to 20% to give 3-d true dryness in each 7. In case of rain, the field should be drained to ensure 3-d dryness per week.

A system of intermittent irrigation with introduction of water on only 1 d in 4 or 5 is successfully used to increase rice yield and as a mosquito control method in 80,000 ha in Henan Province, China, where the soil is sandy and dries quickly. Yield increases of 14% and 74% reduction in mosquito breeding are reported.[41] Intermittent irrigation is used throughout the 100-d rice-growing period (i.e., 20 to 26 floodings in all) apart from the first 2 to 3 weeks after transplanting when water is required not only for the direct benefit of the rice plant, but also to suppress weeds. It is noteworthy that the time of continuous flooding after transplantation, which is universally considered necessary, corresponds to the time of the peak populations of *An. sinensis*.

In India, the first mention of intermittent irrigation was in the Annual Report of the Agricultural Research Station, Aduturai (Tanjore Delta) for 1938—39, which also states that no harm results to the crop if there is no standing water in the field for 2 d between waterings. A detailed series of experiments was carried out over a period of 3 years by Russell and his colleagues in Pattukottai.[40] The soil of Pattukottai taluk is a sandy loam with a low moisture-retaining capacity. Mosquito breeding was successfully controlled when intermittent irrigation, with cycles of 5 wet followed by 2, 3, or 4 dry days, was carried out, and no pupae could be collected during the 5 wet days under any of these regimens. A 5-d wet and 1-d dry cycle did not control breeding. There was no cracking or clod formation in the soil of the paddy fields if the dry periods did not exceed 4 d and the subsoil remained moist, although the surface-water film disappeared. During the period when the northeast monsoon was active, it was not possible to dry out the fields sufficiently to control mosquito breeding, but this was not felt to be a serious defect because the malaria vector *An. culicifacies* did not breed actively at that time. Careful measurements showed that there was no effect on the yield and quality of grain and straw, and that, contrary to popular belief, weed growth did not increase. Subsequently intermittent irrigation was the control method used in paddy fields covering about 800 ha in a successful demonstration of reduction of malaria incidence by various antilarval measures in an area of 18 km^2 which included six villages.[41] This demonstration was never followed up, perhaps because *An. culicifacies* transmitted malaria ceased to be a problem in Pattukottai once the area was fully irrigated and once the wastage of water associated with the early phases of irrigation was minimized.[28]

Russell et al.[40] reported in 1942 that a high-yielding variety of rice (S 705) had been developed which could be grown in sandy soil with only one period of wetting every 7 to 10 d. It is not known what has happened to this strain. There are still areas where a single rain-fed rice crop is raised and where frequently there is not enough water to stand in the fields for long periods — a kind of enforced intermittent irrigation. Elsewhere, where irrigation water is available, the Indian Department of Agriculture presently recommends that a level of 5 cm of water is maintained in nursery fields, and 5 to 10 cm in the main field. In practice these levels are frequently exceeded. In clay soil, hairline cracks appear and rapidly widen on drying out, and this destroys roots of rice seedlings. Continuous flooding of fields is believed to reduce weed growth. Moreover, water supply is carefully regulated in canal systems and must be taken whenever it is made available, particularly by those whose fields are at the ends of canals. All these factors make the individual farmer extremely reluctant to adopt a 5-wet and 2-dry-day cycle. Russell and his colleagues found that an important factor in their success was that they had control of the sluicegates. They doubted whether the farmers, left to themselves, would have carried out any organized community system of irrigation.[40]

Some recent observations suggest that even partial drying out may reduce mosquito breeding. Under the present system fields are partially dried out for a day or so at irregular intervals so that agricultural operations such as deweeding or topdressing with fertilizers can take place. At these times aggregations of mosquito larvae are observed in pools which form in footprints or in other depressions within fields. Preliminary observations suggest that mortality due to overcrowding and predation may be high under such conditions. Bottom-dwelling predators such as dragonfly larvae may be able to capture sizeable numbers of mosquito larvae when the water level is low. In this context, experiments to develop efficient methods of water utilization and scheduling of irrigation, being carried out by the Indian Council of Agricultural Research at 34 centers in different parts of India, are of interest. One such center, in the Lower Bhavani Project command area, in Periyar District (Tamil Nadu), has demonstrated in large-scale on-farm trials that intermittent supply of a 2.5-cm depth of water once in 4 d during the vegetative and ripening phases of rice growth and a 5-cm depth of water once in 4 d during the reproductive phase gave an increased yield of 12.9 and 4.4%, respectively, of the popular rice varieties IR 50 and IR 20, together with 21% saving in water. It is also claimed to have reduced infestation of brown plant hopper *Nilaparvata lugens,* which spreads quickly under conditions of continuous flooding. Such demonstrations are effectively changing traditional attitudes because they result directly in economic gain. The effect of the improved water management systems which are being developed on mosquito-vector populations requires detailed study.

Early ripening varieties have replaced traditional varieties in many rice-growing areas, notably Japan,[39] thus reducing the period during which the fields must be flooded and contributing substantially to the control of paddy-field mosquitoes. The same early-ripening varieties have not found favor with the rice-eating peoples of India, and the most popular of the high-yielding varieties, IR 20 and IR 50, require 135 to 140 d and 105 d, respectively, from transplantation to harvest. Biological control of mosquitoes in rice fields is discussed in Chapter 8.

VI. PLANT POPULATIONS AND VECTOR CONTROL*

A. SHADE CONTROL, MARGINAL PLANTS, AND WATER-AVID TREES

There has been some interest in the use of live plants to inhibit mosquito breeding. The

* J. D. Lines

classic example is the cultivation or removal of shade trees and bushes on the margins of breeding sites, according to the preference of the local vector for shaded or sunlit water.[42] In this regard, a potentially important observation by Reiter[34] does not seem to have been followed up. He noticed that with short-stem rice varieties, a closed canopy is formed by the tillers only a few weeks after transplanting, effectively curtailing the period of intense breeding by *An. gambiae,* which continues for much longer with taller rice varieties.

In India, Bhasker Rao and Ramoo[43] found that in a small stream the density of anopheline larvae was three times greater in clear stretches than where the banks were naturally overgrown with *Ipomea biloba*. Even greater reductions in breeding were artificially achieved by planting *Vitex negundo*. It is not clear whether this was simply an effect of shading or more specifically related to the aromatic leaves of the *Vitex*. In Africa, Goma[44] pursued the observation that while papyrus plants and *An. gambiae* are both common inhabitants of swamps, they are rarely found together. He concluded that this was not due to any direct interaction; it was simply that the areas of the swamp suitable for papyrus were not suitable for the mosquito, and vice versa.

In Gujarat, India, the Malaria Research Centre (which is based in Delhi) has organized a series of community-based forms of environmental management to suppress the breeding of malaria vectors.[45,46] One of the major activities has been the communal planting of over 1 million saplings of trees such as eucalyptus which absorb and transpire large amounts of water. This has had a noticeable effect in drying up the puddles in which the malaria vector *An. culicifacies* breeds. This form of "social forestry" has also created a source of income for the villagers.

The use of floating mats of *Azolla* in paddy fields is described in Chapter 8.

B. VEGETABLE POLLUTION

Plants can be used nonspecifically against *Anopheles* mosquito larvae by the deliberate pollution of water with decaying vegetable matter. A useful review of this and other naturalistic measures against malaria vectors was provided by Hackett et al.[42] at a time when the incorporation of such measures into organized control schemes was at its height and before the advent of DDT caused the naturalistic approach to be neglected. These authors state that "the effect of a fortnight-old brew of cut grass is remarkable. There is no need for the filling to cover the water surface . . . ". Pollution with sisal and sugarcane waste has been used with success in Sri Lanka and the Philippines. There is a danger that the polluted water will become a source of culicine mosquitoes. This may be no problem if these generally bite animals or birds, but could be a fatal objection if they are man-biters. Recently, in Brazil, river pollution by sugarcane waste from the alcohol industry has been a source of nuisance flies.[47] Nevertheless, in most cases vector anophelines will be the greater danger.

Shallow, small-volume, running channels may be packed with a thick and complete layer of herbage over the water. The water is thereby deprived of oxygen, and the mosquito fauna changes from vector anophelines to various culicines.[42] Palm leaves may be used, but in India the best plants were toxin-containing *Cleistanthus* and *Holorrhena* species. It is important that only green-cut vegetation is used because dry straw will only result in a hay-infusion favorable to larval growth. Herbage packing has considerable limitations: it is, for instance, washed away by heavy showers.

C. PLANT SPECIES AS MARKERS FOR MOSQUITO BREEDING

Plant species can serve as markers for conditions suitable, or unsuitable, for breeding of different mosquito species. In the south of France, plant marker species are used in combination with the monitoring of tidal movements to direct chemical larviciding operations, resulting in greater efficiency. More accurate targeting of the insecticide reduces waste and means that breeding can be controlled over a wide area, only 3.3% of which is actually treated in the average year.[48] The method in this case relies on the strong zonation of flora and fauna on tidal shorelines.

In the U.S., control of *An. quadrimaculatus* and other species was achieved mainly through physical means by the Tennessee Valley Authority following the construction of many dams.[49] An important element was knowledge of the plants which encouraged mosquito breeding (some of which were destroyed by stranding or by herbicides) and of those which discouraged it. A useful classification of water plants (microscopic, submerged, floating mat, etc.) emerged from this work.[49] Application of these principals elsewhere could be most useful since the main problem in many larviciding operations is not how to treat areas of mosquito breeding but how to find them all.

REFERENCES

1. **Rajagopalan, P. K. and Das, P. K.,** *The Pondicherry Project,* Vector Control Research Centre, Pondicherry, India, 1987.
2. **Menon, P. K. B. and Rajagopalan, P. K.,** Relative importance of different types of breeding habitats in contributing to the population of *Culex pipiens fatigans* in Pondicherry, *Indian J. Med. Res.,* 71, 725, 1980.
3. **Cairncross, S., Rajavel, A. R., Vanamail, P., Subramanian, S., Paily, K. P., Ramaiah, K. D., Amalaj, D., Mariappan, T., and Srinivasan, R.,** Engineering, mosquitoes and filariasis: a case report, *J. Trop. Med. Hyg.,* 91, 101, 1988.
4. **Menon, P. K. B. and Rajagopalan, P. K.,** Control of mosquito breeding in wells using *Gambusia affinis* and *Aplocheilus blochii* in Pondicherry town, *Indian J. Med. Res.,* 68, 927, 1978.
5. **Rajagopalan, P. K., Panicker, K. N., and Das, P. K.,** Control of malaria and filariasis in south India, *Parasitol. Today,* 3, 233, 1987.
6. **Armitage, P. and Berry, G.,** *Statistical Methods in Medical Research,* 2nd ed., Blackwell, Oxford, 1987, 459.
7. **Bheema Rao, U. S., Panicker, K. N., and Andree, K.,** Malaria persistence in some coastal villages in south India, *Indian J. Med. Res.,* 75, 808, 1982.
8. **Panicker, K. N., Geetha Bai, M., Bheema Rao, U. S., Viswam, K., and Suryanarayana Murthy, U.,** *Anopheles subpictus:* vector of malaria in coastal villages of south-east India, *Curr. Sci. (India),* 50, 694, 1981.
9. **Reuben, R., Kalyana Sundaraman, M., and Suguna, S. G.,** Salinity tolerance of sibling species in the taxon *Anopheles subpictus, Indian J. Med. Res.,* 80, 67, 1984.
10. **Panicker, K. N. and Geetha Bai, M.,** Field release of *Toxorhynchites splendens* in controlling container breeding mosquitoes in a coastal village, *Indian J. Med. Res.,* 77, 339, 1983.
11. **Rajagopalan, P. K. and Panicker, K. N.,** Feasibility of community participation for vector control in villages, *Indian J. Med. Res.,* 80, 117, 1984.
12. **Rajagopalan, P. K. and Panicker, K. N.,** Financial rewards ensure community participation, *World Health Forum,* 6, 174, 1988.
13. **Panicker, K. N. and Rajagopalan, P. K.,** Vector control through integrated rural development, *Indian Counc. Med. Res. Bull.,* 10, 1, 1986.
14. **Kimura, E., Penaia, L., and Spears, G. F. S.,** Epidemiology of sub-periodic bancroftian filariasis in Samoa 8 years after control by mass treatment with di-ethyl carbamazine, *Bull. W.H.O.,* 63, 869, 1985.
15. **Coluzzi, M.,** Heterogenieties of the malaria vectorial system in tropical Africa and their significance in malaria epidemiology and control, *Bull. W.H.O.,* 62, 107, 1984.
16. **Service, M. W.,** Problems of vector-borne disease and irrigation projects, *Insect Sci. Appl.,* 5, 227, 1984.
17. **Snow, W. F.,** Mosquito production and species succession from an area of irrigated rice fields in The Gambia, West Africa, *J. Trop. Med. Hyg.,* 86, 327, 1983.
18. **Chandler, J. A., Highton, R. B., and Hill, M. N.,** Mosquitoes of the Kano Plain. I. Results of indoor collections in irrigated and non-irrigated areas using human bait and light traps, *J. Med. Entomol.,* 12, 504, 1975.
19. **Chandler, J. A. and Highton, R. B.,** The succession of mosquito species (Diptera, Culicidae) in rice-fields in the Kisumu area of Kenya, and their possible control, *Bull. Entomol. Res.,* 65, 295, 1975.
20. **Chandler, J. A. and Highton, R. B.,** The breeding of *Anopheles gambiae* (Diptera: Culicidae) in rice-fields in the Kisumu areas of Kenya, *J. Med. Entomol.,* 13, 211, 1975.

21. **Russell, P. F. and Ramachandra Rao, T.**, On habitat and association of species of anopheline larvae in South Eastern Madras, *J. Malar. Inst. India*, 3, 153, 1940.
22. **Ramachandra Rao, T.**, Behaviour of *Anopheles fluviatilis*. III. Larval habitats in North Kanara District, *J. Malar. Inst. India*, 6, 77, 1945.
23. **Russell, P. F. and Jacob, V. P.**, On the epidemiology of malaria in the Nilgiris District, Madras Presidency, *J. Malar. Inst. India*, 4, 349, 1942.
24. **Sen, P.**, Anopheline breeding in rice fields of Lower Bengal: its relation with culture practices and with growth of rice plants, *Indian J. Malariol.*, 2, 221, 1948.
25. **Chandrahas, R. K., Arunachalam, N., Panicker, K. N., and Rajagopalan, P. K.**, On density measurements of *A. culicifacies* Giles during malaria transmission season in Pattukottai, Thanjavur District, Tamil Nadu, *Indian J. Med. Res.*, 70, 429, 1979.
26. **Tewari, S. C., Appavoo, N. C., Mani, T. R., Reuben, R., Ramadas, V., and Hiriyan, J.**, Epidemiological aspects of persistent malaria along the river Thenpennai (Tamil Nadu), *Indian J. Med. Res.*, 80 (Suppl.), 1, 1984.
27. **Sharma, V. P., Malhotra, M. S., and Mani, T. R.**, Entomological and epidemiological study of malaria in Terai, U. P., in *Facets of Environmental Problems, Five Case Studies*, Krishnamurthy, C. R., Ed., Indian National Science Academy, New Delhi, 1984, 35.
28. **Rao, T. R.**, The Anophelines of India, Monogr. Malaria Research Centre, Delhi, 1984.
29. **Pant, C. P.**, Control of vectors of Japanese encephalitis, mimeographed doc. WHO/VBC/76. 733, World Health Organization, Geneva, 1979.
30. **Kamimura, K. and Katori, K.**, Seasonal occurrence of *Culex tritaeniorhynchus summorosus* in Toyama Prefecture, with special reference to the effect of aerial application of carbaryl dust on rice fields and Japanese encephalitis, *Jpn. J. Sanitary Zool.*, 20, 87, 1969.
31. **Bang, Y. H. and Self, L. S.**, Pesticide spray practices in rice fields and control of culicine mosquitoes in South Korea, mimeographed doc. WHO/VBC/71. 269, World Health Organization, Geneva, 1971.
32. **Rajendran, R.**, Mosquito larval density in relation to agricultural practices, in *Alternatives to Synthetic Insecticides in Integrated Pest Management System*, Proc. Symp., Reuben, R. and Sundara Babu, P. C., Eds., Centre for Research in Medical Entomology, Madurai, India, 1987, 153.
33. **Service, M. W.**, Mortalities of the immature stages of species B of the *Anopheles gambiae* complex in Kenya: comparison between rice fields and temporary pools, identification of predators and effects of insecticidal spraying, *J. Med. Entomol.*, 13, 535, 1977.
34. **Reiter, P.**, The action of lecithin monolayers on mosquitoes. III. Studies in irrigated rice-fields in Kenya, *Ann. Trop. Med. Parasitol.*, 74, 41, 1980.
35. League of Nations, French Indo-China, Rep. of Trop Gen. Med. and Publ. Hlth., Preparatory paper of Inter-Govt. Conf. Far Eastern Countries on Rural Hygiene, League of Nations Health Org. Doc., C.H. 1235, 1937.
36. **Hill, R. B. and Cambournac, F. J. C.**, Intermittent irrigation in rice cultivation, and its effect on yield, water consumption and *Anopheles* production, *Am. J. Trop. Hyg.*, 21, 123, 1941.
37. **Self, L. S.**, Unpublished observations in China.
38. **Service, M. W.**, Vector control in China, *Bull. Soc. Vector Ecologists*, 6, 1988.
39. **Kamimura, K. and Watanabe, M.**, Some aspects of great diminution of *Culex tritaeniorhynchus summorosus* the principal vector of Japanese encephalitis with special reference to the recent advance in agricultural methods, *Botyu-Kagaku*, 38, 245, 1973.
40. **Russell, P. F., Knipe, F. W., and Rao, H. R.**, On the intermittent irrigation of rice fields to control malaria in South India, *J. Malar. Inst. India*, 4, 321, 1942.
41. **Russell, P. F. and Knipe, F. W.**, A demonstration project in the control of rural irrigation malaria by antilarval measures, *J. Malar. Inst. India*, 4, 615, 1942.
42. **Hackett, L. W., Russell, D. R., Scharff, J. W., and Senior White, R.**, The present use of naturalistic measures in the control of malaria, *Bull. Health Org. League of Nations*, 7, 1016, 1938.
43. **Bhasker Rao, R. and Ramoo, H.**, The control of mosquito breeding in canal distributaries by growing certain plants on their banks, *J. Malar. Inst. India*, 4, 409, 1942.
44. **Goma, L. K. H.**, Experimental breeding of *Anopheles gambiae* in papyrus swamps, *Nature (London)*, 187, 1137, 1960.
45. **Sharma, V. P.**, Community based malaria control in India, *Parasitol. Today*, 3, 222, 1987.
46. **Sharma, V. P., Sharma, R. C., and Gautam, A. S.**, Bioenvironmental control of malaria in Nadiad, Kheda district, Gujarat, *Indian J. Malariol.*, 23, 95, 1986.
47. **Laurence, B. R.**, unpublished data, 1981.
48. **Gabinaud, A.**, Ecological mapping to support mosquito control on the French Mediterranean coast, *Parasitol. Today*, 3, 317, 1987.
49. Malaria Control on Impounded Water, Tennessee Valley Authority (with U.S. Public Health Service), U.S. Government Printing Office, Washington, D. C., 1947.

Chapter 8

BIOLOGICAL CONTROL METHODS SUITABLE FOR COMMUNITY USE

R. Reuben, D. Raghunatha Rao, Anthony Sebastian, Philip S. Corbet, Wu Neng, and Liao Guohou

TABLE OF CONTENTS

I. THE APPLICABILITY OF BIOLOGICAL CONTROL

Biological control of mosquitoes has only been seriously attempted on the aquatic stages, and this generally requires that people go and find the breeding sites and release the biological control agent. Perhaps the only exceptions are releases of *Toxorhynchites amboinensis* reported by Focks et al.[1] These mosquitoes have predatory larvae and nonbiting adults which find a considerable number of *Aedes aegypti* breeding sites and lay eggs in them, thus stocking them with predaceous larvae. However, the mass rearing of *Toxorhynchites* is a specialized task because of problems of cannibalism etc.

Biological control attempts against mosquitoes have generally had to be of the augmentative rather than the inoculative type, though there has been a report of persistence for 27 months of the parasitic nematode worm *Romanomermis culicivorax* after two stockings of the breeding sites of *Anopheles albimanus* in the town of El Valle, Colombia.[2] Following the introduction of the worms there was a marked decline in the larval and adult *An. albimanus* populations and in malaria, compared with an untreated control town. However, most experiments with parasites of this type have yielded disappointingly low rates of parasitism and/or short persistence.

Thus, it appears that biological control will require repeated checking of sites and/or restocking with the control agent. It seems unlikely that this would be done on a continuing basis where it required the searching out of mosquitoes breeding in numerous seminatural puddles or seasonally variable marshy pools. The best chances of success would seem to be where mosquitoes breed in wells (see Chapter 7), in clearly definable domestic water containers (see Section II of this chapter), or in paddy (irrigated rice) fields (see Section III), especially where there is an economic incentive for the restocking operations; for example, where the larvicidal agents are edible fish (see Section III.A) or plants or plant products which function as fertilizers of rice crops (see Sections III.D and III.E).

The discovery in the 1970s of bacterial agents specific for and highly toxic to a narrow range of Diptera, including mosquitoes and *Simulium* blackflies, seemed to promise a new era of biorational vector control. However, a decade and a half of intensive research has not yet resulted in extensive use of these agents in operational vector control. The toxins of these bacteria, *Bacillus thuringiensis* ssp. *israelensis* (= serotype H-14 and often called "Bti") and *B. sphaericus* have to be ingested by larvae to kill them, and it has proved difficult to retain the toxins within the feeding range of the larvae for more than a day or two. *Culex* are generally more susceptible than *Anopheles* because they frequently descend to feed on the bottom. Attempts to solve this problem are described in Section III.C.

A trial using *B. thuringiensis* ssp. *israelensis* against *Aedes aegypti* is described in Chapter 6, Section IV. This agent has so far only found large-scale operational use:

1. In the Onchocerciasis Control Program in areas of West Africa where *Simulium damnosum* s.l. has become resistant to organophosphates[3] and where no insecticide is persistent because of the fast flow in the breeding sites
2. In certain mosquito abatement projects in developed countries where there are strong political pressures to avoid damage to nontarget organisms[4]

The effect of *B. sphaericus* is generally more lasting than that of *B. thuringiensis* ssp. *israelensis*, but there has been considerable confusion as to whether this is due to persistence of bacterial spores in the environment or to true recycling, i.e., multiplication of spores in larval cadavers[5] and their subsequent larvicidal effect in the same habitat. The use of *B. sphaericus* in sanitation systems is described in Chapter 10.

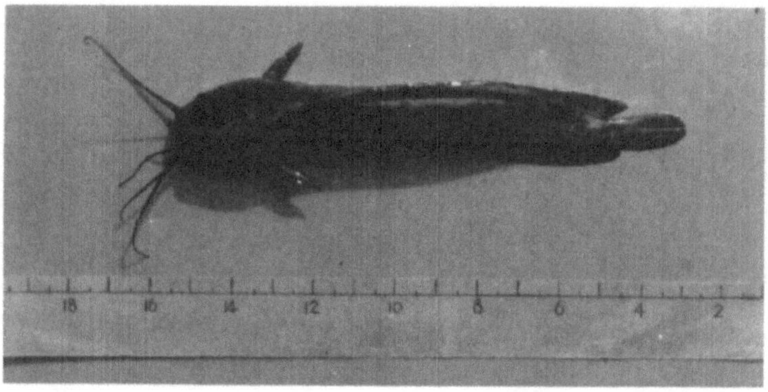

FIGURE 1. Chinese catfish, *Clarias fuscus* (scale in cm). (Photograph by Wu Neng.)

II. WATER-STORAGE CONTAINERS

A. THE USE OF FISH

1. Control of *Anopheles gambiae* s.l. in Water Tanks in Somalia

In the dry parts of northern Somalia underground water tanks ("barkits") provide the only water available for most of the year for breeding of the malaria vector *Anopheles gambiae* s.l. These tanks were stocked with the fish *Oreochromis (= Tilapia) spilurus spilurus* at a density of 0.3 fish per cubic meter.[6] Sharp reductions were reported in larval and adult mosquito populations and in malaria following the stocking in the treated villages, in contrast to the untreated control villages. This is one of the few cases where evidence has been presented for suppression of larval and adult mosquitoes and disease by biological means.

2. Control of *Aedes aegypti* in Guangxi, China*

There was an outbreak of dengue in Guangxi in 1980—1981. The disease was transmitted by *Aedes aegypti* which bred in domestic water-storage containers. In eight villages with a total human population of 66,000, it was decided to use the Chinese catfish *Clarias fuscus* (Figure 1) for larval control following initial emergency use of insecticides to eliminate the virus-infected population of adults. It was required of householders that there was a live fish in every water-storage container at all times. The container infestation rate was recorded in terms of the Breteau index, i.e., the number of infested containers per 100 premises.[7] Initially this index was over 100 (Figure 2), but within a short time of introducing the fish it was greatly reduced and has remained at a low level in the following years.[8] In 1985—1986 there was a dengue epidemic in the neighboring villages, but not in the eight villages in which fish had been stocked. The cost of this use of fish was estimated at 7% that of insecticidal control.

When mature, *C. fuscus* is edible. The young fish are readily obtainable on the market in China. They are resistant to adverse conditions, voracious eaters of mosquito larvae, and can survive in a water container for 6 months to a year provided that they are prevented from jumping out of it. It has to be recognized that the use of fish is a relatively slow-acting method of mosquito control and that if an *Ae. aegypti* population has been allowed to build up so that a dengue epidemic has begun, insecticidal fogging is indispensable for rapid killing of infected mosquitoes.

The effective use of larvivorous fish is not just a biological problem, but a political,

* Wu Neng and Liao Guohou

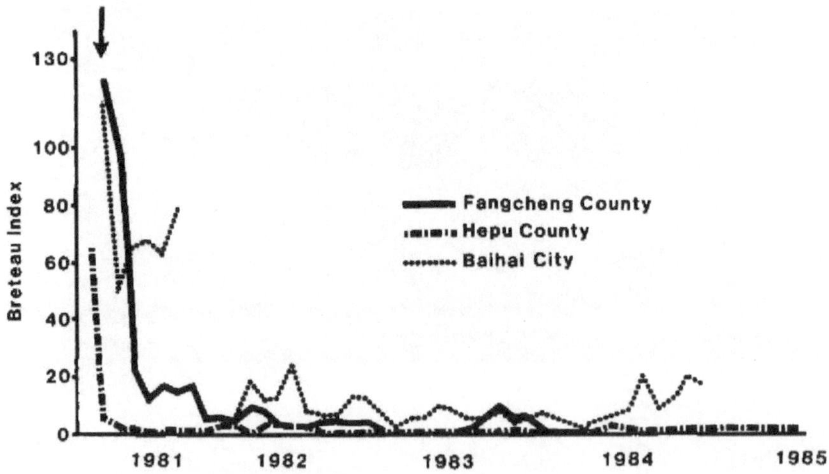

FIGURE 2. The Breteau index (number of containers infested with *Aedes aegypti* per 100 premises) after the introduction of Chinese catfish into containers in villages in Guangxi, China.[8] (Reproduced by permission of the World Health Organization, Geneva.)

legislative, and organizational one. It requires both community participation and a trained professional team for supervision and surveillance.

For the program in Guangxi a working group was set up by the local primary health care personnel (formerly known as "barefoot doctors"), and they were trained in the use of *C. fuscus* and in monitoring *Ae. aegypti* larval infestation. After stocking of the water containers they checked them for the presence of fish and absence of mosquito larvae every 10 to 15 d. In the event of a fish disappearing from a container, the householder was required to replace it within a specified number of days, failing which a fine was levied. The system was maintained throughout the period indicated in Figure 2, and in 1985—1986 when dengue broke out in the surrounding area, the benefits to the treated area became apparent.

B. USE OF DRAGONFLY LARVAE*

1. A Trial In Rangoon

A successful field demonstration of the use of dragonfly (Odonata) larvae for the control of *Ae. aegypti* has been carried out in Rangoon, Burma.[9,10]

Ae. aegypti control in Rangoon is required because it is the vector there of the hemorrhagic form of dengue (DHF). Epidemics with many deaths have recurred annually since a major one in 1970.[9,11] In Rangoon, 98% of *Ae. aegypti* breeding is in domestic water-storage containers, 92% of which are jars, pots, and metal drums.

In 1975, during ecological studies in the Hlaing Study Area in western Rangoon, a water-filled metal drum (capacity 200 l) was being regularly exposed outdoors to collect living larvae of *Ae. aegypti*. In the normal course of events many larvae would have appeared in the drum a few days after its first exposure. However, on one occasion, to the investigator's surprise, the drum contained only newly hatched larvae. Inspection of the contents of the drum revealed the presence of dragonfly larvae of the anisopterous family Libellulidae. The only species of dragonfly which appeared in drums exposed in the field was *Bradinopyga geminata* (Rambur) which, according to Kumar,[12] often oviposits in cement tanks (i.e., water-storage cisterns) and sometimes in large steel vessels.

Subsequent laboratory studies showed that when medium-sized *B. geminata* larvae were confined for 24 h in beakers containing 50 of each developmental stage (the four larval

* A. Sebastian and P. S. Corbet

instars and pupae) of *Ae. aegypti*, each dragonfly larva consumed on average 133 ± 21 larvae and pupae per 24 h — a rate more than three times that achieved by full-grown larvae of the predatory mosquito *Toxorhynchites splendens* when confined under laboratory conditions. The dragonfly larvae consumed most first-instar mosquito larvae and progressively fewer of each successive stage.[9]

In the field, metal drums (as described above) and pegu jars (oval, dark-brown, elongated, glazed jars with a round opening 35 cm in diameter and a capacity of 90 l) in daily use were selected for study. Numbers of larvae and pupae of *Ae. aegypti* in each container were counted before introduction of two medium-sized dragonfly larvae and then daily thereafter. The numbers of larvae and pupae in ten such containers before introduction of dragonfly larvae ranged from 87 to 780 (average 458.8). Under these conditions it required only 4 to 9 d for all *Ae. aegypti* to be eliminated by the dragonfly larvae, which then continued to maintain control during the 1-month study period.

Two study areas were then chosen for a comparative trial. Each featured a high density of *Ae. aegypti* and similar conditions of housing and water supply and storage. The area selected for treatment (about 0.02 km^2) was a squatter settlement of 282 houses, accommodating 447 family units (a total of 2262 adults and children). The houses were single (93%) or double storey with walls of bamboo mat or timber and roofs of thatch or galvanized sheet iron. Members of each family stored their daily requirement of water in their own containers. The control, or untreated, area was closely similar to the test area though smaller (about 0.01 km^2 with 511 adults and children). It may be noted that only for a relatively sedentary mosquito species like *Ae. aegypti* would a trial in such a small area have any chance of avoiding being swamped by immigrants from outside.

All major containers able to support aquatic stages of *Ae. aegypti* were listed in both areas. By public announcements and talks, the local community was encouraged to cooperate in the forthcoming biological control trial. All but one of the residents did so readily, perhaps partly because some of the storage containers already harbored dragonfly larvae. Indeed, enthusiasm was such that householders asked that regular releases of dragonfly larvae continue after the trial had been completed.

Monitoring, undertaken by trained scouts, was conducted fortnightly in both areas, beginning with the onset of the monsoon, 4 weeks before dragonfly larvae were first released. At each visit the number of larvae and pupae of *Ae. aegypti* was recorded in all listed containers, as the Larval Density Index defined by the Service;[7] adult incidence was recorded in a standardized way at certain fixed stations as the Indoor Resting Density, expressed as the average number of females collected per man-hour.[13]

Treatment entailed placing medium-sized dragonfly larvae in each listed container on four occasions at 4-week intervals. For this, large numbers of dragonfly larvae were needed. Despite its natural occurrence in drums and jars in daily domestic use, *B. geminata* proved unsuitable for this purpose, partly because larvae could not readily be reared in captivity and partly because adults (from which to obtain eggs) could not be obtained at all times of year. Instead, another species of libellulid, *Crocothemis servilia* (Drury), was chosen. It was common in garden pools in the neighborhood, easy to rear in captivity, and available as adults throughout the year. A female *C. servilia*, captured in the field, could be induced to lay up to 1500 eggs without interruption when her abdomen was dipped into water.

At 29 to 34°C more than 90% of eggs hatched within a week, and the larvae then fed readily on copepods (initially) and mosquito larvae, becoming large enough to release 3 weeks after hatching. Larvae reared in this way were transported to release sites in large containers furnished with *Elodea*, to provide perches for the larvae, and placed in domestic containers (four larvae per major source of *Ae. aegypti*) using a spoon.

The trial was a success. All criteria used to measure larval numbers showed that (despite the great initial increase of *Ae. aegypti* larvae after the onset of the monsoon) larvae were

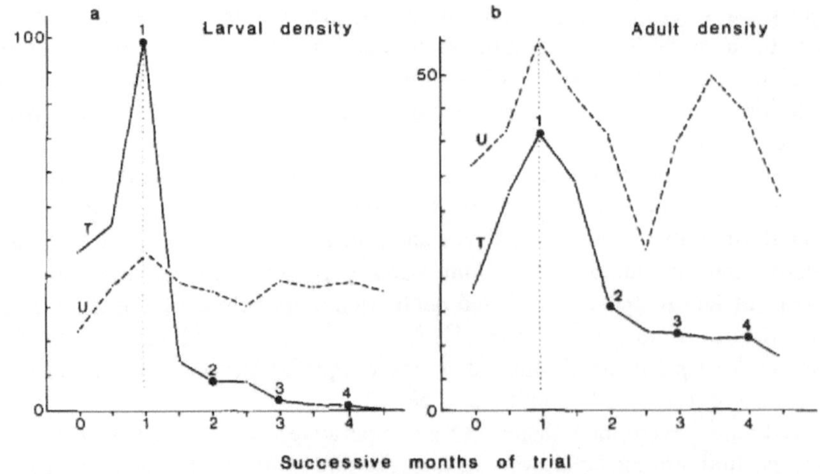

FIGURE 3. Suppression of *Aedes aegypti* by release of dragonfly larvae in Rangoon from mid-May through September 1979. (a) The Larval Density Index; (b) the average adult Indoor Resting Density in treated (T) and untreated (U) settlements during a field trial, the onset of which is indicated by a vertical dotted line. Closed circles mark successive releases of dragonfly larvae.

reduced by more than 85% within the first 2 weeks of the treatment, and then almost eliminated during the following 3 months (Figure 3a). There was also marked and unequivocal reduction of adults (Figure 3b), but it was less abrupt and less extensive than for larvae, presumably because of influx from outside the treated area and because some adults which had emerged before treatment began could have survived for days or weeks afterwards.

2. Conditions for the Applicability of Dragonflies

Dragonflies are generalized, obligate predators which are known from laboratory studies to be able, as larvae, to consume large numbers of larvae and pupae of mosquitoes.[14,15] No hard evidence is available of dragonfly larvae achieving acceptable suppression of mosquito populations in the field without human intervention. As generalized predators, dragonflies in nature are unlikely ever to be sufficiently abundant or selective to keep mosquitoes down to levels which would satisfy human requirements.[16] Even with human intervention in the form of augmentative release, dragonflies are unlikely to exert effective control except in domestic water containers, as in the Rangoon trial.

In this trial, augmentative release of 3-week-old *C. servilia* rapidly achieved virtual elimination of larvae and pupae of *Ae. aegypti* in the treated area.

For use in domestic water containers, dragonflies with the attributes of *C. servilia* have the following advantages: they cling to the sides of the container and so, unlike other predators, are unlikely to be accidentally flushed or scooped out by the householder; their development is relatively slow; they do not leave the container until emergence; and they remain alive in containers for up to 3 months, sometimes with very little food. *C. servilia* has a wide geographical distribution and ecological valency and is often abundant, so it is exceptionally suitable among Anisoptera, although probably not uniquely so. It occurs commonly as an early colonizer in lowlands from Iraq in the west, throughout east Asia, with subspecies in the Papuan region and Australia,[17] and it has recently become established in North America.[18] Four species of *Crocothemis* occur in Africa, one of these being common also in south, east, and central Europe, central and west Asia, and Arabia.[19]

A prerequisite for success in the Rangoon trial was the participation of the local community. Whereas such involvement will always depend partly on the approach adopted by the initiator, it must be recognized that in some places people would object to predators being placed in the water that they use for domestic purposes. It is possible that some predators would contaminate the water with pathogens or parasites harmful to humans. In the case of dragonflies, this is extremely unlikely, unless the householder were to eat a larva without first cooking it, whereupon trematode metacercariae might be ingested.[20]

It remains to be determined whether the scale of a control operation with dragonflies could be stepped up enough to accomplish suppression throughout a city the size of Rangoon. At present, colonization of *C. servilia* has not been achieved, and this could constitute a serious drawback at campaign level. Dragonflies are notoriously difficult to colonize, only one species having so far been reared through several successive generations in captivity.[21] Perhaps the public could be encouraged to use naturally occurring *B. geminata* collected from major sources where it may be found established naturally in large numbers, sometimes attaining about 200 in a concrete tank of 1.9-m^3 volume or 10 per metal drum or pegu jar containing water for daily domestic use. In such places attempts could be made to raise numbers by exposing more receptacles as surrogate oviposition sites. If an adequate source of supply could be assured, it can be concluded that biological control of *Ae. aegypti* is feasible using augmentative release of dragonfly larvae. Where it is possible to use dragonflies in this way, they may constitute the method of choice, especially where, as in Burma, regular treatment of drinking water with an insecticide such as temephos (Abate) is too expensive in foreign exchange or because people object to the taste of the insecticide.

III. BIOLOGICAL CONTROL OF MOSQUITOES IN PADDY FIELDS

A. LARVIVOROUS FISH*

Until recently the most promising larvivorous fish have been considered to be the "top minnows", *Gambusia* and *Aplocheilus* followed by *Oryzias* and *Poecilia*, because they are small, hardy, and carnivorous, can flourish in shallow waters, and are "worthless and insignificant" as food.[22,23] (This last attribute will certainly not protect them in the fish-eating areas of the northeastern Indian states, where the smallest of fish are trapped and fried to provide a tasty addition to the basic diet.)

Gambusia affinis has been tested for mosquito control in paddy fields on numerous occasions. For example, at a stocking rate of 250 to 300/ha in California it gave 82% control of *Culex tarsalis* and *Anopheles freeborni* larvae.[24-26] In Uzbekistan, U.S.S.R., at stocking rates of 2500 to 50,000/ha it gave 96.4 to 100% control of *Culex* larvae,[27] and in Louisiana stocking 12,500 to 50,000/ha gave 80 to 96% control of *Psorophora confinis*.[28] However, it has more recently been shown that if the biomass of *Gambusia* is allowed to become too high (more than 500 g/ha) consumption of invertebrate predators of mosquitoes may lead to increases in the larval density of *Cx. tarsalis*.[29]

Gambusia may eat the eggs of other fish, but the importance to be attached to this is controversial.[30] Its low fecundity and viviparity make mass rearing difficult, and despite the use of solar energy, production costs in the U.S. could not be reduced below $16.5/kg.[31] Thus, it seems unlikely that the hundreds of millions of fish which would be needed to stock the major rice growing areas of the world could be produced economically.

In the Republic of Korea trials with *Aplocheilus latipes* and *Aphyocypris chinensis* (formerly named *Zacco platypus*) at a stocking density of 10,000 to 20,000/ha reduced the density of the larvae of malaria and Japanese encephalitis vectors by 80 to 90% and maintained them at that level for 6 weeks.[32,33]

Low priority was given to large food fishes, such as *Labeo, Catla, Cyprinus, Ctenopharyngodon idella*, etc., whose fry eat mosquito larvae, and it is only recently, since

* Wu Neng, Liao Guohou, R. Reuben, and D. Raghunatha Rao

emphasis has shifted to community participation in vector control, that their potential has been realized. Another possible approach is to encourage the use of a mixture of compatible species, including food fish to provide economic incentive and the classical larvivorous fish which are too small by themselves to provide a worthwhile harvest.

A paddy field, in order to be suitable for pisciculture, should have stronger and higher dikes than usual to retain water up to the desired depth and to prevent escape of the fish during flooding. Inlets and outlets should be screened. An essential feature is the presence of a deeper pond, or trench near the outlet, where fish can shelter when not foraging among the rice plants or when the field dries out. This also makes capture and harvesting easy. In West Bengal, the quantity of manure ordinarily applied is increased by 50 to 100% for fish and rice culture, by addition of cow dung, bonemeal, superphosphate, etc.

According to Nie and Wang,[34] paddy fields in China are richer in nutrients and plankton than conventional fish ponds, and fish grow faster in the fields. However, Jhingran[35] points out that because they are shallow, paddy fields are subject to wide temperature and oxygen fluctuations, and there may be adverse pH and low phosphate. According to Jhingran,[35] in India they are relatively poor in plankton production. Therefore, he considers that they provide a poorer environment than a pond, and species suitable for culture in rice-fields must be capable of withstanding these conditions and of growing to a marketable size in a short period.[35] In India a number of species have been successfully cultured on an experimental scale as well as more extensively, particularly in Kerala and West Bengal. The most interesting experiment, from the point of view of vector control, is one carried out jointly by the Central Inland Fisheries Research Institute and the Central Rice Research Institute, Cuttack, Orissa, in 1958 with *Oreochromis (= Tilapia) mossambica,* the fingerlings (young stages) of which feed voraciously on mosquito larvae. Fingerlings of *O. mossambica* and *Cyprinus carpio* were stocked at 2500/ha, and fish yield, at the end of 3 months, was about 77 kg/ha of which *O. mossambica* constituted 40%.[35] *O. mossambica* has also been found to be suitable for rice-fish culture in the Philippines.[36] *C. idella*, the grass carp, eats weeds as well as mosquito larvae, and its utility for mosquito as well as weed control has been demonstrated in China,[37] U.S.S.R.,[38] and India.[39,40]

Breeding edible fish in paddy fields has a long history in China, especially in the south and east. Recently, the Chinese Ministry of Agriculture and Fisheries has strongly advocated this system, and more than 700,000 ha of paddy fields are stocked with edible fish. The main species are a special strain of *Cyprinus carpio* (whose name may be translated as the "foolish carp") which is particularly well suited to paddy fields and the grass carp, *C. idella*.

Figure 4 shows the decline in larval and adult mosquito densities achieved after introduction of fish into paddy fields early in each season, compared with control areas where fish were not used.[41] Table 1 shows the average levels of control of mosquito larvae (and in some cases adults) reported in various parts of China and the U.S.S.R. after stocking with carp. Figure 5 shows the effect of predation by carp in shifting the age distribution among the aquatic stages toward first and second instars. In the last 8 years the percentage of paddy fields in Quangzhou county, Guangxi, with carp breeding has risen from 0 to 43%, and the fields near villages have been especially heavily stocked. Over the same period annual incidence of malaria per 100,000 population has declined from 11.6 to 0.1. It would seem that predation by the fish on the larvae of the malaria vectors must have been a contributory factor, especially as in other parts of Guangxi where fish rearing is not so popular, malaria incidence has not declined.[41]

The main motive which has stimulated the increase in carp rearing in China is not mosquito control, but production of edible fish for sale, which can add 70% to the farmer's income per hectare.[34,46] In the U.S.S.R. yields of 94 to 274 kg of fish per hectare of paddy field have been reported.[38] During the 90-d growth of a rice crop carp grow to an edible

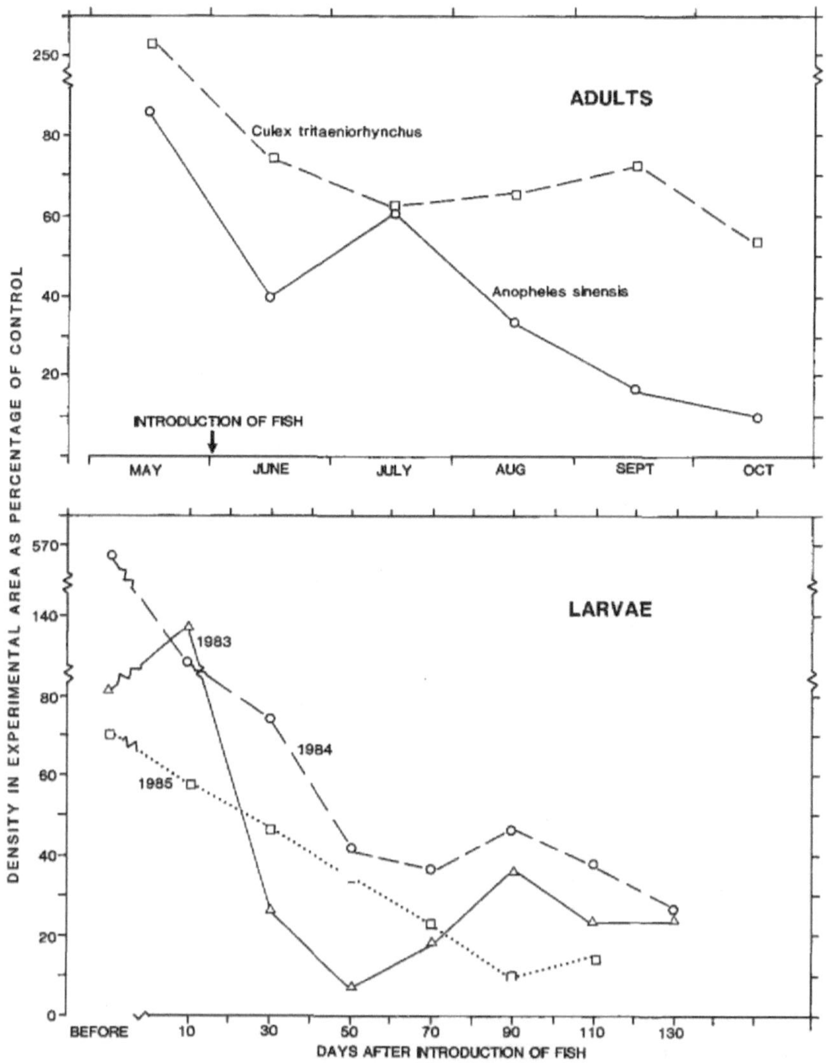

FIGURE 4. Effects on larval and adult mosquito populations following annual stocking of fish in rice fields of Quanzhou county, Guangxi, China. Larval density in experimental and control areas was determined by dipping, and adult density by timed collections in buffalo sheds.[41]

and marketable weight of 50 to 160 g; however, grass carp are better left to continue their growth during the next crop of rice.

The presence of large fish in the fields increases rice yields in Chinese conditions by an average of 10%.[46] This arises because

1. The fish selectively eat weed grasses which may decrease weed populations by 93% and remove the need for the farmer to weed the crop.[47]
2. The fish excrete 60 to 70% of their food intake, and the excrement raises the content of the main plant nutrient elements in paddy fields with fish compared with those without fish.[48]
3. The stirring action as the fish swim in the shallow water in search of food breaks up the film of algae and clay on the soil surface, which otherwise reduces the oxygenation of the soil.[49]

TABLE 1
Mosquito Control by Rearing Edible Fish in Rice Fields in China and the U.S.S.R.

No. fish stocked/ha (size of fish introduced in cm)			Reduction in larval mosquito population (%)			
Carp	Grass carp	Mixture of carp and grass carp	*Anopheles*	*Culex*	Both *Culex* and *Anopheles*	Ref.
2,850 (9—12)	1,590	—	80	99.5	—	42
8,850 (3—3.5)	2,100	—	89	99.5	—	42
—	—	15,600 (5—12)	72	95	—	42
4,500	—	—	50	80	—	43
—	—	12,000—15,000	69	91	—	38
3,150	1,350	—	—	—	57[a]	44
—	—	9,000 (3.5—5)[b]			70[c]	45

[a] 22 to 50% reduction in adult mosquito population.
[b] *Carassias* also included among the fish introduced.
[c] 76% reduction in adult mosquito population.

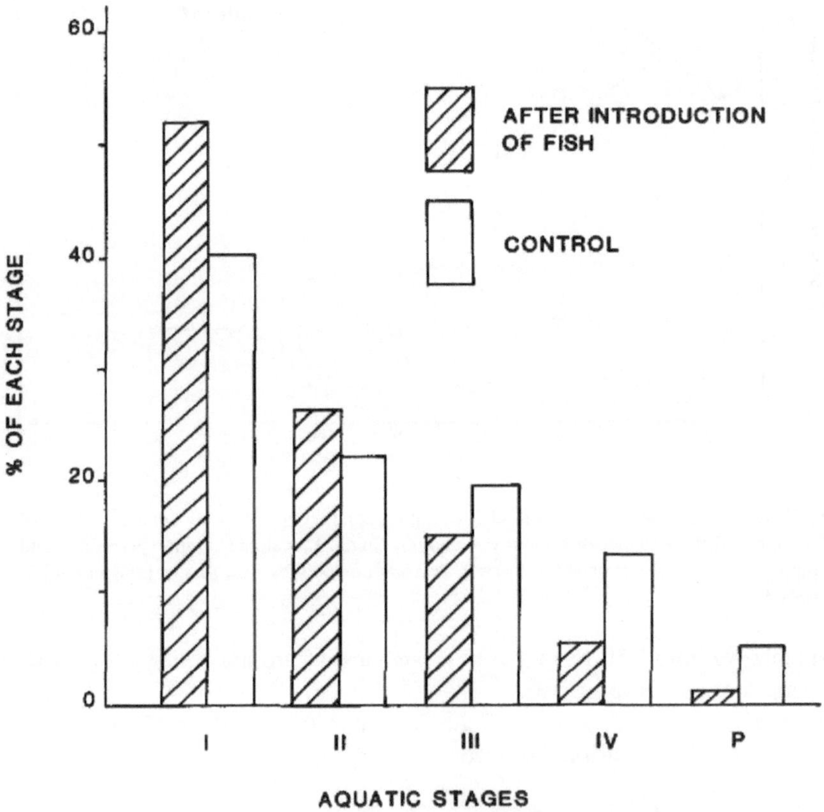

FIGURE 5. Distribution of aquatic stages of mosquitoes between the larval instars and pupae in the presence and absence of fish breeding in rice fields, Guangxi, China.[41]

In contrast to the situation in the U.S.S.R. and China, carp rearing has not been carried out on a large scale in paddy fields in India. Predation may be a problem in some areas. In an experiment with common carp, conducted for 10 weeks from September to December 1958 near Cuttack, the survival of the stocked fish was 0 to 3% due to predation by birds and otters. The average yield of fish was only 6 kg/ha, although the survivors had attained an average of 144 g from an initial weight of 4 g each.[35] The availability of large numbers of fish in shallow water could lead to local congregations of herons and egrets. These birds are known to be amplifying hosts and sources of Japanese encephalitis (JE) virus, and care should be taken not to aggravate the situation in JE-prone areas by causing these birds to flock together in large numbers.

With the introduction of high-yielding rice varieties, the use of insecticides, herbicides, and fungicides toxic to fish has increased, and some authorities feel that fish culture is no longer compatible with rice culture.[35] Even if fish are not killed outright, they would, because of their position at the highest trophic level in the paddy-field ecosystem, bioaccumulate considerable amounts of pesticide residues either directly from the water or through food, and become unfit for consumption. Such considerations have curtailed paddy-field fish culture in Japan, where a greater harvest of rice is preferred to a supplementary crop of fish and a lower yield of rice. However, it is reported from China that, when appropriately applied, chemical fertilizers and insecticides do not kill fish, and there is a deliberate effort to avoid pesticides which have a cumulative toxic effect and can be stored within the body tissues of fish.[40] Newer high-yielding varieties are being developed which are resistant to insect attack and insect-borne diseases; this will decrease the need for protection with pesticides in the future, making rice-fish culture feasible again.[36] For the present, fish culture for food as well as mosquito control in rice fields may be useful in those areas in India where traditional rice varieties with durations from 3 to 8 months are still being grown. These varieties can usually tolerate relatively deeper water and also have stronger roots than the high-yielding varieties, making them suitable for this type of practice.

B. NATURAL PARASITES AND PREDATORS*

There has been a great deal of interest in natural pathogens and predators of mosquito larvae in paddy fields in recent years. In rice fields near Pondicherry natural infestations with a nematode, *Romanomermis iyengari,* ranged from 0.4 to 7.5% in anophelines, and 1.1 to 11.2% in Culicines in different months, whereas infestations with a fungal agent, *Coelomomyces* sp., varied between 0.3 and 12.3% in anophelines and 0.0 to 1.05% in culicines. *Coelomomyces* appeared to be of little importance as a mortality factor for culicine larvae, being totally absent during 5 months of a year-long study.[50] The aquatic bug *Anisops bouvieri* was an important predator in the same area. However, natural levels of parasitism and predation have reached equilibrium with prey densities which permit sufficient adult emergence to maintain significant levels of disease transmission. Mass rearing these agents for inundative releases is not at present being considered for practical application. However, it should not be forgotten that natural mortality factors may account for as much as 97.6% to 99.1% of larval mortality in rice fields, most of which has been shown to be due to predation.[51] Thus, it is important that this natural advantage should be retained and that technology for control of mosquito larvae in rice fields should as far as possible be compatible with natural predators and parasites. Since there are no comparable data from the period when insecticides were not so extensively used, it is impossible to quantify the long-term effect of destruction of natural enemies on the increase in mosquito populations which has occurred in recent years, but short-term increases in anopheline populations following the spraying of Kenyan paddy fields with Dimecron are reported in Chapter 7.

C. BACTERIAL AGENTS*

One of the main obstacles to the effective use of bacterial agents against *Anopheles* larvae is the settling out of the spores from the feeding zone of the larvae which filter food particles at the air-water interface. Particles sinking lower than 1 to 2 mm are not ingested,

* R. Reuben and D. Raghunatha Rao

but surface-bound particles such as wheat flour are rapidly consumed, suggesting that a bait formulation with a spreading agent might be very effective.[52] Corn oil was chosen as a spreading agent, and a combination of this with wheat flour and 2% of *Bacillus thuringiensis* ssp. *israelensis* primary powder proved to be about 50 times more effective against several *Anopheles* species than an aqueous suspension of equivalent strength of the primary powder.[52] The sensitivity of the toxin to ultraviolet light remains as a problem to be solved.

Formulations which release the toxins at a sustained rate have been devised to provide this agent with the persistence it otherwise lacks. Floating briquette formulations have controlled anopheline and culicine larvae in concrete cisterns, wells, and casuarina pits in India for 26 to 40 d.[53] Such briquettes might be tried in paddy fields, but they would have to be anchored in some way as otherwise they tend to drift with the wind resulting in incomplete coverage.

Granular formulations help to penetrate dense vegetation and then float within the feeding zone of *Anopheles* larvae. Granular formulations using corn grit as a core and incorporating one or other of the commercially available preparations of *B. thuringiensis* ssp. *israelensis* were tested in dense stands of maturing rice.[54] Complete mortality of early instars of *Anopheles* was obtained and 92 to 100% mortality of late instars confined in floating cages. Laboratory studies showed release of the active ingredient from the granules, and some of it floated on the surface for up to 2 d.

Another major problem to be solved is that of application over extensive areas. It was shown that mosquito breeding in small rice plots in the U.S. could be controlled using a wettable powder formulation of *B. thuringiensis* ssp. *israelensis* at a dosage of 4.4 ppm.[55] Over 99% control of breeding in rice plots was obtained with two commercial formulations of this bacterium and one locally produced at the V.C.R.C., Pondicherry, India (Bactimos, Teknar, and Deltox at 0.5 to 0.75 kg/ha), and with a strain of *B. sphaericus*, VCRC B-64, at 1.5 kg/ha. However, none of them showed any persistence, with the larval populations showing recovery by the fourth day.[56] Another formulation of *B. sphaericus* (Biocide-S) produced at Anna University, Madras, gave sustained control of culicine vectors of JE for 15 d in a small rice plot treated at 1.1 kg/ha.[57] However, if bacterial agents are to be used operationally in such habitats, methods must be developed to treat extensive rice-growing tracts. Spraying from the air has been used for this purpose, but the cost is exorbitant. A more feasible method is to introduce a formulation into an irrigation system and to allow water movement to distribute the toxin over the entire flooded area. The first attempt to do this employed rice hulls, as carriers for *B. thuringiensis* ssp. *israelensis*, mixed with dissolved gelatin as a sticker. These were broadcast with a manually operated device at one end of a field, but after flooding they were swept into one corner, with the result that larval control was not achieved elsewhere.[58]

Better results were obtained by point source introduction of a liquid flowable concentrate of the bacterium using a simple constant flow device.[59] An operational control trial was carried out over a period of 41 d in 72 separate rice-fields, 52 of which (735 ha) were treated and 20 (425 ha) were untreated controls.[60] The area was quite variable, including some fields with difficult access and others belonging to farmers who flooded very slowly, which reduced the efficiency of introduction of the insecticide. The sites for dispenser placement were selected on the basis of physical and hydrologic characteristics, and dispensing sites were shifted downfield daily to suit the flooding schedule. The amount of formulation delivered was 467 ml/ha, and it was estimated that at full flooding a concentration of 0.2 ppm had been achieved. Because of operational problems, some pans remained untreated or were inadequately treated, but overall, larval counts were reduced to less than 0.5 per dip, and up to 76% control was achieved. This appears to be a simple and effective approach to the treatment of habitats of all types which are irrigated by flowing water from a restricted number of sources, which were hitherto virtually unmanageable. However, the amount of the formulation used was considerable, and cost-benefit analysis has yet to be worked out.

One other approach remains which may turn out to be a simple and elegant solution to the problems of application. While genetic engineering has chiefly been confined to selecting and cloning highly virulent strains of bacteria with multiple copies of the DNA sequence responsible for toxin production, it has recently been reported that the toxin gene of *B. thuringiensis* ssp. *israelensis* has been transferred to blue-green algae, and studies are underway to introduce it into algae that normally grow in the feeding zone of the target mosquito larvae.[61] If successful, due consideration will have to be given to environmental safety before introduction. Although resistance of *Culex quinquefasciatus* to *B. thuringiensis* ssp. *israelensis* did not develop under laboratory selection pressure,[62] the presence of a self-proliferating insecticidal agent in the environment might well succeed in precipitating resistance.

Because of the safety of *B. thuringiensis* ssp. *israelensis* and *B. sphaericus* to man and other nontarget organisms, several workers have favored their production as a cottage industry, pointing out that locally available cheap materials, ranging from broths of boiled meat scraps or bones, sugarcane bagasse, or even dung, can be used as fermentation substrate.[63] Jaggery (molasses) medium has been successfully used in India.[64] Cottage industry production under uncontrolled conditions would result in a product of relatively low potency, and there may be a risk of overgrowth of the bacterial cultures by dangerous bacteria such as those causing anthrax. However, proponents of this approach point out that packaging and distribution problems would be eliminated, and large quantities of the product could be applied by unskilled workers using buckets.

Perhaps more practical is the decision to develop indigenous products in developing countries and to encourage local industry to take these up. This would permit fermentation parameters to be standardized to give a high-potency product subject to normal quality control checks and determination of shelf life and stability, while still retaining the advantage of cheapness. This is being vigorously pursued in India.

D. WATER PLANTS*

There have been reports that the water fern *Azolla imbricata* which grows in rice fields and can completely cover the water surface, causes enhanced mortality among larvae of *Anopheles peditaeniatus* and *An. sinensis.*[65] Species of *Azolla* grow very abundantly in China, and a project is under way there to see if the larval population of rice fields can be reduced by growing this plant.[66] In India, species of *Azolla* are being recommended, because of their ability to fix nitrogen, for cultivation in rice fields and later utilization as organic fertilizer. *Azolla pinnata* was found to inhibit egg laying by *Culex quinquefasciatus* and *Anopheles culicifacies* in the laboratory, but no effect on hatchability or larval survival was observed.[57] Recent field trials in India showed that once a complete mat of *Azolla* is formed this suppresses all breeding. However, this takes about 10 d, and peak culicine pupal production occurs during the first fortnight after flooding of the fields.

E. NEEM*

Many plants produce chemicals with antifeedant or insecticidal properties, which could potentially be exploited to provide pesticides which are safer and considerably less environmentally damaging than synthetic chemicals. Among the most promising is the neem tree *(Azadirachta indica)*. The insecticidal and medicinal properties of neem have been known for centuries. Among other things, the leaves are traditionally used to protect clothes and stored grain from insect attack, and are plowed into the soil as manure. Powdered neemcake, i.e., the residue after the extraction of oil from the seed kernel, was applied to rice fields as early as 1932 at about 45 g/m^2. This was reported to protect the crop from rice stemborer and termites and to serve as manure.[67] The Indian Agricultural Department today

* R. Reuben and D. Raghunatha Rao

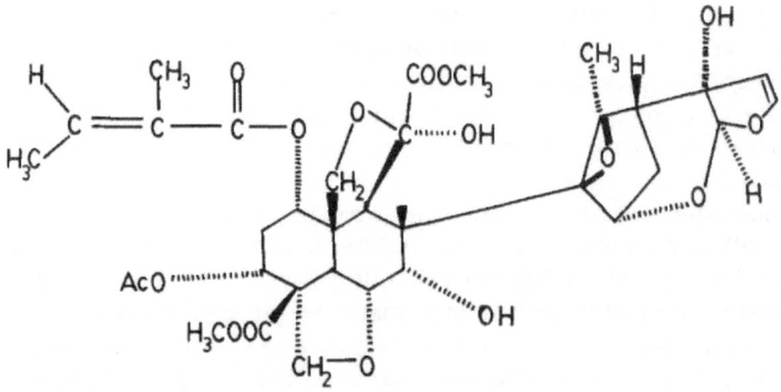

FIGURE 6. Structure of azadirachtin.

recommends the use of neem-coated urea in rice fields, based on the finding that neem-cake inhibits the growth of nitrifying bacteria in the soil and, thus, favors better utilization of urea with resultant better crop yield. This is being widely practiced by farmers. Recent reports that incorporation of neem-cake in the soil can inhibit sucking pests of rice,[67] soil nematodes, and mosquito larvae in the standing water in the fields[68] have raised hopes that low-cost technology can be developed to control simultaneously mosquito vectors and rice pests, and to increase grain yield as well.

1. Chemical Nature of Neem

At least 35 secondary plant metabolites have been identified in neem, and new compounds are still being added to the list. The majority of these compounds are triterpenoids, and there are also some diterpenoids and flavonoids.[69] Not all of them have been studied from the point of view of insecticidal activity, but it is likely that there are several active principles, particularly in view of the range of effects which have been reported on various insects. These include repellency, antifeedant properties, mortality when ingested, reduced growth and abnormal development, production of deformed and miniature adults, and sterilization.[70] The most potent antifeedant and growth regulatory compound is azadirachtin,[71] a triterpenoid which occurs in all parts of the plant, but which can be obtained in higher concentrations from the seed kernel along with other active compounds. The structure of azadirachtin is shown in Figure 6.

2. Effect of Neem on Mosquito Larvae

There are several laboratory studies of the effect of neem products on mosquitoes. Chavan et al.[72] tested various extracts against *Cx. quinquefasciatus*, and Gill found them effective against *Anopheles stephensi*.[73] Zebitz[74] found that exposure of fourth-instar larvae of *Aedes aegypti* to water extracts resulted in conspicuous disruption of growth, whereas exposure of the first-instar larvae caused a prolongation of the larval period and eventually about 90% mortality. Exposure of fourth-instar larvae of *Cx. quinquefasciatus* to water-soaked neem-cake powder resulted in a 50% mortality and 50% of deformed larvae at a concentration of 0.5%, and total mortality at higher concentrations.[68] Neem-cake powder was incorporated in the soil in rice-field simulation trays at concentrations of 12.5, 25, 50, and 100 g/m^2. One set of trays was exposed to full sunlight throughout the day, while a replicate set was partly shaded by trees. Egg rafts of *Cx. quinquefasciatus* were introduced into the trays daily. In the trays kept in full sunlight hatching occurred normally, but there was heavy mortality of larvae at all concentrations, in comparison with the control tray without neem. The effect persisted up to 40 d after transplanting of the seedlings, after which the experiment had to be terminated. Deformities were particularly common at 12.5 g/m^2. In the partially shaded trays there was no effect on larval populations. This was interpreted to mean that the neem-cake powder did not putrefy rapidly at lower temperatures to release the active principles into the water. Observations were also made in farmers' fields in which an experimental plot was treated at the rate of 50 g/m^2 and sampled periodically until harvesting. The predominant species in the field were *Cx. pseudovishnui* and *Cx. tritaeniorhynchus*. There was low culicine density in the experimental plot and high larval density in the control for the first 4 weeks, after which there was low larval incidence in both. There was a sudden rise in the larval population of both plots after about 2 months, and the high level persisted until harvesting. Clearly neem was not effective 2 months after application. Throughout the observations the density of anopheline larvae was high in both plots. The reason for this may lie in the different feeding habits and behavior of anophelines and culicines. Neem-cake powder settles on the bottom where it decomposes to release active compounds. The culicine larvae frequently come down to the bottom to browse and probably ingest the released terpenoids, which show insect growth regulatory (IGR) activity. The anophelines are purely surface feeders and are therefore unaffected.[68]

A second experiment[57] was carried out in a 0.6-ha field with good bunds (channel banks) and separate inlets and outlets for irrigation and drainage. The field was treated with a total of 300 kg of neem-cake, and the results are shown in Figure 7. There was a marked and sudden reduction in numbers of both early- and late-instar larvae in the treated compared with the untreated site. However, reduction in number of pupae was only about 40%. This disappointingly small effect on the pupae was presumably due to density-dependent regulation improving the chances of survival of those larvae which survived the neem. Thus, neem does not seem to give adequate control on its own but might contribute to an integrated pest management system, particularly as the method is highly acceptable to farmers. In the experiment just described the grain yield was enhanced by 25% in the neem-treated plot, resulting in a profit equivalent to about $100 to the farmer.

3. Formulations of Neem

The use of crude extracts of neem, such as neem-cake powder for culicine control in rice fields, shows considerable promise. Crude extracts are natural slow-release formulations containing a wide range of chemicals with differing modes of action, and insects would need to develop a battery of enzymes to become resistant to all of them. Thus, the likelihood of resistance developing appears to be remote, although critical experiments subjecting insect populations to selection pressure with neem products do not appear to have been carried out. However, if this technology is to be widely used, there will be problems of production and supply. One neem tree is estimated to produce about 50 kg of seed in a season, and 1000 kg of seed yields 139.5 kg of neem oil (used in soap manufacture and in Ayurvedic medicine) and 241.8 kg of neem cake.[75] There are about 14 million neem trees in India, of which only about 25% are utilized at present.[76] Even if they could be fully exploited, neem cake is bulky and rots or grows *Aspergillus* if improperly stored. One would have to envisage a "social forestry" approach, with each village growing trees and grinding seeds for its own use. In fact, there are already moves in the Philippines to grow and distribute neem seedlings on a large scale.

The alternative approach is to use formulations of neem. Isolation or synthesis of azadirachtin is expensive and therefore impracticable.[77] Moreover, the pure chemical is photo- and thermolabile. However, it is possible to obtain, by selection of solvents and a few steps of purification, azadirachtin-enriched fractions which also contain most other active compounds.[78] Thus, a relatively stable lyophilized water extract of neem-seed "bitters" has proved to be very potent in preliminary tests with agricultural pests,[79] and four new extracts known as Vemidin, Nemidin, Nemol, and Vepcidin are extremely effective against both

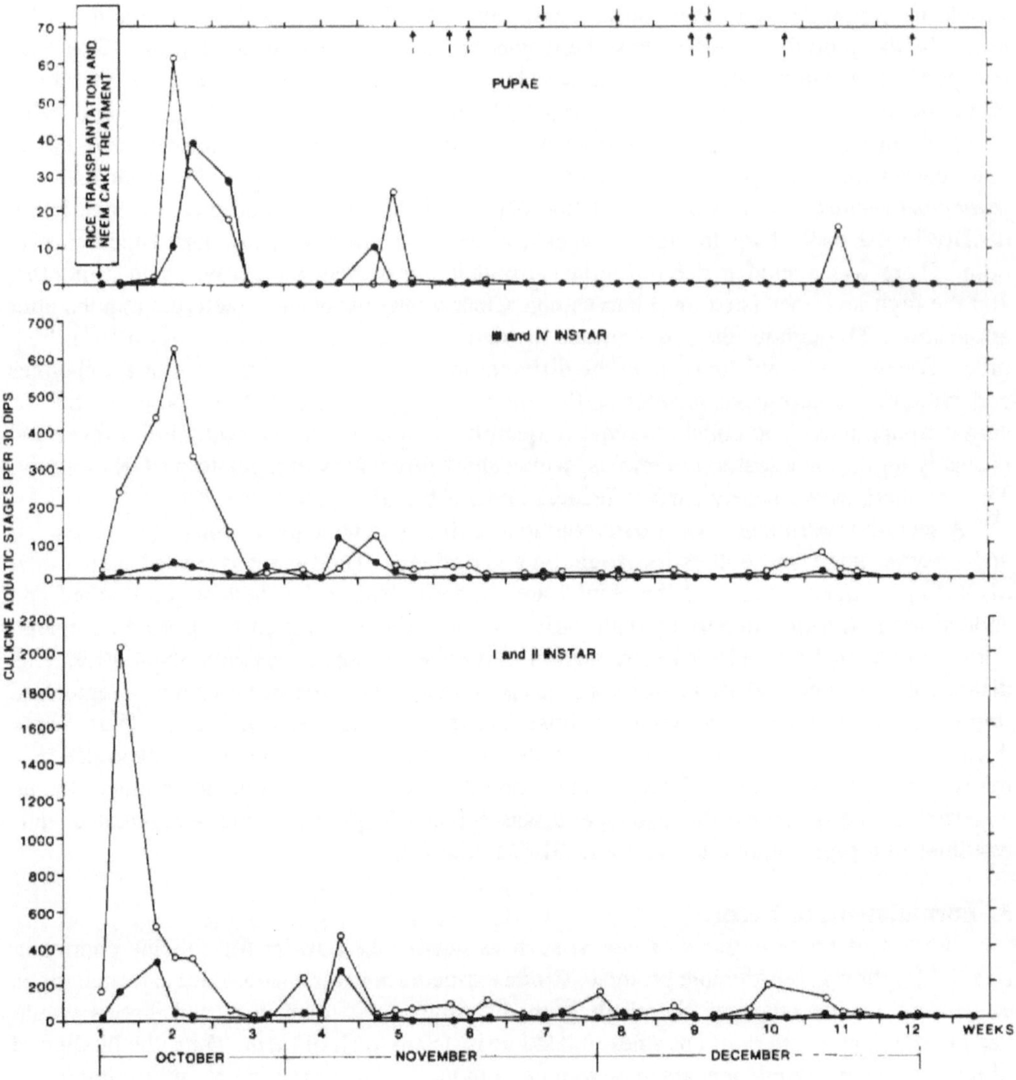

FIGURE 7. Results of a field trial[57] with neem cake (480 kg/ha) for mosquito larval and pupal control in a paddy fields. The graphs show the density, as determined by dipping, of pupae (top graph), third- and fourth-instar larvae (middle), and first- and second-instar larvae (bottom). Closed circles refer to the field treated with neem cake, and open circles refer to the control. Occasions when the fields were dried are indicated at the top of the diagram by solid arrows (control field) and dotted arrows (treated field).

Culex and *Anopheles* larvae.[57] The advantages of these fractions are that they can be standardized and formulated, their shelf life can be defined, and they become susceptible to the usual quality-control procedures. This is a considerable advantage because neem grown in different places and climates differs in azadirachtin content, the African neem being more potent than the Indian, and within India a higher azadirachtin content is associated with dry climates. It is claimed that the extraction processes are simple and cost effective.[80] If this is so, an extremely potent new range of natural products would become available. A granular formulation has been developed and will soon be tested in rice fields.

4. Effect on Nontarget Organisms of Neem

While neem products can be expected to affect all insects, there are several reports that predatory arthropods, such as mirid bugs and wolf spiders, and various parasites of agricultural pests have been relatively unaffected by treatments which had a significant effect on the pests themselves.[81] This suggests that active principles of neem are rapidly broken down into nontoxic compounds within the tissues of the target pests, thus adding partial selectivity to their other advantages. However, the presence of neem seems to have an antagonistic effect on the growth of *Azolla* (Section D above), and the effects on pathogens and predators of mosquito larvae in the paddy field ecosystem needs to be examined, particularly their effect on larvivorous fish such as *Gambusia*. Infants given oral doses of up to 5 ml of neem oil for minor ailments developed vomiting, drowsiness, and encephalopathy within hours of ingestion. Injection of an aqueous extract of the leaves caused appreciable contraction of guinea pig ileum. However, in smaller doses and when applied externally neem products are generally safe and are widely used in traditional medicine in India and in East Africa, where its Swahili name *(mwarobaini)* refers to the fact that it is considered to cure 40 diseases.

REFERENCES

1. **Focks, D. A., Sackett, S. R., Dame, D. A., and Bailey, D. L.,** The ability of *Toxorhynchites amboinensis* to locate and oviposit in artificial containers in an urban environment, *Environ. Entomol.*, 12, 1073, 1983.
2. **Rojas, W., Northrup, J., Gallo, O., Montoua, A. E., Montoua, F., Restrepo, M., Nimnich, G., Arango, M., and Echavarria, M.,** Reduction of malaria prevalence after introduction of *Romanomermis culicivorax* in larval *Anopheles* habitats in Colombia, *Bull W.H.O.*, 65, 331, 1987.
3. **Kurtak, D. C.,** Insecticide resistance in the Onchocerciasis Control Program, *Parasitol. Today*, 2, 20, 1986.
4. **Becker, N. and Ludwig, H. W.,** Mosquito control in West Germany, *Bull. Soc. Vector Ecologists*, 8, 85, 1983.
5. **Mulla, M. S., Darwazeh, H. A., Davidson, E. W., Dulmage, H. T., and Singer, S.,** Larvicidal activity and field efficacy of *Bacillus sphaericus* strains against mosquito larvae and their safety to non-target organisms, *Mosq. News*, 44, 336, 1984.
6. **Alio, A. Y., Isaq, A., and Delfini, L. F.,** Field trial on the impact of *Oreochromis spilurus spilurus* on malaria transmission in northern Somalia, mimeographed doc. WHO/MAL/85.1017, World Health Organization, Geneva, 1985.
7. **Service, M. W.,** *Mosquito Ecology: Field Sampling Methods*, Applied Science, London, 1976.
8. **Wu, N., Wang, S., Han, G., Xu, R., Tang, G., and Qian, C.,** Control of *Aedes aegypti* larvae in household water containers by Chinese catfish, *Bull. W.H.O.*, 65, 503, 1987.
9. **Sebastian, A., Thu, M. M., Kyaw, M., and Sein, M. M.,** The use of dragonfly nymphs in the control of *Aedes aegypti*, *Southeast Asian J. Trop. Med. and Public Health*, 11, 104, 1980.
10. **Corbet, P. S.,** Using dragonflies to suppress mosquitoes in domestic water-storage containers, *Waterlines*, 4, 10, 1986.
11. **Bang, Y. H.,** Integrated management of urban mosquito vectors of human diseases, *J. Commun. Dis.*, 17, 1, 1985.
12. **Kumar, A.,** The life history of *Bradinopyga geminata* (Rambur) (Odonata: Libellulidae), *Gurukul Kangri Vishwa Vidayalya, J. Sci. Res.*, 5, 50, 1973.
13. Technical Guide for Diagnosis, Treatment, Surveillance, Prevention and Control of Dengue Haemorrhagic Fever, World Health Organization, Geneva, 1975.
14. **El Rayah, E. A. and Abu Shama, F. T.,** The predatory capacity of nymphs of the dragonflies: *Trithemis annulata scortecii* Nielsen (Odonata: Zygoptera), *Rev. Zool. Afr.*, 91, 381, 1977.
15. **Lien, J. C. and Matsuki, K.,** [On the larvae of two species of the genus *Lyriothemis* on Taiwan (Libellulidae; Odonata)] (in Chinese), *Nat. Insects*, 14, 57, 1979.
16. **Corbet, P. S.,** *A Biology of Dragonflies*, Witherby, London, 1962.
17. **Lieftinck, M. A.,** Odonata from Ceylon, *Entomol. Scand.*, Suppl. 1, 188, 1971.
18. **Paulson, D. R.,** Additional record of *Crocothemis servilia* (Drury) from Florida (Anisoptera: Libellulidae), *Not. Odonatol.*, 1, 29, 1978.
19. **Pinhey, E.,** *A Descriptive Catalogue of the Odonata of the African Continent (up to December 1959)*, Companha de Diamantes de Angola, Lisbon, 1962.
20. **Manning, G. S., and Lertprasert, P.,** Studies on the life cycle of *Phaneropsolus bonnei* and *Prosthodendrium molenkampi* in Thailand, *Ann. Trop. Med. Parasitol.*, 67, 361, 1973.

21. **Grieve, E. G.,** Studies on the biology of the damselfly, *Ischnura verticalis* Say, with notes on certain parasites, *Entomol. Am.*, 17, 121, 1937.
22. **Covell, G.,** *Malaria in Bombay*, Government Central Press, Bombay, 1928.
23. **Hora, S. L. and Mukerjee, D. D.,** Table for the Identification of Indian Fresh Water Fishes, with Description of Certain Families and Observations on the Relative Utility of the Probable Larvivorous Fishes of India, Health Bull. No. 12, Government of India Press, New Delhi, 1937.
24. **Hoy, J. B. and Reed, D. E.,** Biological control of *Culex tarsalis* in a California rice field, *Mosq. News*, 30, 222, 1970.
25. **Hoy, J. B. and Reed, D. E.,** The efficacy of mosquitofish for control of *Culex tarsalis* in California rice fields, *Mosq. News*, 31, 567, 1971.
26. **Hoy, J. B., O'Berg, A. G., and Kauffman, E. E.,** The mosquitofish as a biological control agent against *Culex tarsalis* and *Anopheles freeborni* in Sacramento Valley rice fields, *Mosq. News*, 31, 146, 1971.
27. **Zainiev, S. A. and Muminov, M. S.,** [The larvicidal effectiveness of *Gambusia* in rice fields in the Uzbek S.S.R.] (in Russian), *Med. Parazitol. Parazit. Bolezni*, 61, 74, 1983; *Rev. Appl. Entomol. Ser. B.*, 72 (Abstr.), 211, 1983.
28. **Craven, B. A. and Steelman, C. D.,** Studies on a biological and a chemical method of controlling the Dark Rice Field mosquito in Louisiana, *J. Econ. Entomol.*, 61, 1333, 1968.
29. **Bence, J. B. and Murdoch, W. W.,** Some indirect effects of *Gambusia* upon *Culex tarsalis* survival in rice paddies, *Bull. Soc. Vector Ecologists* 8, 37, 1983; *Rev. Appl. Entomol. Ser. B.*, 74 (Abstr.), 2391, 1985.
30. **Gerberich, J. B. and Laird, M.,** Larvivorous fish in the biocontrol of mosquitoes with a selected bibliography of recent literature, in *Integrated Mosquito Control Methodologies*, Vol. 2, Laird, M. and Miles, J. W., Eds., Academic Press, London, 1985, 47.
31. **Hoy, J. B.,** Experimental mass-rearing of the mosquitofish, *Gambusia affinis*, *J. Am. Mosq. Control Assoc.*, 1, 295, 1985.
32. **Yu, H. S.,** Biological control of mosquito larvae by the use of indigenous larvivorous fishes in South Korea, *Korean J. Entomol.*, 13, 85, 1983.
33. **Yu, H. S., Lee, D. K., Na, J. O., and Ban, S. J.,** Confined field release of native fish *(Aphyocypris chinensis)* and biological control of encephalitis vector *(Culex tritaeniorhynchus)* in rice paddies of endemic area in South Korea, *Korean J. Entomol.*, 14, 94, 1984.
34. **Nie, D. S. and Wang, J. K.,** [Studies on the economic value of the rice-fish commensal ecosystem] (in Chinese), *Tech. Sci. Inf. Aquat. Prod. China*, 6, 1, 1983.
35. **Jhingran, V. G.,** in *Fish and Fisheries of India*, Hindustan Publishing Corporation of India, Delhi, 1983, chap. 16.
36. **Anon.,** Larvicidal effects of pisciculture in paddy fields, *Chin. Med. J.*, 3, 276, 1977.
37. **Luh, P. L.,** The present status of biocontrol of mosquitoes in China, in *Biocontrol of Medical and Veterinary Pests*, Laird, M., Ed., Praeger, New York, 1983, 54.
38. **Shumkov, M. A., Motenkov, Y. M., Starchikov, A. V., and Shilenko, Y. V.,** [Production evaluation of the use of fry of the Grass Carp in biological control of mosquitoes in rice-growing areas] (in Russian), *Med. Parazitol. Parazit. Bolezni.*, 50, 75, 1981; *Trop. Dis. Bull.*, 79 (Abstr.), 722, 1982.
39. **Panicker, K. N., Srinivasan, R., Viswam, K., and Rajagopalan, P. K.,** Larvivorous potential of some cypriniformes fishes, *Indian J. Med. Res.*, 82, 517, 1985.
40. **Grover, J. H.,** Rice-fish culture and the green revolution, in Advances in Aquaculture, papers presented at the FAO Tech. Conf. on Aquaculture, Kyoto, Japan, Pillay, T. V. R. and Dill, W. A., Eds., 1976, 223.
41. **Wu Neng,** Guangxi, China, unpublished data.
42. **Cheng, J. S. and Gau, K.,** [The evaluation of mosquito control by fish breeding in rice fields] (in Chinese), *J. Zool.*, 3, 237, 1959.
43. Henan Malaria Control Program, [Observations on mosquito control by pisciculture in rice fields] (in Chinese), *Chin. Med. J.*, 3, 142, 1976.
44. **Li, C. M.,** personal communication from Anti-Epidemic Station, Chengolo, Sichuan, China.
45. **Li, P. S. and Xu, R. M.,** [Observations on the efficacy of mosquito control by pisciculture in rice fields] (in Chinese), *Bull. Biocontrol China*, 2, 135, 1986.
46. **Nie, D. S.,** [The theory of rice-fish commensal ecosystem] (in Chinese), *Sci. Man*, 1, 10, 1984.
47. **Yang, Z. H.,** [Pisciculture in the rice field in relation to increasing rice yield] (in Chinese), Symp 2nd Meet. Pisciculture in Rice Fields, Hubei Provincial Society for Aquatic Products, 1983, 24.
48. **Nie, D. S.,** [Pisciculture in rice fields] (in Chinese), *Freshwater Fishery China*, 1, 28, 1985.
49. [Pisciculture in Paddy Fields] (in Chinese), Aquatic Products Dept., Agricultural Bureau of Sichuan Province, China, 1979.
50. **Chandrahas, R. K. and Rajagopalan, P. K.,** Mosquito breeding and the natural parasitism of larvae by a fungus, *Coelomomyces* and a mermithid nematode, *Romanomermis*, in paddy fields in Pondicherry, *Indian J. Med. Res.*, 69, 63, 1979.

51. **Mogi, M., Miyagi, I., and Babrera, B. D.,** Development and survival of immature mosquitoes (Diptera: Culicidae) in Philippine rice fields, *J. Med. Entomol.*, 21, 283, 1984.
52. **Aly, C., Mulla, M. S., Schetter, W., and Xu, B.,** Floating bait formulations; increase effectiveness of *Bacillus thuringiensis* var. *israelensis* against *Anopheles* larvae, *J. Am. Mosq. Control Assoc.*, 3, 583, 1987.
53. **Balakrishnan, N., Pillai, P. K. G. K., Kalyanasundaram, M., and Balaraman, K.,** Efficacy of a slow releasing formulation of *Bacillus thurigiensis* H-14 against mosquito larvae, *Indian J. Med. Res.*, 80, 580, 1984.
54. **Lacey, L. A. and Inman, A.,** Efficacy of granular formulations of *Bacillus thuringiensis* (H-14) for the control of *Anopheles* larvae in rice fields, *J. Am. Mosq. Control Assoc.*, 1, 38, 1985.
55. **Hembree, S. C., Meisch, M. V., and Williams, D.,** Field test of *Bacillus thuringiensis* var. *israelensis* against *Psorophora columbiae* larvae in small rice plots, *Mosq. News*, 40, 67, 1980.
56. **Kramer, V.,** Evaluation of *Bacillus sphaericus* and *B. thuringiensis* H-14 for mosquito control in rice fields, *Indian J. Med. Res.*, 80, 642, 1984.
57. Centre for Research on Medical Entomology, Madurai, India, unpublished data.
58. **McLaughlin, R. E. and Billodeaux, J.,** Effectiveness of *Bacillus thuringiensis* var. *israelensis* against *Psorophora columbiae* breeding in rice fields, *Mosq. News*, 43, 30, 1983.
59. **McLaughlin, R. E.,** An improved constant flow device for dispensing liquid mosquito larvicides to flowing water, *Mosq. News*, 43, 232, 1983.
60. **McLaughlin, R. E. and Vidrine, M. F.,** Point source introduction system for control of *Psorophora columbiae* (Diptera: Culicidae) by *Bacillus thuringiensis* serotype H-14: results of a large-area operational test, *Environ. Entomol.*, 13, 366, 1984.
61. **Marsac, N. T., Torrec, F., and Szulmajster, J.,** Expression of the larvicidal gene of *Bacillus sphaericus* 1593M in the cyano-bacterium *Anacystis nidulans* R2, *Mol. Gen. Genet.*, 209, 396, 1987.
62. The potential for developing resistance to *Bacillus thuringiensis* H-14 toxin, in Resistance of Vectors and Reservoirs of Disease to Pesticides, *W.H.O. Tech. Rep. Ser.*, No. 737, World Health Organization, Geneva, 1986, 43.
63. **Hertlein, B. C., Hornby, J., Levy, R., and Miller, T. W.,** Prospects of spore-forming bacteria for vector control with special emphasis on their local production potential, mimeographed doc. WHO/VBC/80.791, World Health Organization, Geneva, 1980.
64. **Balaraman, K., Bhatia, M. C., Tripathi, S. C., and Hoti, S. L.,** Large scale multiplication of *Bacillus thuringiensis* H-14 asporogenic mutant and *B. sphaericus* strains for mosquito control, *Indian J. Med. Res.*, 85, 266, 1987.
65. **Mogi, M., Okazawa, T., Miyagi, I., Sucharit, S., Tumrasvin, W., Deesin, T., and Khamboonruang, C.,** Development and survival of anopheline immatures (Diptera: Culicidae) in rice fields in northern Thailand, *J. Med. Entomol.*, 23, 244, 1986.
66. **Anon.,** Managing the environment for the control of disease vectors, *WHO Chron.*, 40, 226, 1986.
67. **Saxena, R. C., Justo, H. D., and Epino, P. B.,** Evaluation and utilization of neem cake against the Rice Brown Planthopper, *Nilaparvata lugens* (Homoptera: Delphacidae), *J. Econ. Entomol.*, 77, 502, 1984.
68. **Rao, D. Raghunatha,** Assessment of neem (*Azadirachta indica* A. Juss) cake powder as a mosquito larvicide, in *Alternatives to Synthetic Insecticides in Integrated Pest Management Systems*, Reuben, R. and Sundara Babu, P. C., Eds., Centre for Research in Medical Entomology, Madurai, India, 1987, 163.
69. **Rao, K. N. and Parmar, B. S.,** A compendium of chemical constituents of neem, *Neem News Lett.*, 1, 39, 1984.
70. **Singh, R. P.,** Extraction of Neem seed kernel and its significance in insect pest management, in Mid-Term Review Workshop on Botanical Pest Control in Rice-Based Cropping Systems, June 1—6, Coimbatore, India, 1987.
71. **Schmutterer, H.,** Ten years of neem research in the Federal Republic of Germany, in Proc. 1st Int. Neem Conf., Rottan-Egern, 1980, 21.
72. **Chavan, S. R., Deshmukh, P. B., and Renapurkar, D. M.,** Investigations of indigenous plants for larvicidal activity, *Bull. Haffkine Inst.*, 7, 23, 1979.
73. **Gill, J. S.,** Studies on Insect Feeding Deterrents with Special Reference to the Extracts of the Neem Tree, *Azadirachta indica* A. Juss., Ph.D. thesis, University of London, 1972.
74. **Zebitz, C. P. W.,** Effect of some crude and azadirachtin-enriched neem *Azadirachita indica* seed kernal extracts on larvae of *Aedes aegypti*, *Entomol. Exp. Appl.*, 35, 11, 1984.
75. **Kandaswamy, A. and Raveendran, N.,** Neem production and marketing system in Tamil Nadu — a preliminary study, in Mid-Term Review Workshop on Botanical Pest Control in Rice-Based Cropping Systems, June 1—6, Coimbatore, India, 1987.
76. **Ketkar, C. M.,** *Utilization of Neem,* (Azadirachta indica *A. Juss) and Its By-products*, Directorate of Non-Edible Oils and Soap Industry, Khadi and Village Industries Commission, Bombay, India, 1976.
77. **Kubo, I. and Nakanishi, K.,** Some terpenoid insect antifeedants from tropical plants, in *Advances in Pesticide Science*, Geissbihler, H., Ed., Pergamon Press, Oxford, 1979, Pt. 2, 284.

78. **Feuerhake, K. and Schmutterer, H.,** Development of a standardized and formulated insecticide from a crude neem kernel extract, *Z. Pflanzenkr. Pflanzenschutz*, 92, 643, 1985.
79. **Saxena, R. C.,** Neem seed "Bitters" for management of planthopper and leafhopper pests of rice, in Mid-Term Review Workshop on Botanical Pest Control in Rice-Based Cropping Systems, June 1—6, Coimbatore, India, 1987.
80. **Srimannarayana, G.,** personal communication, 1987.
81. **Jacobson, M.,** Pharmacological and toxicological effects of neem and Chinaberry on warm-blooded animals, *Neem News Lett.*, 3, 39, 1986.

Chapter 9

GENETIC CONTROL BY TRAPPING

Toshiaki Ikeshoji, Peter Langley, and Ludvik Gomulski

TABLE OF CONTENTS

I. THEORETICAL CONSIDERATIONS*

The term ''genetic control'' may be used broadly to cover a variety of control techniques dependent on mating and usually interfering with reproduction.[1] They exploit the mate-seeking behavior of males and are species specific, thus avoiding the damage to nontarget organisms which may result from insecticidal control. Where these nontarget organisms are relatively slow-breeding predators of the pest species, their destruction can lead to surges of pest density soon after insecticide application (see chapter 7).

There have recently been successes in Mexico against agricultural pests: the screw worm fly[2] and Mediterranean fruit fly.[3] Among vectors there have been several successful field trials of genetic control of tsetse flies by sterile male release, including eradication over an area of several hundred square kilometers in Nigeria[4] and some promising smaller-scale trials with mosquitoes.[5] However, all of these required the establishment of sizeable rearing plants staffed by specialists, which make these techniques inappropriate for local use by a community.

Another potential drawback of methods dependent on the release of reared insects is the possible loss of mating competitiveness of sterile males reared in factory conditions, as a result either of the rearing environment or of selection for a factory-adapted genotype which may lead to assortative mating between released males and females.[6] A further possible objection to male releases arises with the tsetse flies where these could, in the short term, add to the vector population.

It may be possible to enjoy the advantages of genetic control without the disadvantages if, instead of rearing, insects are trapped from the wild population, chemically sterilized, and released back into the wild population to mate there with other wild insects. In most medically important species of mosquito, males form mating swarms over contrasting features in the landscape. Females fly into the swarms, and the males use their elaborate antennae to locate the females by the sound of their wingbeats.[7] It is, therefore, possible to trap mosquito males with a combination of artificial swarm markers and acoustic traps. Female mosquitoes have a different antennal structure and show little responsiveness to sound — they are certainly not repelled by buzzers which are advertized as simulating the sound of male mosquitoes (see chapter 4). There are other ways of trapping female mosquitoes using suction traps baited with carbon dioxide and/or visible or UV light. These are much used in monitoring mosquito populations and have been considered as a control technique in China.

In contrast to mosquitoes, both sexes of tsetse fly have similar feeding behavior, and both can be caught in the same type of trap (see Chapter 3).

Several attempts have been made to model the effects of trapping one or both sexes and sterilizing or killing them.[8-10] These have been either simple models with nonoverlapping generations in which all the events of one generation are supposed to occur at one instant, or more complex models which take account of events on a daily basis. These models have not considered the catching and killing of males only, i.e, the ''male annihilation'' technique as used in the control of certain fruit fly species.[11] A model[12] to take into account this possibility as well as the other possible trapping strategies has been constructed, based on tsetse biology, i.e., a 30-d pupal life, monandrous females capable of mating from their fourth day of adult life onward and able to produce a pupa every 10 days from day 17 of adult life. Each male is capable of mating up to six times from its seventh day onward. Figure 1 shows some results obtained with this model when 2% of one or both sexes are trapped daily for 100 d. The survival parameters were such that without trapping, the population size remained constant. Trapping or killing 2% of the males per day temporarily

* L. M. Gomulski

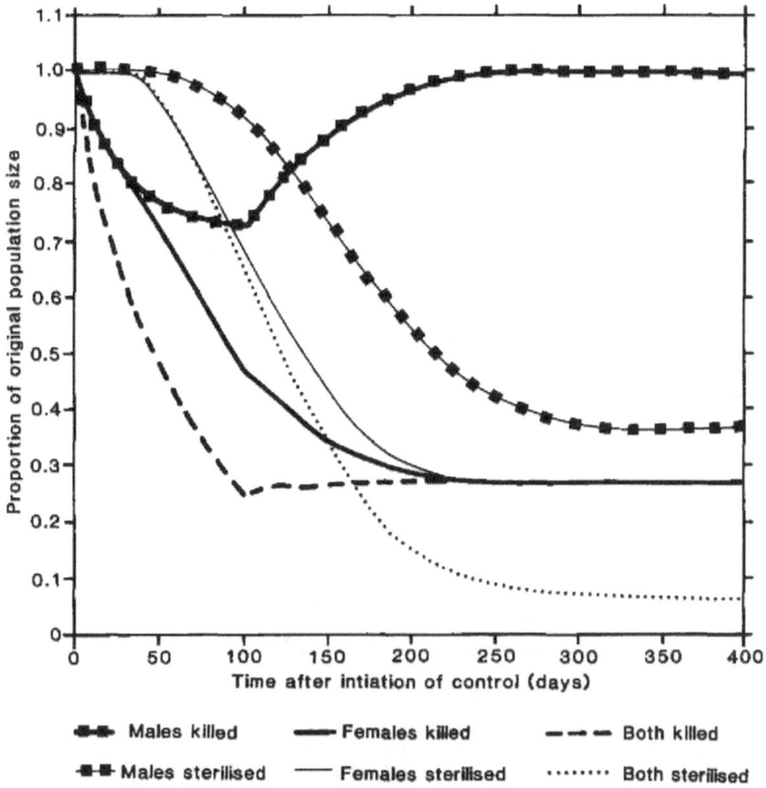

FIGURE 1. Simulated effects of trapping 2% of the males and/or females of a tsetse fly population per day.[12] The trapped flies are either killed or sterilized and trapping continues for 100 d only. Without trapping the population size remains constant.

lowers the adult vector population (which in tsetse includes males) but has no reproductive effect as the males that survive are sufficient to inseminate all of the emerging females. Thus, when trapping stops the vector population soon recovers. Only by greatly increasing the trapping rate does one reach the point at which some females are left uninseminated and a lasting impact is made on the population (Figure 2).

Killing females is reproductively equivalent to sterilizing them, and killing males in addition to females would have no additional reproductive effect because there would still be sufficient males to inseminate all the females. Therefore, the long-term consequences of these three strategies are the same. However, the immediate impact on the population is greater if females are killed (and even greater if both sexes are killed).

It might be expected that sterilizing males only would be equivalent to sterilizing females only. However, Figure 1 shows that this is not so. The difference appears to be connected with the facts that (1) in the case of female chemosterilization, they are at risk from the traps every day for 17 d before their first chance to larviposit whereas males are at risk for only 7 d before their first chance to mate; and (2) in the case of male sterilization, one has to "dilute" both the pool of fertile males and the pool of females already inseminated with fertile sperm before one can make an impact on the population, whereas when females are sterilized, only the pool of fertile females has to be "diluted".

Sterilizing both sexes has the greatest long-term effect because in addition to eliminating the reproduction of trapped females, the trapped males will sterilize some females which were not trapped — only matings between untrapped females and untrapped males will be fertile. In Figure 1 sterilizing 2% of both sexes daily for 100 d reduces the population by 94%. To achieve the same effect by sterilizing females or killing females or both sexes

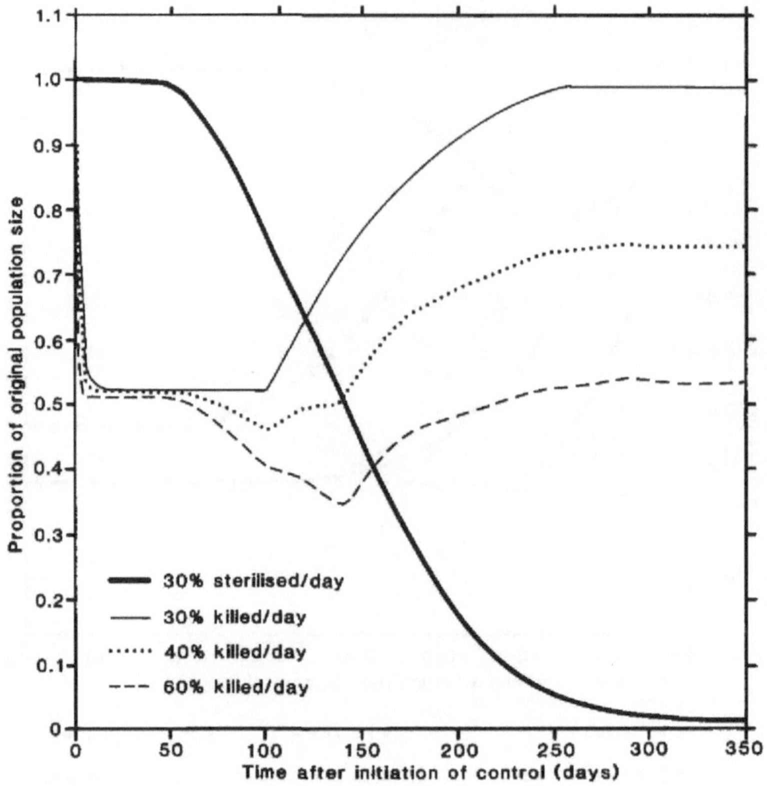

FIGURE 2. Simulated effect of killing or sterilizing large percentages of males in a tsetse population, using the same model as Figure 1.[12]

would require raising the trapping rate to 4%/d. However, the initial rate of population suppression is greatest when flies are killed, and in some circumstances this would be an important consideration. Males sterilized in traps are expected to disperse to find mates. This has the potential advantage over killing females in traps since one can rely on the males to penetrate into terrain where human access to erect traps would be difficult. Also, the selection pressure for trap avoidance behavior might not be so strong where males are sterilized that they would prevent the reproduction of some females which had not been attracted to the traps.

There have been a number of experimental studies on sound traps for male mosquitoes and on sterilization methods to be used in conjunction with the tsetse traps described in Chapter 3.

II. EXPERIMENTS WITH MOSQUITOES*

A. SOUND FOR THE ATTRACTION OF MOSQUITOES

Sound frequency is the most important aspect of the female wingbeat for male attraction. These frequencies are in the range 300 to 500 Hz[13-15] The frequency of the female wingbeat of a given species depends on its wing length according to an equation of the form $y = -100.2x + 765.9$ where y is the frequency in Hz and x is the wing length in mm.[16] The frequency is affected by temperature, and at any one temperature the frequency to which males respond is not precise and includes frequencies about 50 Hz on either side of the wingbeat-frequency range of females of their species.

* T. Ikeshoji

The intensity of sound is another important parameter. From electrophysiological measurements, one would expect that *Anopheles subpictus* could respond to 10 dB of sound at a frequency of 380 Hz.[17] However, to be attractive the sound intensity must be much higher. Wishart and Riordan[18] observed that 100 dB at 1.2 cm above a speaker induced maximal seizing and clasping behavior in male *Aedes aegypti*, whereas 68 to 85 dB showed the best attractancy. Ikeshoji[19] obtained the maximal attractancy of *Ae. aegypti* with 100 dB and *Culex pipiens* var. *molestus* with 110 and 120 dB in the laboratory. However, a less intense 50- to 60-dB sound may be applicable without losing much of the attractancy to swarming males in the field.

Since the response of the mosquito's auditory sense becomes habituated to a continuous sound, temporal modulation of sound intensity is essential for an efficient trap. Usually, sound is emitted for a few seconds at intervals of a few seconds. Furthermore, an intensity-modulated sound shows somewhat better attractancy than a similar but unmodulated sound. When the waveform of sound of female *Cx. quinquefasciatus* was emitted in a 5-s pulse at 5-s intervals with a 1-s fading out phase at each pulse, it was twice as attractive as the same sound turned off abruptly with no fading out phase.[20] The gradually damped sound may mimic a female flying away from a chasing male.

B. SOUND TRAPS FOR MOSQUITOES

Forty years ago Khan and Offenhauser[21] attracted a large number of *Anopheles albimanus* males by replaying a tape-recorded sound of the female wingbeat beside a swamp. Twenty years ago Belton[22] attracted *Aedes stimulans* from a swarm into a suction trap baited with sound and suggested its use for trap-sterilize-release procedures. Recently, Ikeshoji[23] and Ikeshoji et al.[24,25] have further developed the sound-trapping technique for mosquito control. They caught *Cx. tritaeniorhynchus*, *Cx. tarsalis*, and *Cx. quinquefasciatus* with a cylindrical sound trap. It consisted of a 9.3-cm diameter, 8-Ω dynamic speaker wrapped around with a 30 $\times$ 32-cm transparent, colorless, polyethylene sheet to form a cylinder. The inside wall was sprayed with adhesive. The speaker was connected to a small audiooscillator which was powered by four 1.5-V batteries and emitted a 370- or 400-Hz 100-dB sinusoidal sound for 5 s at intervals of 5 s. The battery lasted for several months when operated for 20 min every night. The trap was placed at a height of 0.5 to 1.5 m on a tripod standing at the center of a swarm marker. Figure 3 shows the trap, Figure 4 shows mosquitoes being attracted to it, and Figure 5 shows mosquitoes stuck to the inside of the trap.

Ikeshoji and Ogawa[26] tested two other types of sound traps — board and cup traps — for *Cx. tritaeniorhynchus* and *Ae. albopictus*. The board trap consisted of a 53 $\times$ 73 $\times$ 1-cm polyethylene foam board in the center of which a 5-cm-diameter, thin piezo-electric, plastic sheet speaker was embedded. This produced uniform sound emission over the whole front and back surfaces of the board. The board was sprayed with black paint and adhesive on both sides and was tested standing vertically. The cup trap was formed from half of a 30- to 40-cm-diameter hollow polyethylene foam ball. On the apex of the inside of the cup was taped a 2-cm-diameter, piezo-electric, metal disk of a type which is commonly used as a door buzzer. The outer surface of the cup was sprayed with black paint and adhesive. It was tested with the convex side upward on a tripod (Figure 6).

Kanda et al.[27] compared the effectiveness of a cylinder trap with that of a guinea pig or dry ice or both for catching *Mansonia* near a breeding place in the swamp. A tape-recorded sound of 375 Hz (the female wingbeat frequency) attracted more males and females than a silent trap with the olfactory baits.

C. ARTIFICIAL SWARM MARKERS

Since males are attracted to sound only at close range,[28] other stimuli are needed to attract them from a distance. Visual or chemical stimuli can be employed for this purpose.

FIGURE 3. Cylinder sound trap.[23] (Photograph by T. Ikeshoji.)

FIGURE 4. *Culex tritaeniorhynchus* males responding to a cylinder trap from a swarm.[23] (Photograph by T. Ikeshoji.)

Bidlingmeyer and Hem[29] attracted blood-seeking females of various species with 1.5 × 2.4-m dark pieces of plywood as visual markers and estimated the range of response of most species as 15.5 to 19 m. If these figures apply to the attraction of males also, the appropriate size of a swarm marker and the distance between the markers can be calculated. In fact, when 1 × 2-m markers were experimentally placed at a 20- to 50-m distance from each other in a paddy field, all swarms of *Cx. tritaeniorhynchus* were observed at the markers, and none were between them. In practice, the size of a marker should be about 1 × 2 m for an open field, but smaller where visibility is obstructed as in residential areas or woods.

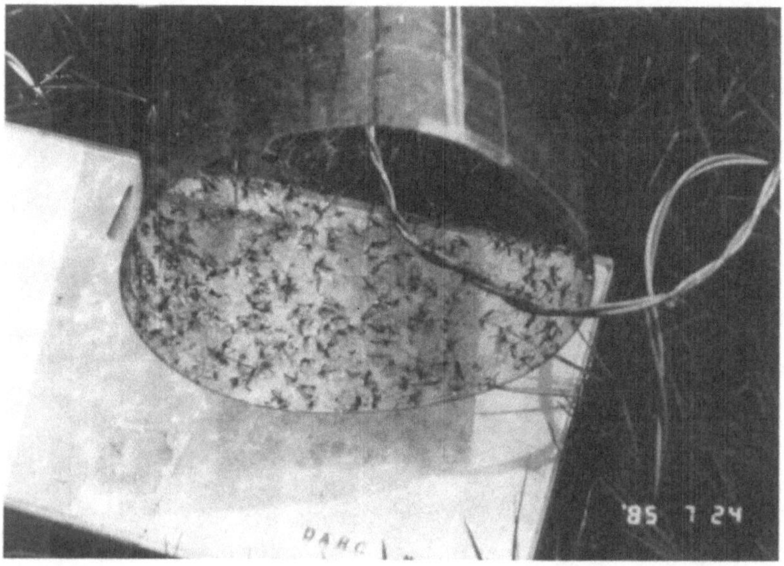

FIGURE 5. Males stuck on the inside wall of a trap.[23] (Photograph by T. Ikeshoji.)

In an initial survey to find the potential swarming sites of a given species one should first set the markers at likely sites or distribute them at random over the area and then shift unattractive markers to adjacent untested sites. Through such trial and error tactics the best sites can be found for a limited number of markers.

The color of a good swarm marker is somewhat controversial. Out of 26 species for which there are reports, 12 species were attracted to a black marker, 10 species to a white one, and 4 species were attracted to both black and white.[30] Males probably respond to a sharp color contrast of the marker against the background. In this regard it should be noted that *Cx. tritaeniorhynchus* males swarming at a height of a few meters descend toward a black marker as daylight fades. The contrast between the light reflected from the marker and the ground decreases with decreasing illumination, but this can be compensated by descending toward the marker, and a very compact swarm results. A piece of cloth, paper, or polyethylene sheet may be used temporarily and later replaced by a semipermanent painted marker, since males swarm at fixed sites throughout the season.

D. EFFECTS OF SOUND TRAPPING AND CHEMOSTERILIZATION OF MALES

1. Impact of Male Killing

Around a 25 × 50-m stand of marsh fleabane, *Pulchea purpusascens,* in Bakersfield, CA, 12 cylinder sound traps were operated for 13 nights with the intention of removing males from a *Cx. tarsalis* population. The mean age of the resting male population decreased from 3.9 d during the first half of the test to 3.1 d during the second half (Figure 7).[24] The mean age of both the trapped and resting male populations decreased as the 3- and 4-day-old males were continuously removed. The percentage insemination of females resting in the fleabane gradually decreased from 36% at the beginning to 14% at the end. This trend was more pronounced in the unfed females, among which the rate decreased from 22 to 0%. Meanwhile, in females caught for comparison in a red box trap 500 m away, insemination remained at an average of 55%. Hence, it appears that in this case so many males were killed that there were not enough to inseminate all the females.

FIGURE 6. Cup sound trap.[26] (Photograph by T. Ikeshoji.)

2. Decrease of an *Aedes albopictus* Population

To trap *Ae. albopictus* one cylinder, two board, and two cup traps were distributed in a 15 × 100-m area of woods inside a temple compound in Tokyo. They were operated for 20 min every day for 8 d between 1700 and 1800 h. The maximum daily catch of 45 males and 53 females on the third day of the experiment decreased to 1 and 23, respectively, on the last day.[26] The proportion of males in the catch decreased from 0.85 to 0.04. Such a marked decrease of the male population in a short period may be attainable in other sedentary species of *Stegomyia* and *Finlaya* breeding at low densities.

3. Chemosterilization of a *Culex quinquefasciatus* Population

Cylinder traps were tested at Butterworth, Malaysia, in a 2.2-ha area of a village where there were 40 houses. The inside wall of the cylinder traps were treated with the chemosterilant hempa at a dose of 0.7 mg/cm^2. The traps were set at 13 sites in the field for 20 d. The hatch of egg rafts collected from three ditches was checked every day (Figure 8). As 95% of egg rafts in the pretreatment period showed less than 7.5% unhatched eggs, an egg raft with 7.5% or more unhatched eggs was classified as sterilized.

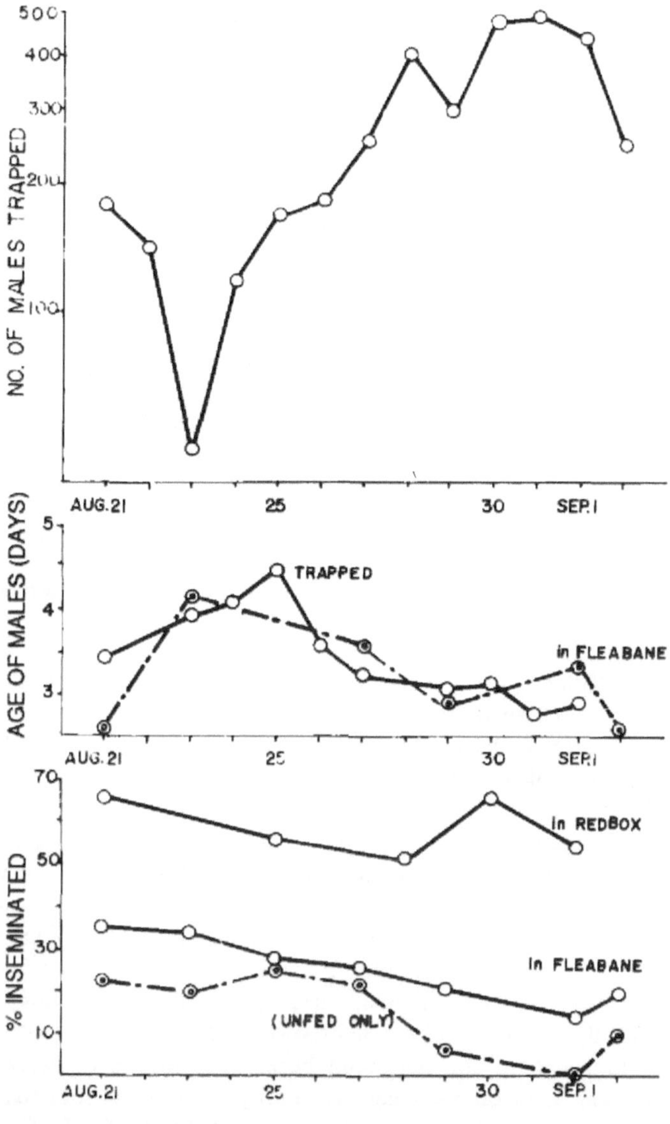

FIGURE 7. Impact of male removal by sound trapping on the *Culex tarsalis* population in a patch of marsh fleabane in Bakersfield, CA.[24]

When the results for all the ditches were averaged, the egg raft sterility was 17.1 ± 4.7% and the egg sterility was 7.8 ± 3.4%. These values were significantly different from the corresponding values of 4.2 ± 4.8% and 1.8 ± 0.4% found in the pretreatment period.[31] The disappointing proportion of sterilized egg rafts may be attributed to both migration of mosquitoes from outside the area and insufficient attractancy of the sound trap. The ineffectiveness of hempa and the formulation used were also blamed for the poor results. Subsequently, a synergistic formulation of metepa in a mixture with lactic acid was found[25] and will be tested in the near future.

III. EXPERIMENTS WITH TSETSE*

A. CHEMOSTERILANT TRAPS OR TARGETS FOR TSETSE

The concept of sterilization of members of wild tsetse populations[32,33] arose from earlier

* P. A. Langley

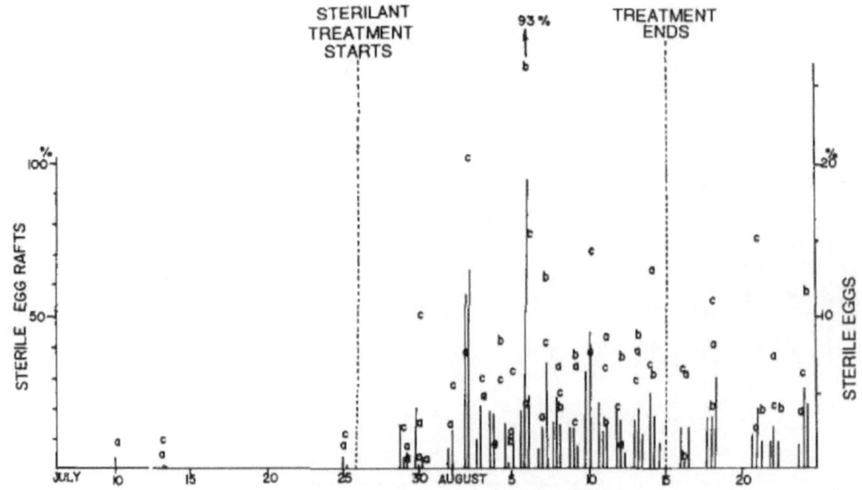

FIGURE 8. Chemosterilization of *Culex quinquefasciatus* males by sterilant-treated sound traps. The vertical lines show the percentages of the egg rafts collected from three ditches and classified as sterile. The letters a, b, and c show the percentages of all the eggs from each ditch which were sterile.[31]

observations of the efficiency of odor-baited traps of the kind described in Chapter 3. Flies entered a trap and were chemically sterilized using an aerosol system before being automatically released. However, insect chemosterilants are highly toxic, and the apparatus for their delivery was so complex, requiring frequent servicing, that the idea was abandoned. The discovery of the sex pheromone of *Glossina morsitans morsitans*[34,35] and its use on decoys mounted on dark cloth screens[35,36] provided new possibilities for delivering sterilizing doses of chemosterilants. Male flies attracted visually to such screens would alight on decoys composed of small pieces of nylon cloth of approximately the same shape and size as females. The presence of pheromone on the decoys induced males to engage in copulatory behavior, while the combination of pheromone with the chemosterilant bisazir, *p,p*-bis(1-aziridinyl)-*n*-methylphosphinothioic amide, ensured that they were sterilized in the attempt.[37,38] This was also thought to be a safer way of deploying the chemosterilant in the environment. For the purposes of sterilizing both sexes of tsetse, bisazir can be made relatively safe by using it in traps[39] where its ability to sterilize both males and females following short exposure to its vapor[40] can be exploited (Figure 9). Unfortunately, bisazir is not only volatile, but it is somewhat unstable. Therefore, a sterilizing trap baited with bisazir would require frequent servicing. Also, bisazir could not be used safely on a screen of the type described in Chapter 3.

B. JUVENILE HORMONES AS TSETSE STERILANTS

Accepting that a sterilization technique for tsetse must be as cost effective as an insecticidal technique and ought to provide at least the same degree of safety to other life forms (including the operators), there seems little future for the conventional insect chemosterilants.

For this reason attention has turned in recent years to compounds which interfere with the normal functioning of the insect endocrine system and which are referred to generally as insect growth regulators. Such compounds should be specific to insects and not present a hazard to other organisms. They should disrupt certain aspects of the reproductive or developmental process in the insect, and for this reason compounds allied to insect hormones have received much attention.

The mode of reproduction in tsetse is unusual and complex in that ovarian development,

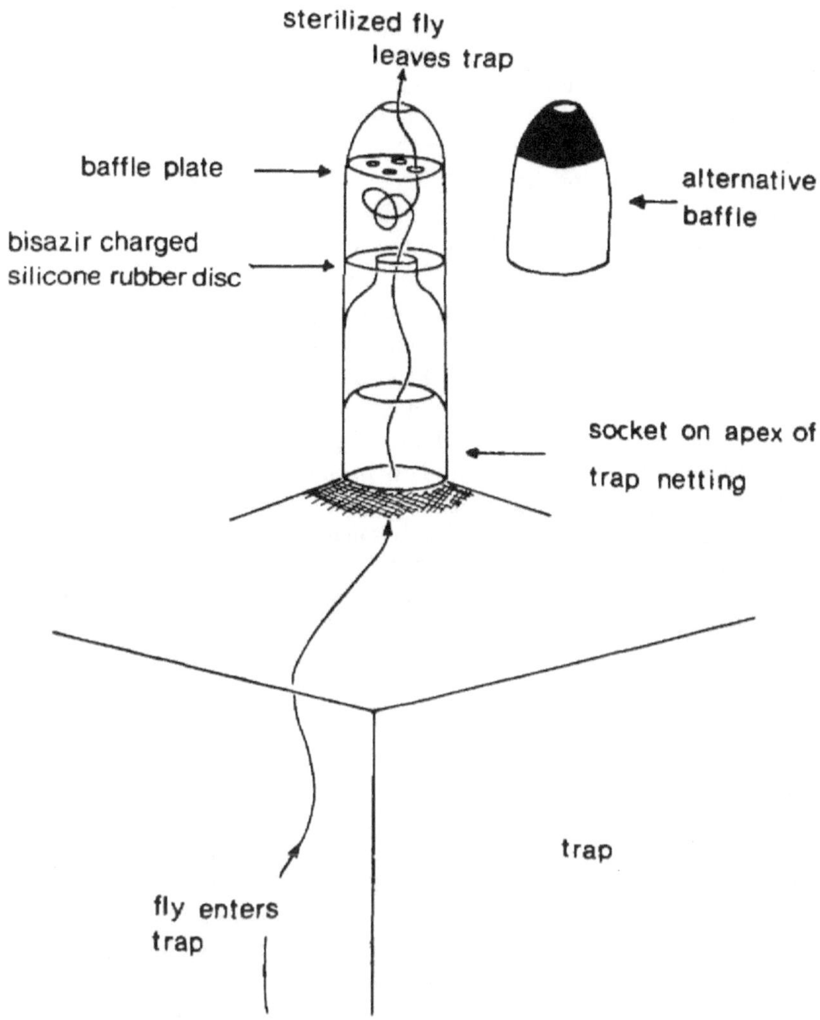

FIGURE 9. Sterilizing system for tsetse consisting of chambers made from transparent plastic bottles which push-fit together on top of the trap.[40] A fly entering the trap passes through the chambers and can be sterilized on contact or by the vapor phase of a suitable sterilant before leaving from the top chamber. The complexity of the baffle system will determine the length of time the fly is exposed to the sterilant. (Diagram reproduced by permission of *Parasitology Today*.)

ovulation of a single egg, larval nourishment, and parturition of the fully grown third-instar larva are timed to ensure an unbroken reproductive sequence and a lifetime of larval production at regular intervals of 9 or 10 d. This suggests perhaps a greater involvement of the tsetse endocrine system in the process of reproduction than in those insects which mature and lay successive batches of eggs. Whether this is true or not, a disappointingly large number of the hormones involved in tsetse reproduction have proved to be proteinaceous in nature and, being highly polar, are only soluble in water.[41-47] This means that they cannot be considered as agents for tsetse control because of the impossibility of delivering them to the insect through its waxy cuticle. The role of the ecdysteroids has yet to be evaluated, but they also are likely to be too polar to be formulated easily for cuticle penetration. However, the untimely presence of juvenile hormone in the pupa has the classic effect of disrupting metamorphosis,[48] and it seemed possible that this phenomenon could be exploited with tsetse in a manner which would not be possible with an egg-laying insect.

The effects of a relatively new compound, 2-[1-methyl-2-(4-phenoxy-phenoxy)ethoxy]pyridine, produced by the Sumitomo Chemical Company and coded S-31183, have been compared with those of the well-known synthetic juvenile hormone analogue *s*-methoprene. The results show that S-31183 is several orders of magnitude more potent in its effect upon tsetse than is *s*-methoprene.[48,49]

Unfortunately, treatment of the pupal stage with juvenile hormones is not a practical proposition for most insects, including tsetse. However, the viviparity of tsetse ensures that treatment of the adult female with appropriate formulations of S-31183 results in transfer of lethal amounts to her larva *in utero*, thus effectively sterilizing her for life. Recent experiments have shown that topical applications of only 20 ng of S-31183 in acetone or oil induce total sterility in the sense that all the offspring produced by such females fail to develop beyond the 20th day of their 30-d intrapuparial life. Furthermore, tarsal contact with treated filter papers (1 cm^2) or black cotton cloth disks (2 cm^2) dosed with 5 or 10 μl of oil per square centimeter containing 0.1 mg of S-31183 leads to the pickup of between 0.004 and 0.12% of the material present. This means that exposure of a fly for 1 min to the lower dose of oil leads to the pickup of about 8 ng, while a 5-min exposure to the higher dose leads to pickup of around 240 ng of active ingredient.

Further experiments have shown that a 20 to 25% solution of oil in acetone will deposit between 7 and 10 μl of oil per square centimeter onto a black cloth simply by dipping and hanging up to dry. The cloth does not remain visibly oily, and a single solution can be used repeatedly without any consistent change in the amount of oil deposited until around 90% of the solution has been used.

Further experiments are in progress to determine the extent to which male tsetse contaminated with S-31183 can transfer sterilizing doses to females during mating. Initial results look promising.

S-31183 has a high specificity for insects and has extremely low mammalian toxicity (acute oral LD_{50} in rats more than 5 g per kg body weight). It has high thermal stability and does not decompose appreciably in UV light. We believe the use of S-31183 on black cloth screens or in traps used for tsetse control should represent a significant improvement over the use of insecticides on such devices. The safety of the chemical involved, combined with its great stability and ease of application, could provide the means of a worthwhile self-help campaign to control tsetse in Africa.

REFERENCES

1. **Curtis, C. F.,** Genetic control: growth industry or lead balloon?, *Biol. J. Linn. Soc.*, 26, 359, 1985.
2. **Krafsur, E. S.,** Screwworm eradication in North and Central America, *Parasitol. Today*, 3, 131, 1987.
3. **Patton, P.,** Programe contra la mosca Mediterraneo en Mexico, in *Sterile Insect Technique and Radiation in Insect Control*, I.A.E.A., Vienna, 1981, 25.
4. **Takken, W., Oladunmade, M. A., Dengwat, L., Feldmann, H. U., Onah, J. A., Tenabe, S. O., and Hamann, H. J.,** The eradication of *Glossina palpalis palpalis* using traps, insecticide impregnated screens and the sterile insect technique in Central Nigeria, *Bull. Entomol. Res.*, 76, 225, 1986.
5. **Lofgren, C. S., Dame, D. A., Breeland, S. G., Weidhaas, D. E., Kaiser, J. R., Ford, H. R., Boston, M. D., and Baldwin, K. F.,** Release of chemosterilized males for the control of *Anopheles albimanus* in El Salvador. III. Field methods and population control, *Am. J. Trop. Med. Hyg.*, 23, 288, 1974.
6. **Reisen, W. K.,** Male mating competitiveness: the key to some problems associated with the genetic control of mosquitoes, in *Ecology of Mosquitoes*, Lounibos, L. P., Rey, J. R., and Frank, J. H., Eds., Florida Medical Entomology Laboratory, Vero Beach, FL, 1985, 345.
7. **Roth, L. M.,** A study of mosquito behavior; an experimental laboratory study of the sexual behavior of *Aedes aegypti*, *Am. Midl. Nat.*, 40, 265, 1948.
8. **Knipling, E. F.,** *Basic Principles of Insect Suppression and Management*, Agriculture Handbook No. 512, U.S. Department of Agriculture, Washington, D. C., 1979.

9. **Weidhaas, D. E. and Haile, D. G.,** A theoretical model to determine the degree of trapping required for insect population control, *Bull. Entomol. Soc. Am.*, 24, 18, 1978.
10. **Langley, P. A. and Weidhaas, D. E.,** Trapping as a means of controlling tsetse, *Glossina* spp. (Diptera: Glossinidae); the relative merits of killing and of sterilization, *Bull. Entomol. Res.*, 76, 89, 1986.
11. **Koyama, J.,** The Japan and Taiwan projects on the control and/or eradication of fruit flies, in *Sterile Insect Technique and Radiation in Insect Control*, I.A.E.A., Vienna, 1982, 39.
12. **Gomulski, L. M.,** unpublished data, 1988.
13. **Clements, A. N.,** *The Physiology of Mosquitoes*, Pergamon Press, London, 1963, 293.
14. **Belton, P. and Costello, R. A.,** Flight sounds of the females of some mosquitoes of Western Canada, *Entomol. Exp. Appl.*, 26, 105, 1979.
15. **Ikeshoji, T.,** Acoustic attraction of male mosquitoes in a cage, *Jpn. J. Sanit. Zool.*, 32, 7, 1981.
16. **Ogawa, K. and Kanda, T.,** Wingbeat frequencies of some anopheline mosquitoes of East Asia (Diptera: Culicidae), *Appl. Entomol. Zool.*, 21, 430, 1986.
17. **Tischner, H.,** Über den Gehörsinn von Stechmücken, *Acustica*, 3, 335, 1953.
18. **Wishart, G. and Riordan, D. F.,** Flight responses to various sounds by adult males of *Aedes aegypti* (Diptera: Culicidae), *Can. Entomol.*, 151, 181, 1959.
19. **Ikeshoji, T.,** Attactive sounds for auto-chemosterilization of the male mosquitoes, *Jpn. J. Sanit. Zool.*, 33, 41, 1982.
20. **Ikeshoji, T., Yamasaki, Y., and Yap, H. H.,** Attractancy of various waveform sounds in modulated intensities to male mosquitoes, *Culex quinquefasciatus*, in the field, *Jpn. J. Sanit. Zool.*, 38, 249, 1987.
21. **Khan, M. C. and Offenhauser, W., Jr.,** The first field tests of recorded mosquito sounds used for mosquito destruction, *Am. J. Trop. Med.*, 29, 811, 1949.
22. **Belton, P.,** Trapping mosquitoes with sound, in Proc. 35th Conf. Calif. Mosq. Control Assoc., 1967, 98.
23. **Ikeshoji, T.,** Distribution of the mosquitoes, *Culex tritaeniorhynchus*, in relation to disposition of sound traps in a paddy field, *Jpn. J. Sanit. Zool.*, 37, 153, 1986.
24. **Ikeshoji, T., Sakakibara, M., and Reisen, W. K.,** Removal sampling of male mosquitoes from field populations by sound trapping, *Jpn. J. Sanit. Zool.*, 36, 197, 1985.
25. **Ikeshoji, T.,** Synergistic effect of lactic acid with a chemosterilant metepa to the sound-attracted male mosquitoes, *Jpn. J. Sanit. Zool.*, 38, 323, 1987.
26. **Ikeshoji, T. and Ogawa, K.,** Field catching of mosquitoes with various types of sound traps, *Jpn. J. Sanit. Zool.*, 39, 119, 1988.
27. **Kanda, T., Cheong, W. H., Loong, K. P., Lim, T. W., Ogawa, K., Chiang, G. L., and Sucharit, S.,** Collection of male mosquitoes from a field population by sound trapping, *Trop. Biomed.*, 4, 161, 1987.
28. **Charlwood, J. D. and Jones, M. D. R.,** Mating behaviour in the mosquito *Anopheles gambiae s.l.* I. Close range and contact behaviour, *Physiol. Entomol.*, 4, 111, 1979.
29. **Bidlingmeyer, W. L. and Hem, D. G.,** The range of visual attraction and the effect of competitive visual attractants upon mosquito (Diptera: Culicidae) flight, *Bull. Entomol. Res.*, 70, 321, 1980.
30. **Ikeshoji, T.,** Swarming and mating behaviour in evolution of mosquitoes, presented at the 18th Int. Congr. Entomol., Vancouver, July 3 to 9, 1988.
31. **Ikeshoji, T. and Yap, Y. Y.,** Monitoring and chemosterilization of a mosquito population, *Culex quinquefasciatus* (Diptera: Culicidae) by sound traps, *Appl. Entomol. Zool.*, 22, 474, 1987.
32. **Bursell, E.,** Chemosterilization of tsetse flies using a pressurized metepa aerosol, *Trans. Rhodesian Sci. Assoc.*, 58, 43, 1977.
33. **House, A. P. R.,** Chemosterilization of *Glossina morsitans moristans* Westwood and *G. pallidipes* Austen (Diptera: Glossinidae) in the field, *Bull. Entomol. Res.*, 72, 6, 1982.
34. **Langley, P. A., Pimley, R. W., and Carlson, D. A.,** Sex recognition pheromone in tsetse fly *Glossina morsitans, Nature (London)*, 254, 51, 1975.
35. **Langley, P. A., Huyton, P. M., Carlson, D. A., and Schwarz, M.,** Effects of *Glossina morsitans morsitans* Westwood (Diptera: Glossinidae) sex pheromone on behaviour of males in field and laboratory, *Bull. Entomol. Res.*, 71, 57, 1981.
36. **Langley, P. A., Coates, T. W., Carlson, D. A., Vale, G. A., and Marshall, J.,** Prospects for autosterilization of tsetse flies *Glossina* spp. (Diptera: Glossinidae), using sex pheromone and bisazir in the field, *Bull. Entomol. Res.*, 72, 319, 1982.
37. **Coates, T. W. and Langley, P. A.,** Laboratory evaluation of contect sex pheromone and bisazir for autosterilization of *Glossina morsitans, Entomol. Exp. Appl.*, 31, 276, 1982.
38. **Hall, M. J. R.,** The orientation of males of *Glossina morsitans morsitans* Westwood (Diptera: Glossinidae) to pheromone-baited decoy 'females' in the field, *Bull. Entomol. Res.*, 77, 487, 1987.
39. **Hall, M. J. R. and Langley, P. A.,** Development of a system for sterilizing tsetse flies, *Glossina* spp., in the field, *Med. Vet. Entomol.*, 1, 201, 1987.
40. **Langley, P. A. and Hall, M. J. R.,** Tsetse control by auto-sterilization, *Parasitol. Today*, 2, 125, 1986.
41. **Pimley, R. W.,** Neuroendocrine stimulation of uterine gland protein synthesis in the tsetse fly, *Glossina morsitans, Physiol. Entomol.*, 8, 429, 1983.

42. **Pimley, R. W.,** Effects of biogenic amines on fat cells of *Glossina morsitans in vitro, J. Insect Physiol.,* 30, 587, 1984.
43. **Pimley, R. W.,** Chromatographic separation of some *corpora cardiaca* peptides that influence fat cell activity in female *Glossina morsitans, Insect Biochem.,* 14, 521, 1984.
44. **Pimley, R. W.,** Cyclic AMP and calcium mediate the regulation of fat cell activity by octopamine and peptide hormones in *Glossina morsitans, Insect Biochem.,* 15, 293, 1985.
45. **Pimley, R. W. and Langley, P. A.,** Hormonal control of lipid synthesis in the fat body of the adult female tsetse fly, *Glossina morsitans, J. Insect Physiol.,* 27, 839, 1981.
46. **Pimley, R. W. and Langley, P. A.,** Hormone stimulated lipolysis and proline synthesis in the fat body of the adult tsetse fly, *Glossina morsitans, J. Insect Physiol.,* 28, 781, 1982.
47. **Pimley, R. W. and Stafford, K. A.,** Octopamine distribution of the tsetse fly *Glossina morsitans, Comp. Biochem. Physiol.,* 81C, 297, 1985.
48. **Langley, P. A. and Pimley, R. W.,** A role for juvenile hormone and the effects of so-called anti-juvenile hormones in *Glossina morsitans, J. Insect Physiol.,* 32, 727, 1980.
49. **Langley, P. A., Felton, T., and Oouchi, H.,** Juvenile hormone mimics as effective sterilants for the tsetse fly *Glossina morsitans, Med. Vet. Entomol.,* 2, 29, 1988.

Chapter 10

INSECT PROOFING OF SANITATION SYSTEMS

C. F. Curtis, P. R. Morgan, J. N. Minjas, and Caroline A. Maxwell

TABLE OF CONTENTS

I. INSECTS IN SANITATION SYSTEMS AND INSECTICIDAL CONTROL*

The separation of people from their feces is of major importance in the prevention of many infectious diseases. For the forseeable future a large proportion of the population of the world cannot be provided with water-borne sewerage because this is expensive and requires in-house piped water. Various forms of pit latrines are a more feasible alternative,[1] but these tend to provide breeding sites for

1. *Culex quinquefasciatus*, vector of Bancroftian filariasis in many areas and the major urban nuisance mosquito[2]
2. Blowflies, especially *Chrysomya putoria* and *C. megacephala*, which may carry fecal pathogens to food,[3] and when present in large numbers are generally considered a repulsive feature of pit latrines

One solution has been regular programs of insecticidal spraying. Oil has been used[4] but is now considered less cost effective than organophosphate insecticides.[5] A program[6] using the organophosphate, chlorpyrifos (= Dursban), was carried out by the City Council in Dar es Salaam, Tanzania, for more than 10 years. It was shown[6] that because of the persistence of chlorpyrifos, 6 cycles of spraying per year were sufficient to give permanent control when a dose of about 1 mg/l was applied. In 1980 $135,000 was spent on the import of insecticide to carry on this program in over 100,000 pit latrines and about 60,000 cess pits, septic tanks, etc. More recently, because of foreign-exchange shortage there has been insufficient insecticide to spray all the pits. Furthermore, by 1980, some 10 years after the program began, a high level of organophosphate resistance was detectable by standard tests on the Dar es Salaam population of *Culex quinquefasciatus*.[7] Studies on the impact of this on control in the field[8] showed that the resistance did not prevent the larvae being killed when the insecticide was freshly applied, but the resistant larvae could survive in the insecticide residue left in a typical pit 2 weeks after spraying, whereas susceptible larvae would continue to be killed for at least 9 weeks after spraying. The stepping up of the spraying frequency to once every 2 weeks would have been prohibitively expensive and could not have been managed with the existing spraymen, each of whom could visit 40 houses per day and required 37 working days to complete a round of spraying.

All the pits were sprayed, but in fact only a minority contained water and were infested with mosquitoes[9,10] or flies.[9] Spraying only the infested pits would allow more frequent treatment without an increase in the foreign-exchange cost, but regular checking of all pits for insect breeding would require a large and well-supervised labor force. Furthermore, laboratory selection shows that *Cx. quinquefasciatus* has the potential to evolve higher and higher levels of organophosphate resistance, apparently by selection for duplications of the gene coding for the esterase which causes the resistance.[11] Thus, increasing the frequency of organophosphate spraying would probably only be a temporary solution.

II. *BACILLUS SPHAERICUS***

The toxin of *Bacillus sphaericus* is lethal to the larvae of *Cx. quinquefasciatus* after ingestion.[12] The insecticidal effect can persist for many weeks after introduction of *B. sphaericus*[13] (in contrast to the very short persistence of the effect of *B. thuringiensis* ssp. *israelensis* — see Chapter 8). *B. sphaericus* tolerates pollution well and is effective against larvae resistant to conventional insecticides.

* C. F. Curtis and J. N. Minjas
** J. N. Minjas

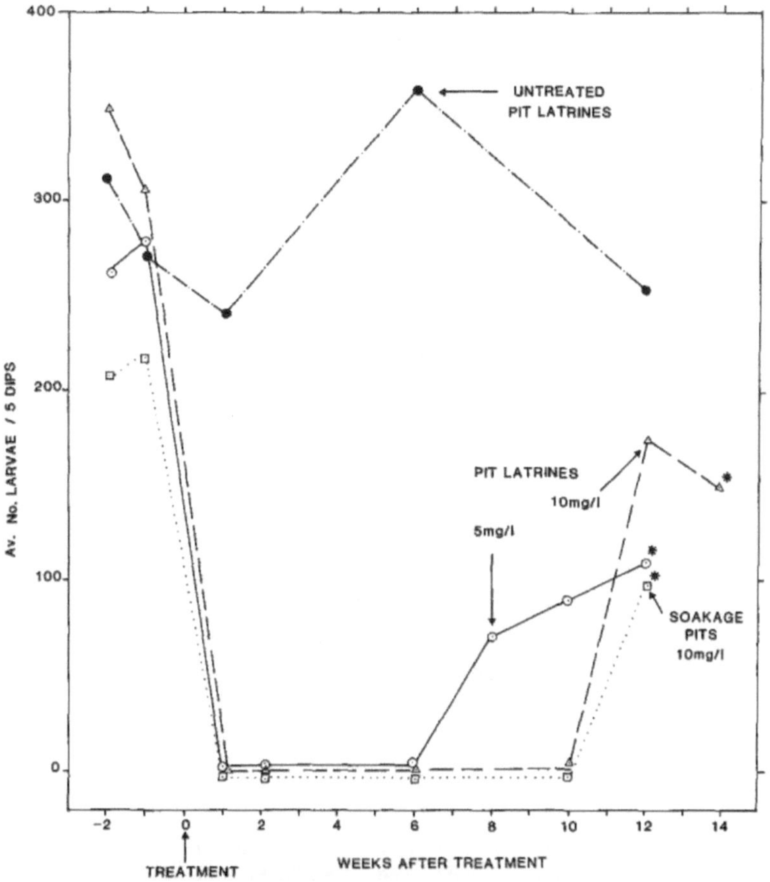

FIGURE 1. Results of a trial of *B. sphaericus* in 17 pit latrines and 2 soakage pits in Dar es Salaam, Tanzania.[14] The number of pit latrines given the high dose, the low dose, and no dose were 10, 5, and 2, respectively. Stars indicate when all of the pits in each category became positive for larvae.

In a trial in pit latrines and soakage pits in Dar es Salaam, *Cx. quinquefasciatus* breeding was prevented for 6 to 10 weeks after application of *B. sphaericus*, (Figure 1),[14] i.e., for about as long as chlorpyrifos used to be effective before the onset of resistance. In this trial the target dosages in the pits were 5 or 10 mg/l, i.e., very much greater than the LC_{90} of 0.05 mg/l observed in laboratory tests. The high target dosages were chosen to compensate for the large dilution factors expected over several weeks in the pits into which much washing water is poured.

III. POUR-FLUSH LATRINES AND CONCRETE LIDS*

Several physical solutions have been proposed for the problems of mosquitoes and flies in latrines, as well as the problem of latrine odor.

One of these solutions is a water seal (S-bend) or pour-flush latrine.[15] The water seal prevents the escape of odor from pit latrines, just as it does from sewers, and the seal also prevents the entry or exit of insects, provided that there are no cracks in the masonry at the top of the pit. Such latrines are appropriate where people use water and nothing else for anal cleansing, so they are accustomed when going to the latrine to taking water, about a

* C. F. Curtis

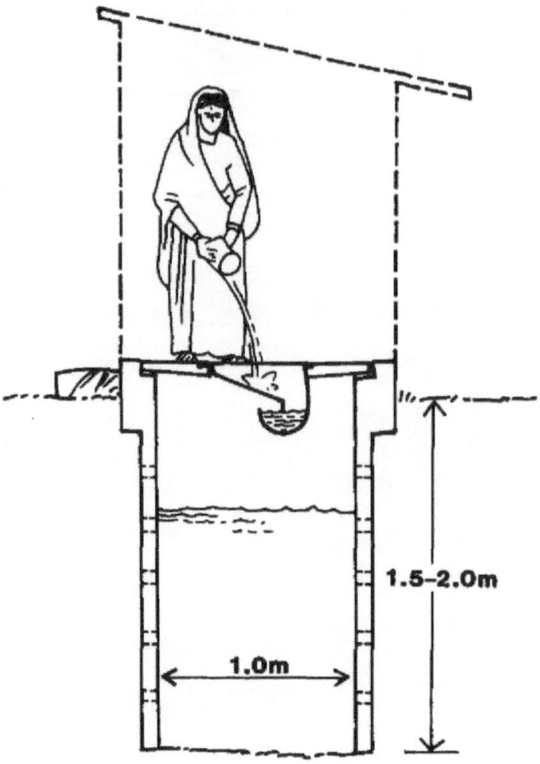

FIGURE 2. A latrine slab incorporating a pour-flush system.[15] (From Winblad, W. and Kilama, W., *Sanitation without Water,* Macmillan, New York, 1986, 25. With permission.)

liter of which can be used for flushing. In Asia latrine slabs incorporating a pour-flush system are now produced on a large scale (Figure 2). However, where paper, stones, or maize cobs are used for anal cleansing the S-bend tends to become blocked, and in trying to clear it, it is often broken.

Another solution is a latrine with a tightly fitting lid. Wooden or sheet-metal lids are usually ineffective because they warp or bend so as not to be insect proof. However, in Mozambique a system has been developed for casting a concrete lid in the very hole into which it is to fit.[16] The sides of the hole are beveled to assist the fitting, the latrine slab is domed for greater strength with minimal thickness, and no wire reinforcement is used (Figure 3). The fitting of the lids is reported to be so tight that flies, cockroaches, and odor cannot leave the pit, except when the latrine is actually in use. The great advantage of this system is that it is economical of building materials.[16] However, it is perhaps not an ideal solution because odor will inevitably be emitted while the latrine is in use, and insects may take this opportunity, while the lid is lifted, to enter and exit. Also, one would expect that small children might have difficulty in lifting the concrete lid. Furthermore, if the lid became chipped, insect and odor proofing would be lost.

For these reasons a latrine with an insect-screened vent pipe and no lid over the latrine hole is preferable when it can be afforded.

IV. THE CONTROL OF FLIES IN VENTILATED PIT LATRINES*

The development of the ventilated pit latrine in Zimbabwe demonstrated that a well-

* P. R. Morgan and C. F. Curtis

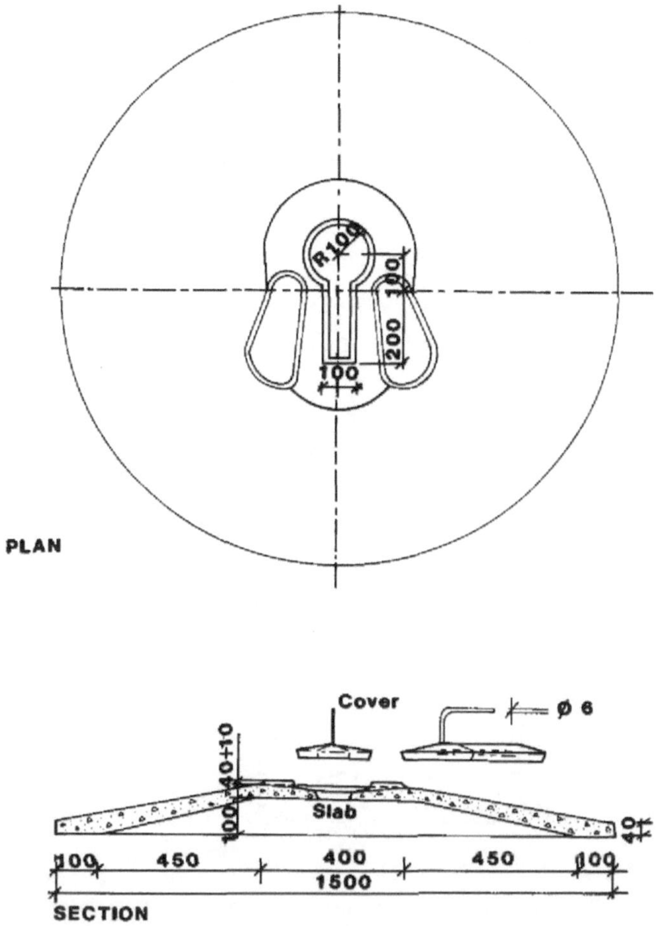

FIGURE 3. A conical unreinforced concrete latrine slab with cast concrete insect-proof lid made in Mozambique.[16] Dimensions are in millimeters. (From Cairncross, S. and Feachem, R. G., *Environmental Health Engineering in the Tropics*. Introductory Text, John Wiley & Sons, New York, 1983, 116. With permission.)

screened and efficient ventilation pipe, when fitted to a pit latrine, can be very effective in controlling flies.[17] This was later confirmed by the work of Curtis and Hawkins[9] and Evans.[18,19]

Although the principle of the ventilated improved pit (VIP) is simple, it is often poorly understood, and this has led to frequent examples of poorly functioning VIPs throughout the world.

Ventilated pit latrines are characterized by the presence of a ventilation pipe which draws odors away from the pit when an air current passes across the apex of the pipe (Figure 4). Air is thus drawn in through the squat hole or pedestal to replace air passing up the pipe. If the latrine building is fitted with a roof and is fairly dark within, most of the light entering the pit will do so via the pipe and not the latrine hole. Flies approaching the latrine from outside are attracted to odors emanating from the head of the pipe. Flies within the pit, being positively phototactic, are attracted to the light from the pipe when they emerge. When the pipe is screened these two groups of flies never meet. Flies cannot enter the pit or escape from it through the pipe. Some flies may enter the pit via the latrine hole, but these will be trapped in the pipe when they emerge.

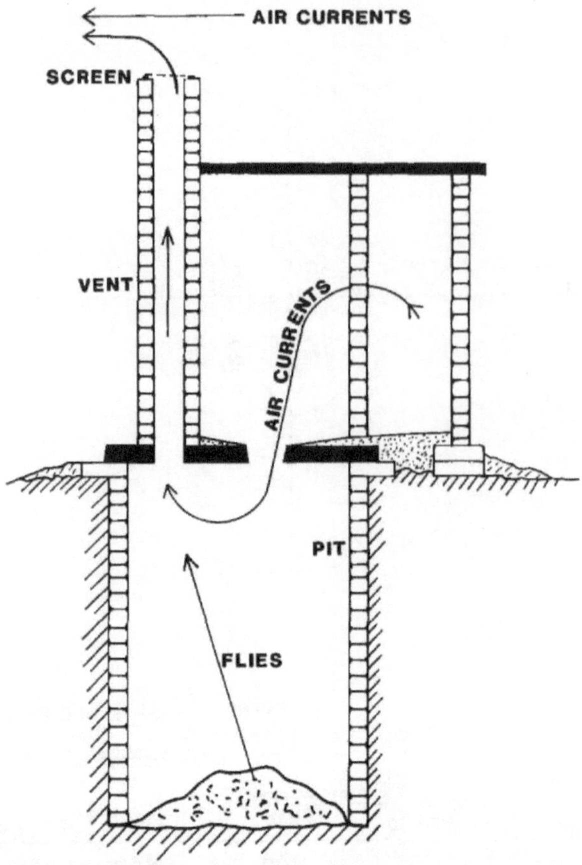

FIGURE 4. Air currents and the movement of flies from the pit in a Blair VIP latrine.

The efficiency of this sytem was demonstrated by a series of experiments using insect exit traps over the latrine holes.[17] In one experiment four identical latrines were built in a row, but only two were fitted with pipes; the other two, being unvented, behaved like normal pit latrines. After 6 months use without insect trapping, a series of weekly counts of fly output were taken from one pair of latrines (vented and unvented). The traps were moved from one pair of latrines to the other at monthly intervals. A total of 13,953 flies were trapped from the unvented latrines, while only 146 were trapped from the ventilated latrines. The structures differed only in the presence or absence of the ventilation pipe. The reliability of this system has been tested many times since. Over 150,000 VIP (Blair) latrines have been built in Zimbabwe alone since that time.

At least four features of the VIP latrine are essential to effective fly control.[20] The pipe must be efficient and should be fitted with a corrosion-resistant screen. Semi darkness within the structure is essential, and a hygienic interior greatly reduces its attractiveness to flies. The construction of a Blair ventilated pit latrine is illustrated in Figure 5.

The cost in Zimbabwe of the rural type of VIP is the equivalent of about U.S. $57, including labor. The family who are to be the pit's owner usually provides labor, bricks, stone, and sand. The Ministry of Health provides the fly screen, and donor agencies frequently give a subsidy of five or six bags of cement. A more expensive version for periurban areas, incorporating a tank (which can easily be desludged) and a soakaway, has been designed, and about 1000 of these are in use.

Studying the effectiveness of ventilation pipes of different diameters Evans[18] demon-

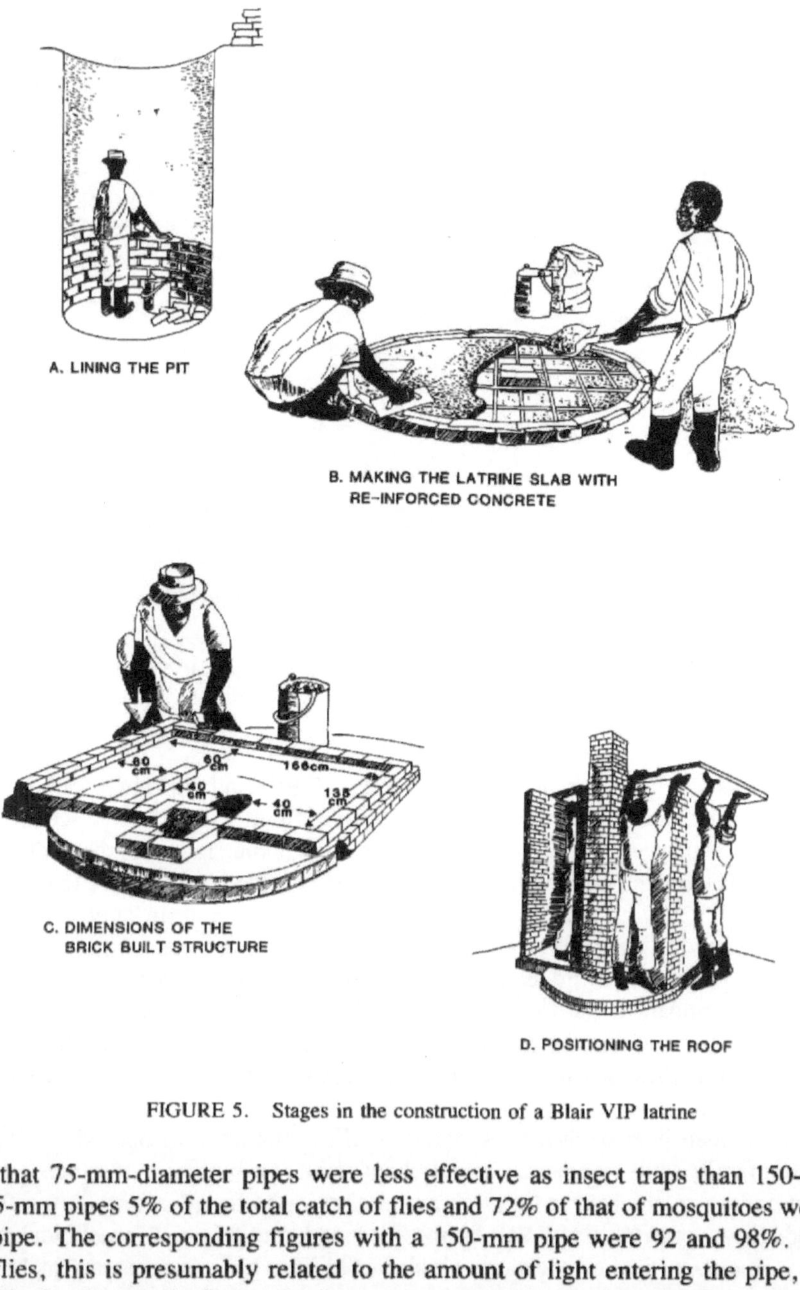

FIGURE 5. Stages in the construction of a Blair VIP latrine

strated that 75-mm-diameter pipes were less effective as insect traps than 150-mm pipes. With 75-mm pipes 5% of the total catch of flies and 72% of that of mosquitoes were trapped in the pipe. The corresponding figures with a 150-mm pipe were 92 and 98%. In the case of the flies, this is presumably related to the amount of light entering the pipe, since flies are positively phototactic. It may be that more mosquitoes are drawn up large diameter pipes by the relatively large masses of air which flow up them.[21] In Zimbabwe, brick ventilation pipes having an internal cross-sectional area of 225 × 225 mm are the most commonly used. The more efficient but more expensive 110-mm-diameter PVC vent pipes are designed with a coned apex to increase the area of the screen which is fitted at the head of the pipe. The PVC pipe is considered more efficient because its walls are smoother and cause less resistance to the air current than do brick pipes.

A corrosion-resistant flyscreen fitted at the head of the pipe is also essential: insect control cannot be achieved if the screen is broken or absent. In Zimbabwe most screens used in rural sanitation programs are made of stainless steel which last the life of the latrine, but aluminum is now being introduced as a cheaper, though still noncorroding, alternative.

FIGURE 6. Polystyrene beads spreading over the surface of a soakage pit. Note the mosquito trap made from a paint tin which was used to monitor emergence from the pit. (From Curtis, C. F., *Parasitol. Today*, 1(1), 36, 1985. With permission.)

Unfortunately, less care has been taken in other areas. Curtis and Hawkins[9] found that none of the vents in Dar es Salaam and few in Gaborone were fitted with effective screens. When screens were fitted, they were often rusty. Significantly higher rates of infestation of flies were noted in VIP latrines fitted with screenless pipes compared with screened pipes.

Semidark conditions within the latrine structure are also essential. In Zimbabwe semidark conditions are ensured in doorless structures with a spiral ground plan (Figure 5). In many other countries, superstructures are rectangular in plan with a door in one wall. Such structures cannot guarantee semidark conditions because the door is often left open by the users. The proportion of flies from within the pit choosing to fly up the vent pipe, and thus becoming trapped, is considerably less when the door is open.[9] Thus, many of these units are ineffective at controlling insects.

Similarly, a clean interior is less attractive to flies compared to one which is fouled. For this reason family units, which are usually kept clean, operate far better than communal units, where odors can emanate from urine and other wastes found on the latrine floor.

The inadequate construction of VIP latrines has made programs of insecticidal spraying necessary in some countries, a procedure which would have been unnecessary had the units been built correctly in the first place.

V. EXPANDED POLYSTYRENE BEADS*

Oil has long been used to make the water surface in wet pit latrines and cess pits an unsuitable place for mosquito larvae to respire.[4] However, oil does not spread well on heavily polluted water, and it is destroyed by bacteria. Layers of expanded polystyrene beads[22-25] are better in both these respects — they spread to fit exactly the water surface if sufficient beads are applied (Figure 6), and persistence of a layer for 4 years has been observed in Dar es Salaam.[26] In fact, it seems that such bead layers would have an indefinite life unless

* C. F. Curtis, P. R. Morgan, C. A. Maxwell, and J. N. Minjas

the pit flooded and the beads were swept away. The likelihood of frequent flooding rules out the use of polystyrene-bead layers in outdoor sites where the water is not confined within four walls. Apart from sanitation systems, polystyrene-bead layers are effective in disused wells and may be applicable to water tanks where the water is drawn off from below and not dipped for from above.

Apart from making it difficult for larvae to respire, thick layers of polystyrene beads prevent emergence of adults from pupae, and the layers present a dry surface unsuitable for egg laying. The effectiveness of bead layers in preventing mosquito emergence from wet pits has been repeatedly demonstrated by placing exit traps (Figure 6) over latrine holes before and after application of bead layers. Catches ranging up to 13,000 *Cx. quinquefasciatus* in 1 night have been recorded before application of polystyrene layers, though the catch from one pit fluctuates considerably from night to night.[27] The catches generally drop to zero soon after application of bead layers and remain there subsequently. In Zanzibar the catch of newly emerged mosquitoes consistently dropped to zero 2 nights after application of a bead layer.[27] This delay until the second night is presumably because some mosquitoes which had already emerged from pupae remained resting on the pit walls for a night and a day before flying out. However, in Zimbabwe[28] and India[25] delays of a week or more have been observed before complete suppression of mosquito emergence was achieved. In Zanzibar beads of 2-mm diameter are used, and these seem optimal[29] as they give a compact "carpet" on the water surface. In Zimbabwe beads of 3 to 4-mm diameter are used, and it may be that these do not completely prevent larval survival and adult emergence and that the final suppression of breeding is due to the inhibition of egg laying.

The prevention of egg laying by a bead layer may divert gravid females to other sites which cannot be, or have not been, treated. On larger pits with several holes, exit traps placed over one of the holes caught *increased* numbers of gravid females after application of a bead layer.[27,30] This was presumably because they entered through the holes not covered by the trap, could not oviposit because of the polystyrene, and were trapped while exiting in search of another site. In most cases the catch of gravid females declines after about 5 d, presumably because the bead layer suppresses the odor from the pit, thus making it unattractive to gravid mosquitoes. The suppression of latrine odor by a polystyrene layer was much appreciated by householders in Dar es Salaam.[26]

A polystyrene-bead layer immediately reforms after feces etc. have dropped through it. If a pit becomes dry and the polystyrene is buried under feces, the extreme buoyancy of polystyrene ensures that the layer reforms on the surface when water returns to the pit.[24]

Unexpanded polystyrene beads are produced by the petrochemical industry with pentane in solid solution in each bead. Heating the beads to about 100°C softens the plastic, and the pentane rapidly expands the beads by up to 30 times in volume. The process does not use chloro-fluorocarbons which are notorious for their effect on the ozone layer of the earth. Industrially, the beads are heated with steam to make expanded polystyrene beads, which are then heated and pressed to form blocks or slabs for packaging, insulation, etc. In or near cities it is probably best to have beads for mosquito control expanded by steam in a factory. They must then be placed in sacks or drums to take them to the pits where they are needed. If piled loose in an open truck, they would blow away before reaching their destination.

For places remote from an appropriately equipped factory it is possible to expand beads in boiling water in a cooking pot over a wood fire (Figure 7). The unexpanded material is added a cupful at a time, stirred, sieved off the water surface, and poured into the pit. The beads spread satisfactorily when still wet from the "cooking" process. In boiling water there is only a 15- to 20-fold expansion in the volume of the beads, presumably because of uneven heating. It is important to keep the unexpanded material sealed in the containers in which it is delivered until one is ready to "cook" it, otherwise the pentane expanding agent may leak away.

FIGURE 7. Expanding polystyrene beads in boiling water over a wood fire in Zanzibar, Tanzania. (Photograph by C. F. Curtis.)

There is virtually no change in mass during the expansion process, so the weight of the polystyrene required per pit is the same whether it is assessed before or after expansion. The relative merits of factory or on-site expansion of the beads depend on the ease and cost of transport of small volumes of high density or large volumes with very low density. With most forms of transport the cost would probably depend mainly on the volume to be transported. For example, the owner of a dhow plying between Dar es Salaam and Zanzibar assessed the cost of transporting sacks of expanded polystyrene at almost as much as for sacks of coconuts, making only a small reduction for the lightness of expanded polystyrene.[27] In those rural areas where firewood is in short supply, purchasing and/or transporting it into the area could add significantly to the costs of a program.

A 2-cm-thick layer of 2-mm beads is sufficient to eliminate mosquito breeding. Thus, 20 l. of expanded beads are needed per square meter of water surface or 30 l. for a typical pit latrine. This weighs about 1.25 kg if expanded in boiling water and costs about $2.50. Expansion in boiling water uses about 0.05 m^3 of firewood to produce 240 l of expanded beads, and this takes 2 to 3 h from lighting the fire.[27]

Using beads expanded with steam in Zimbabwe 1.5 tonnes of polystyrene were sufficient for the treatment of 3000 VIPs each with a cross-sectional area of 1.1 m^2.[28] Such treatment is now applied routinely to VIPs on completion as a precaution against mosquito breeding during the first few months of the life of the pit. After that time a floating layer of scum usually forms which suppresses any further chance of mosquito breeding. With larger traditional pits one cannot rely on such a scum layer forming, especially where the number of people using a pit is not very large.[9]

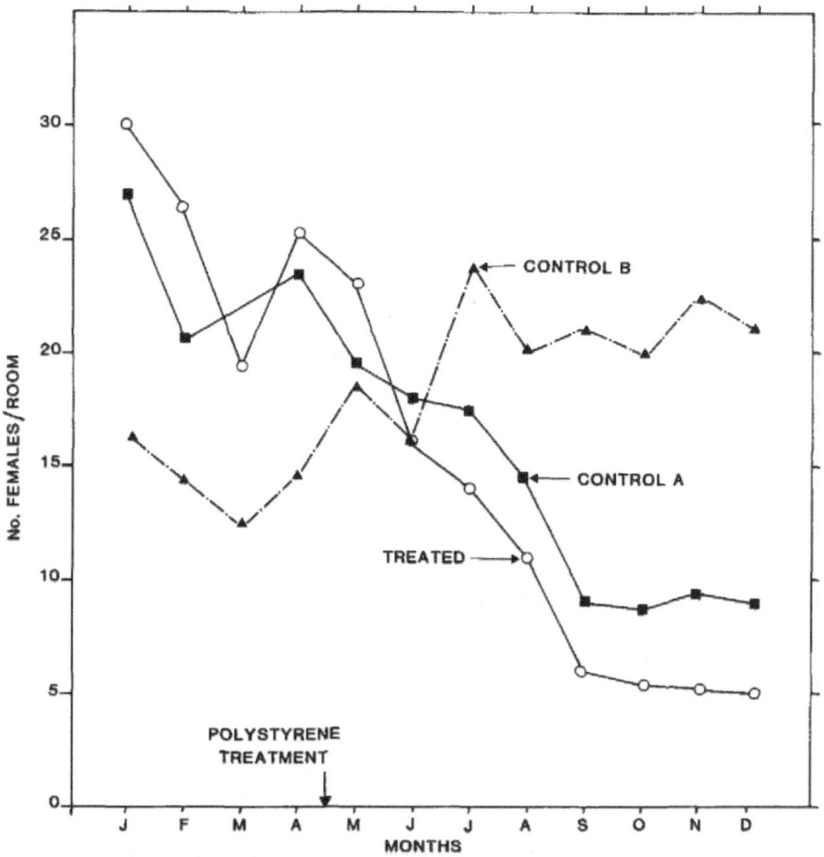

FIGURE 8. The adult female population of *Culex quinquefasciatus* assessed by morning pyrethrum spray catches in houses in a sector of Dar es Salaam in which the pits were treated with expanded polystyrene beads, and in two control sectors of the city.[26] Control A was adjacent to the treatment sector and control B was remote from it.

VI. CONTROL OF FILARIASIS-VECTOR POPULATIONS WITH POLYSTYRENE BEADS*

In attempting to control mosquito vectors of diseases by larvicidal methods, it is generally more of a problem to find all the sites which contribute adults to the vector population concerned, than it is to kill the larvae once they have been found. The real test of the effectiveness of a larvicidal program is not the killing of the larvae in a given site, but whether the adult biting population can be suppressed by searching out and dealing with all the breeding sites which can be found and which could be contributing adults to the population.

An attempt was made to suppress the population of *Cx. quinquefasciatus* in a densely populated sector of Dar es Salaam with an area of about 1.5 km^2. There were hundreds of pit latrines in the area, but only a minority,[10,26] 120 in all, contained water and mosquito larvae. The adult mosquito population in this and two control sectors of the city were assessed by pyrethrum spray catches made by the City Council mosquito control organization in houses before and after treatment of all the wet pits in the experimental sector with polystyrene beads (Figure 8).[26] The experimental area showed a population decline of about 80%, and the adjacent control area (control A in Figure 8) showed about a 60% decline. Control B,

* C. A. Maxwell, J. N. Minjas, and C. F. Curtis

TABLE 1
Mean Catches of *Culex quinquefasciatus* in Light Traps in Bedrooms in the Four Districts of Makunduchi, Zanzibar, before and after Application of Expanded Polystyrene Beads to the Pit Latrines in April—May 1988[27]

	Districts				
	Ki	Ka	Mz	Ng	Total
July—August 1987					
Mean	65.6	97.6	44.9	32.2	63.3
SE	17.6	24.1	10.4	9.1	9.7
N	24	16	11	12	63
July—August 1988					
Mean	0.50	0.17	4.1	0.33	1.39
Range	0—5	0—2	0—51	0—2	0—51
N	22	18	21	15	76

Note: Standard errors of the means or ranges and number of nights of trapping (N) are also shown

in another part of the city, showed a slight increase in population over the same period. It seems likely that the declines in the first two sectors were directly or indirectly due to the polystyrene treatment. The lack of a clearer difference between these two sectors may be explained by movements of adult mosquitoes and by the fact that after the polystyrene treatment the spray team, which was routinely responsible for both these sectors, had more time and more of its limited supply of insecticide to devote to the control A sector. For a more convincing test of the widespread use of polystyrene beads it was decided to make the next trial in a whole community without any other mosquito-control program and surrounded by several kilometers of almost uninhabited dry country in which *Cx. quinquefasciatus* is unlikely to be able to exist.

Makunduchi, in Zanzibar, fitted these requirements. It has a population of 12,000 people with a very high endemicity of Bancroftian filariasis, and *Cx. quinquefasciatus* appears to be the only significant vector; an average of 2.5% of the female mosquitoes carried infective stage larvae.

Apart, perhaps, from a short period of heavy rain each year, virtually the only breeding sites appear to be in pit latrines and cess pits. A survey of all 1741 pits found only 323 to be wet, but some others appeared to be seasonally wet. After a year of baseline data collection on the adult biting population, using light traps in bedrooms, beads were applied to all pits for which there was any report of their ever becoming wet. Table 1 shows the mean light trap catches in a 2-month period shortly after completion of the polystyrene application compared with the same period in the previous year.[27] There has been a massive reduction in the vector population. Where mosquitoes are still found they have been mostly traced to newly dug pit latrines; periodic checks for, and treatment of, such pits will continue to be necessary.

It was recognized that vector control alone would take a long time to reduce the prevalence of a chronic disease such as Bancroftian filariasis (see Chapter 7). Therefore, it was decided to integrate the vector control with mass treatment of the whole population with diethyl carbamazine. The drug was distributed via the Tanzanian system of Balozis, each of whom is in charge of 10 households. Table 2 shows the level of microfilarial infection before the drug treatment and in the same individuals after it.[27] Table 2 also shows the percentage of mosquitoes with infective stage (L3) filarial larvae before and after the mass drug-treatment

TABLE 2
The Effect of the Mass Drug-Treatment Program in January—May 1988 in Makunduchi, Zanzibar, on the Infection Rate in the Human and Mosquito Populations[27]

	No. people surveyed	Microfilariae +ve (%)	William's mean	No. mosq. dissected	% mosq. infective
1987	987	46.7	3.41	5556	2.45
1988	432	14.8	0.24	1716	0.41

Note: In humans the rates are expressed by the percentage found positive for microfilarial infection in night blood samples and the geometric (William's) mean of the numbers of microfilariae per 0.1 ml of blood (including all the people sampled, not just the positives, in the denominator). The mosquito infective rate is the percentage of the light trap catch with L3, infective stage, larvae. The mosquito data were collected from after the first month of drug treatment until the mosquito population was suppressed with polystyrene beads.

program. The combined effect of this reduction in the infective rate and in the biting density (Table 1) indicates that the integrated control program reduced the number of infective bites per night to less than 1% of its pretreatment level. It is hoped that long-term vector control via the polystyrene treatment will prevent the disease returning and avoid the need for repeated rounds of drug treatment as have been necessary in other countries where drug treatment has been used[31,32] without the support of a major vector control effort.

REFERENCES

1. **Mara, D. and Feachem, R. G.,** Technical and public health aspects of low cost sanitation programme planning, *J. Trop. Med. Hyg.*, 83, 229, 1980.
2. **Curtis, C. F. and Feachem, R. G.,** Sanitation and *Culex pipiens* mosquitoes: a brief review, *J. Trop. Med. Hyg.*, 84, 17, 1981.
3. **Greenberg, B.,** *Flies and Disease, Vol. 2: Biology and Disease Transmission*, Princeton University Press, New Jersey, 1973.
4. **Burton, G. J.,** Observations on the habits and control of *Culex pipiens fatigans* in Guyana, *Bull. W.H.O.*, 37, 317, 1967.
5. **Graham, J. E., Abdulcader, M. H. M., Mathis, H. L., Self, L. S., and Sebastian, A.,** Studies on the control of *Culex pipiens fatigans*, *Mosq. News*, 32, 399, 1972.
6. **Bang, Y. H., Sabuni, I. B., and Tonn, R. J.,** Integrated control of urban mosquitoes in Dar es Salaam using community sanitation supplemented by larviciding, *East Afr. Med. J.*, 52, 578, 1975.
7. **Curtis, C. F. and Pasteur, N.,** Organophosphate resistance in vector populations of the *Culex pipiens* complex, *Bull. Entomol. Res.*, 71, 1253, 1981.
8. **Curtis, C. F., Keto, A., Ramji, B. D., and Iosson, I.,** Assessment of the impact of chlorpyrifos resistance in *Culex quinquefasciatus* on a control scheme, *Insect Sci. Appl.* 5, 263, 1984.
9. **Curtis, C. F. and Hawkins, P. M.,** Entomological studies of onsite sanitation systems in Bostswana and Tanzania, *Trans. R. Soc. Trop. Med. Hyg.*, 76, 99, 1982.
10. **Minjas, J. N.,** Control of *Culex quinquefasciatus* in pit latrines: reducing costs by selective larviciding, *Trans. R. Soc. Trop. Med. Hyg.*, 78, 847, 1984.
11. **Mouchès, C., Pasteur, N., Bergé, J. B., Hyrien, D., Raymond, M., Robert, B., de Silvestri, M., and Georghiou, G. P.,** Amplification of an esterase gene is responsible for insecticide resistance in a California *Culex* mosquito, *Science*, 233, 779, 1986.
12. **Davidson, E. W.,** Microbiology, pathology and genetics of *Bacillus sphaericus:* biological aspects which are important to field use, *Mosq. News*, 44, 147, 1984.
13. **Des Rochers, B. and Garcia, R.,** Evidence for persistence and recycling of *Bacillus sphaericus*, *Mosq. News.*, 44, 160, 1984.

14. **Minjas, J. N.,** unpublished data, 1987.
15. **Winblad, U. and Kilama, W.,** *Sanitation Without Water*, Macmillan, London, 1986.
16. **Brandberg, B.,** Why should a latrine look like a house?, *Waterlines*, 3, 24, 1985.
17. **Morgan, P. R.,** The pit latrine — revived, *Cent. Afr. J. Med.*, 23, 1, 1977.
18. **Evans, A. C.,** Pit latrines and vent pipes: public health tools in rural areas, *S. Afr. J. Sci.*, 80, 107, 1984.
19. **Evans, A. C.,** Effect of screened vent pipes on the egression of mosquitoes and flies from pit-type latrines, *S. Afr. J. Sci.*, 83, 144, 1987.
20. **Morgan, P. R. and Mara, D. D.,** Ventilated Improved Pit Latrines; Recent Developments in Zimbabwe, Tech. Paper No. 3, World Bank, Washington, D. C., 1982.
21. **Ryan, B. A. and Mara, D. D.,** Ventilated Improved Pit Latrines: Vent Pipe Design Guidelines, T. A. G., Tech. Note No. 6, World Bank, Washington, D. C., 1983.
22. **Reiter, P.,** Expanded polystyrene balls: an idea for mosquito control, *Ann. Trop. Med. Parasitol.*, 72, 595, 1978.
23. **Curtis, C. F. and Minjas, J. N.,** Expanded polystyrene for mosquito control, *Parasitol. Today*, 1, 36, 1985.
24. **Reiter, P.,** A field trial of expanded polystyrene balls for the control of *Culex* mosquitoes breeding in pit latrines, *J. Am. Mosq. Control Assoc.*, 1, 519, 1985.
25. **Sharma, R. C., Yadav, R. S., and Sharma, V. P.,** Field trials on the application of expanded polystyrene (EPS) beads in mosquito control, *Indian J. Malariol.*, 22, 107, 1985.
26. **Minjas, J. N., Keto, A., and Curtis, C. F.,** unpublished data, 1986.
27. **Maxwell, C. A. and Curtis, C. F.,** unpublished data, 1988.
28. **Morgan, P. R.,** unpublished data, 1986.
29. **Curtis, C. F., Minjas, J. N., and Hill, N.,** Selection of the optimum grade of polystyrene beads for control of mosquito breeding in pits, *Trans. R. Soc. Trop. Med. Hyg.*, 80, 841, 1986.
30. **Curtis, C. F. and Myamba, J.,** unpublished data, 1988.
31. **Kimura, E., Penaia, L., and Spears, G. F. S.,** Epidemiology of subperiodic bancroftian filariasis in Samoa 8 years after control by mass treatment with diethylcarbamazine, *Bull. W.H.O.*, 63, 869, 1985.
32. **Laigret, J., Kessel, J. F., Malarde, L., Bambridge, G., and Adams, H.,** La lutte contre la filariose lymphatique apériodique en Polynésie Francaise, *Bull. Soc. Pathol. Exot.*, 58, 895, 1965.

Chapter 11

THE ROLE OF HOUSE DESIGN IN LIMITING VECTOR-BORNE DISEASE

C. J. Schofield, R. Briceño-León, N. Kolstrup, D. J. T. Webb, and G. B. White

TABLE OF CONTENTS

I. THE HOUSING PROBLEM

The definition of adequate and inadequate housing is often a source of confusion. What is adequate for one family may be inadequate in the eyes of a city architect or town planner. Here, inadequate housing implies houses that do not meet the minimum requirements of low-income families, do not satisfy local building regulations, and predispose to pest infestation and illness among the householders.[1] With current rates of human population growth, the provision of adequate housing is expected to lag far behind demand, especially in the poorer sectors of tropical developing countries.[2,3] Although precise data are unavailable, current estimates indicate that over 100 million people have no shelter at all, while around 1000 million lack adequate shelter.[4,5] On present trends, these figures could be expected to more than double by the end of this century.

Attention is now being focused on this need, which represents a major challenge to planners, architects, engineers, and social scientists. The United Nations declared 1987 as "International Year of Shelter for the Homeless", with a series of national and international projects designed to increase public awareness of the problem and raise funds for specific projects related to low-cost housing. In addition, the World Health Organization — through its Rural and Urban Development program (RUD) — has stressed the health risks associated with inadequate shelter, emphasizing a series of principles related to housing that include protection against communicable diseases, protection against injury and chronic disease, promotion of mental health, and improving health education, community participation, and land use planning in relation to improved standards of human shelter.[5]

Such attention is welcome. In most cities in tropical developing countries, 30 to 60% of the population live in grossly inadequate conditions, including large numbers who live in precarious, poorly planned, or illegal squatter settlements on marginal areas around the cities. Such settlements, often containing hundreds of thousands of people, may be declared illegal and denied access to basic services such as clean water and sanitation. As a result, they show a wide range of health defects, particularly those associated with stress, malnutrition, and water-borne pathogens such as typhoid, cholera, hepatitis, polio, and pathogenic gastrointestinal bacteria and protozoa. In these slum areas, overall child mortality rates and the rate of infection for specific diseases such as TB, acute respiratory diseases, diarrheal diseases, and gastrointestinal helminths are typically more than double those of more affluent residents of these cities.[6]

To a lesser degree, attention has also been focused on the plight of rural areas, where subsistence farmers and small villages generally lack the resources needed to improve their living standards. As one of the principles of development since World War II has been to emphasize industrialization, which has necessarily been concentrated in urban areas, one result has been a consistent neglect of rural development in most tropical developing countries. This has encouraged the steady migration of rural peoples to swell the urban squatter settlements, exacerbating the already grave problems of overcrowding and poor living conditions of these areas.[7]

The housing challenge encompasses a wide range of issues. First is committment to the principle of improving the quality of housing and shelter for the less privileged sectors of the community. This alone is difficult in many countries, where a fragile economic base associated with a limited taxable surplus leads to an acute shortage of resources to finance public expenditure,[8] while at the same time, provision of housing and health for the poorest is generally seen to offer a lower rate of return than comparable expenditure on, say, roads, bridges, industrial development, or even armies. Moreover, a commitment to low-cost housing also implies a commitment to provide suitable land for such development, which often raises difficult questions about land tenure, especially in rural areas.[7,9,10]

There are two further key issues concerning the type of housing and how it is to be

TABLE 1
Housing Costs and Income Levels in Latin America
(1973 U.S. $)

Country	Lowest-cost dwelling	Minimum income for purchase	Average annual income for manufacturing worker
Argentina	3000	1440	1406
Bolivia	4000	1920	731
Colombia	5135	2465	858
Ecuador	1000	480	915
Guatemala	5500	2640	975
Mexico	3440	1651	2114
Venezuela	5700	2736	3050

Note: Data compiled by Agarwal[11] from World Bank and United Nations estimates for 1970—1973. The minimum cost of a dwelling represents the cheapest urban housing unit then available (including toilet and other services), while the estimated income required for purchase is based on an average 20-year payback period at 10% annual interest.

financed. These are closely interrelated. To place greater reliance on the private sector and on local authorities for a national house improvement program will require the state to open appropriate lines of credit to those involved, while allowing much of the decision-making about how, where, and when to build or improve housing to rest with those directly involved at the local level. If not, housing programs will risk being unacceptable to people whom they are meant to serve, either by the position or style of the housing or by being too expensive to benefit those most in need (e.g., Table 1).[11]

II. CONSTRAINTS ON HOUSE DESIGN

To satisfy the constraints of low cost, ease of construction, permanence, comfort, and acceptability to the householders concerned, housing materials and design must make best use of nonimported materials coupled with techniques and designs that are already familiar in the area. Rather than capital-intensive programs, for example, using prefabricated units which impose unacceptable design constraints, the construction techniques should emphasize labor-intensive systems making best use of the local labor force and using locally available materials. In developing countries where the marginal cost of imported capital and materials may be particularly high, emphasis on labor-intensive systems can be a key factor in minimizing costs as well as increasing the local acceptability of the techniques. The use of imported materials such as cement, steel, and glass must be minimized in order to keep costs low, and energy-expensive materials such as fired clay bricks and concrete should also be discouraged since they can add considerably to material production and distribution costs.

Frequently, the constraints of cost, materials, technique, design, and cultural acceptability lead to an optimum solution in the form of existing traditional styles of housing.[12,13] Generally, traditional styles of housing use locally available materials, constructed with proven local techniques, and follow a style already compatible with the cultural aspirations and social requirements of the householders (even though they may not follow legally defined building regulations). This is often true of traditional rural housing, but less apparent for more recent urban "shanty towns" where shelters may be prepared using the most precarious of materials — packing cases, cardboard boxes, old planking, corrugated iron, and so on. In most cases, however, given land tenure and adequate services such as drainage, water, vehicular access, and, if possible, electricity, the poorer urban settlements can be upgraded using techniques developed for rural areas.

In addition to their technical, material, and cultural aspects, traditional styles of housing usually offer other advantages that may be difficult to replicate using other designs.[13] The most important of these are their suitability for the local climate and local climatic fluctuations (Figure 1). The elegant studies of Kroeger[14] on rural housing in Ecuador demonstrated the advantages of traditional earth-walled, thatch-roofed houses in maintaining a cooler daytime climate within the house. Similarly, traditional houses in Africa (Figure 1b) and the Middle East are designed to provide daytime shade and, in drier areas, to preserve atmospheric humidity,[15] while raised houses of slatted bamboo in humid costal regions of Ecuador (Figure 1c) and Southeast Asia allow welcome cooling ventilation, as do the pole-walled houses of Mayans in Central America (Figure 1d). In any attempt to improve traditional styles of housing it is, therefore, important to identify and preserve these useful characteristics.

In both tropical and temperate areas, however, traditional and modern housing offers a range of habitats that can be exploited by arthropod pests and vectors of disease (Figure 2; Table 2). Of the vast range of vector-borne pathogens — arboviruses, parasitic protozoa, helminths, etc. — there are relatively few for which the domestic and peridomestic environment does not offer an important site for transmission to man.[16] Indeed, for many vectors it is their close association with man's domestic environment that has altered them from agents of zoonotic transmission to be vectors of significant medical importance.

Good house design and use of appropriate materials can minimize the risk of pest infestation. In most cases, however, designers and builders both of traditional and contemporary houses, either ignore the possibility of pest infestation or concentrate only on those pests that threaten the structural integrity and function of buildings and their services. The risk of direct structural damage due to rodents, ants, and termites, or indirect damage due to birds, for example, by their droppings or by blockage of drains and gutters by their nests, is fairly well recognized.[17] However, even in developed countries the risk to human health of pest contamination of catering facilities, foodstores, and hospitals is generally ignored at the design and construction stage, while even subsequent remedial action may be lacking or grossly inadequate.[17-19] In temperate and tropical countries alike, appropriate house design to minimize pest problems is generally given even less priority.[16]

III. THE DOMESTIC VECTOR PROBLEM

Man's domestic and peridomestic environment offers an important habitat for a wide range of arthropods. Obviously, such habitats (i.e., houses) are common and widely distributed throughout temperate and tropical areas. They offer shelter from extremes of climate, often allowing their inhabitants to survive in regions that would otherwise be inhospitable. For arthropods, these habitats also offer an abundant source of potential food in the form of man, his animals, and products. Moreover, the domestic environment tends to be relatively stable over time. Even the temporary shelters of nomadic peoples offer a degree of stability compared to much of the outside environment, while more permanent homes, even of the more precarious types, may last for 5 to 10 years or more.

Together, these characteristics make the domestic and peridomestic environment an attractive habitat for colonization by a wide range of arthropods, many of which are important as carriers or vectors of human pathogens (Table 2). Among the best-known examples are the ubiquitous houseflies and cockroaches, carriers of a range of viral and bacterial pathogens, the triatomine bugs of Latin America that live in cracks and crevices of poor-quality rural houses and transmit *Trypanosoma cruzi* to cause Chagas disease, and the domestic *Ornithodoros* ticks, vectors of a range of rickettsial fevers especially in parts of Africa. Other domestic arthropods are better known for their direct effects, such as the widespread cimicid bedbugs implicated in chronic iron deficiency anemia and possibly hepatitis-B transmission, or the housedust mites (*Dermatophagoides* spp.) which are important causes of asthma and

FIGURE 1. Styles of traditional rural housing developed according to availability of materials, cultural and climatic requirements. (A) 18th Century rural cottages in Cambridgeshire, U.K.; earth walls and thatch roofs provided well-insulated houses suitable for a humid climate, which now command high prices. (B) Masai house, Kenya; the design, formed of mud and sticks, provides a cool shaded interior allowing a platform from which a warrior could keep guard. However, the cracked walls and dark interior can favor domestic arthropods such as ticks, bedbugs, fleas, and cockroaches. (C) House of bamboo slats, coastal Ecuador. The underfloor space allows ventilation, but can also habor pests. Triatomine bugs *(Triatoma dimidiata)* were also found resting between the bamboo slats. (D) Mayan house, Central America; open pole walls allow cooling ventilation and a relatively well-lit interior, unfavorable to domestic pests, although the thick thatch roof may provide refuges for insects. (E) Earth and wood house, central Brazil; a construction style adapted from Iberian origins provides ample refuge in the cracked walls for triatomine bugs *(T. infestans)*, cimicid bedbugs, fleas, cockroaches, and spiders. Grain stored at the side of the house attracts rodent pests that may be disease reservoirs; they and domestic animals such as dogs will offer additional blood sources to boost populations of blood-sucking bugs. (F) House of split palm logs in the humid chaco of Paraguay; the fibrous core of the logs provides an excellent refuge for domestic arthropods. (Photograph B by D. J. T. Webb; photographs A and C—F by C. J. Schofield.)

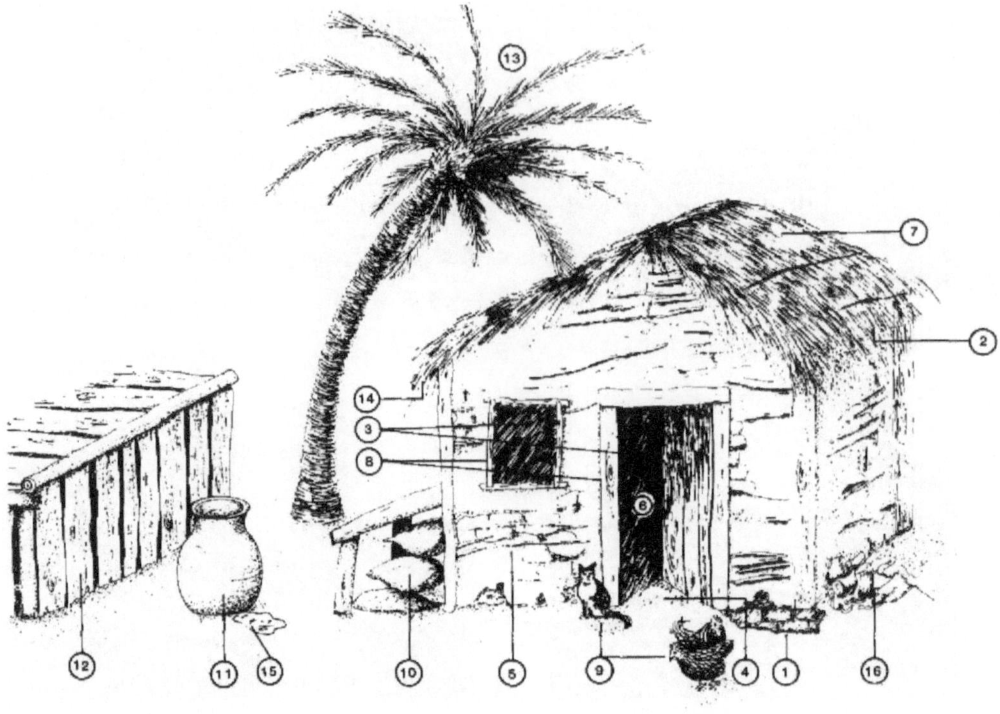

FIGURE 2. Factors predisposing to colonization of the domestic and peridomestic environment by arthropods of medical importance. (1) Inadequate foundations, and inadequate protection of the base of the wall against water erosion, causing damage to footings of wall, allowing entry of crawling pests. (2) Open, unscreened eaves, with no internal ceiling, allowing entry of flying insects into living space. (3) Open, unscreened window and door allowing entry of flying insects. (4) Cracked uneven floor providing refuge for ticks, flea larvae, and other insects. (5) Unrendered, cracked walls providing refuge for cockroaches, bedbugs, triatomine bugs, etc. (6) Dark, poorly ventilated interior favoring arthropods that rest during the day. (7) Roof of thatch allowing resting places for insects. (8) Shrinkage cracks at junction between wooden frames and walls of other material. (9) Domestic animals offering food source for blood-sucking arthropods, and may also act as reservoirs for parasites and pathogens. (10) Stored products in or near the house attracting rodents and other pests. (11) Unscreened water-storage vessels providing breeding sites for mosquito larvae. (12) Animal enclosures near house providing additional refuge and food source for insect pests. (13) Shade tree maintaining roof fabric cool and more suitable for insects, and also offering refuge for some peridomestic insects. (14) Inadequate roof overhang does not protect wall from rain, causing erosion and cracking. (15) Water seepage providing moist breeding grounds for, e.g., psychodids and ceratopogonids. (16) Organic refuse providing breeding ground for houseflies etc.

other allergies. In some areas, venomous spiders such as *Atrax robustus* (Australian funnel-web spider), *Latrodectus mactans* (the "black widow"), and *Loxosceles reclusa* (the "brown recluse spider") are important domestic pests.

Some domestic arthropods go further, colonizing man himself or his domestic animals. Well-known examples include head and body lice, scabies and follicle mites, and various species of fleas whose larvae live in the accumulated debris around the house while adults spend much of their time on the surface of their chosen hosts. Other medically important insects, particularly mosquitoes and certain species of sand fly, might best be regarded as adventitious insects, entering the domestic environment to feed but making use of breeding sites that may be close or distant from the house (Table 2).

With the exception of houseflies and diurnally active *Aedes* mosquitoes, most domestic arthropods are active at night when their vertebrate hosts and potential predators are resting. During the day these arthropods rest, seeking dark undisturbed crevices that offer protection from predators and climatic extremes. In the case of *Triatoma infestans* for example, studies

TABLE 2
Summary of Domestic and Peridomestic Arthropods of Medical Importance

Breeding site	Cause or carrier of
Breeding on the body	
Head lice[a]	
Pediculus capitis	Louse infestation of the head
Body lice[a]	
P. humanus	Louse infestation of clothing Typhus: *Rickettsia prowazeki* Trench fever: *R. quintana* Relapsing fever: *Borrelia recurrentis*
Pubic lice[a]	
Phthirus pubis	"Crab" louse infestation
Scabies mites[b]	
Sarcoptes scabiei	Scabies
Follicle mites[b]	
Demodex folliculorum	*Corynebacterium acnes*?
Breeding in the house	
Fleas[a]	
Pulex irritans, Ctenocephalides spp.	Bites only: not vectors
Xenopsylla spp.	Plague: *Yersinia pestis* Typhus: *R. mooseri*
Cockroaches[c] *Blatta orientalis* *Blattella germanica* *Leucophaea maderae* *Periplaneta* spp. *Supella* spp. etc.	Shigellosis: *Shigella dysenteriae* Salmonellosis: *Salmonella* spp. and other enteric infections Viruses: hepatitis A, poliomyelitis, etc. Legionnaires' disease: *Legionella*, etc.
Bedbugs[a]	
Cimex hemipterus *C. lectularis*	Viral hepatitis B?
Triatomine bugs[a]	
Panstrongylus megistus *Rhodnius prolixus* *Triatoma infestans*	Chagas' disease, *Trypanosoma cruzi*
Soft ticks[a]	
Ornithodoros spp.	Relapsing fever, *Borrelia duttoni*
Argas spp.	Bites only; not a vector of human disease
Dust mites[b]	
Dermatophagoides spp.	House dust allergy
Breeding peridomestically and/or in water storage and sewerage	
Jiggers[a]	
Tunga penetrans	Jiggers
Midges	
Culicoides furens *C. phelbotomus* etc.	*Mansonella ozzardi, M. perstans*
C. austeni *C. grahami*	*M. perstans, M. streptocerca*
Mosquitoes[a]	
Aedes aegypti *Ae. albopictus*	Dengue, chikungunya
Ae. aegypti *Haemagogus equinus*	Yellow fever
Ae. scutellaris group	Subperiodic Bancroftian filariasis
Culex pipiens group	Urban periodic Bancroftian filariasis
Cx. tritaeniorhynchus *Cx. gelidus*, etc.	Arboviruses, e.g., CHIK, JE, WN
Anopheles stephensi	Malaria: *Plasmodium* spp.

TABLE 2 (continued)
Summary of Domestic and Peridomestic Arthropods of Medical Importance

Breeding site	Cause or carrier of
Armigeres subalbatus	Bites only; not a vector of human disease
Sandflies[a]	
Lutzomyia spp. *Phlebotomus* spp.	Sand fly fever; Visceral leishmaniasis: *Leishmania donovani* complex Cutaneous/mucocutaneous leishmaniasis: *L. braziliensis*, *L. tropica*
Moth flies[c]	
Psychoda spp.	Allergy to moth flies
Telmatoscopus spp., etc.	
Blowflies[c]	
Calliphora *Lucilia* *Chrysomya*	Intestinal myiasis; Dysentery: *Clostridium*, *Salmonella*, *Shigella*, etc.
Houseflies[c]	
Musca domestica	Dysentery: *Clostridium*, *Entamoeba*, *Salmonella*, etc. Viruses: hepatitis A, poliomyelitis, etc.
Latrine flies[c]	
Fannia spp.	Dysentery?
Face flies[c]	
Musca sorbens	Yaws: *Treponema pertenue* Trachoma: *Chlamydia trachomitis* Septicemia: *Clostridium*, *Staphylococcus*, *Streptococcus*, etc. Infective dermatitis: *Candida*, *Staphylococcus*, etc.
Stable flies[a]	
Stomoxys spp.	Bites only; not a vector of human disease
Poisonous spiders[c]	
Atrax, *Latrodectus* *Loxosceles*, *Lycosa* etc.	Spider bites
Adventitious — entering the house to feed	
Mosquitoes[a]	
Aedes spp.	Arbovirus, e.g., CE, JE, RVF, VEE Subperiodic Bancroftian filariasis
Anopheles spp.	Arboviruses: e.g., JE, ONN Malaria: *Plasmodium* spp. Periodic Brugian filariasis Rural periodic Bancroftian filariasis
Coquillettidia spp.	Arboviruses, e.g., BSQ, COT, CVO, MAY, MUR, TEN, WN
Culex spp.	Arboviruses, e.g., CHIK, CE, EEE, GMA, JE, RVF, SIN, WN
Mansonia spp.	Arboviruses, e.g., ILE, RVF, VEE Subperiodic Brugian filariasis
Sand flies[a]	
Phlebotomus spp.	Cutaneous leishmaniasis: *L. aethiopica* *L. major*, *L. tropica* Visceral leishmaniasis: *L. donovani*, *L. infantum*; Sand fly fever
Scorpions[c]	
Androctonus, *Buthus*, *Centruroides*, etc.	Scorpion stings

[a] Blood sucking.
[b] Dermatophagous.
[c] Scavening.

From Schofield, C. J. and White, G. B., *Trans. R. Soc. Trop. Med. Hyg.*, 78, 285, 1984. With permission.

of their vertical distribution in the cracked walls of houses in central Brazil show that they tend to congregate in crevices at the upper levels, which is interpreted to represent avoidance of the predatory behavior of domestic chickens.[20] Moreover, since these bugs depend exclusively on a diet of vertebrate blood and their population size is regulated by access to that blood, their nocturnal feeding behavior is interpreted to be an important adaptation to minimize disturbance while feeding from their resting hosts. In contrast, some other species of Triatominae that may feed on nocturnal hosts tend to be more active during the day.[21] Similar adaptations are shown by other blood-sucking insects, representing a balance between minimizing their risk of predation, avoiding climatic extremes, and acquiring adequate blood meals.[22]

In terms of vector-borne diseases, the greatest problems are apparent in tropical areas, where higher temperatures unrelieved by extreme winters allow vector populations to reproduce more quickly and often permit year-round transmission. Some problems are particularly associated with cities, where blocked drains and inadequate sewage disposal can provide breeding sites for mosquitoes such as the culicine vectors of filariasis (Chapter 10), accumulated refuse promotes housefly breeding, and an abundance of water containers can encourage the breeding of nuisance and vector mosquitoes such as *Aedes aegypti, Ae. albopictus,* and *Armigeres subalbatus*. Such problems can best be dealt with by improved urban services aided by appropriate legislation, as, for example, in Bombay,[23] Singapore,[24] and urban areas of Malaysia and Cuba where fines can be imposed on householders who allow mosquito-breeding sites on their property (Chapter 6). Local authorities that refrain from such legislation and its effective enforcement do a disservice to the community. For example, in the U.S. lax legislation concerning the control of *Aedes aegypti* infestations has led to recent outbreaks of dengue in Texas and has paved the way to the current spread of *Ae. albopictus* through importation of used-car tires harboring larvae (Chapter 6).[24] In addition, many features of the design, construction, situation, and maintenance of individual houses — both in urban and rural areas — can also help to minimize these problems as well as inhibiting vectors that actually breed in the domestic and peridomestic environments.

There are three key features of the domestic environment that promote access or colonization by arthropod pests and vectors of disease. The first relates to the position of the house within areas where a given pest species is known to occur. Siting a house close to known or potential breeding sites of mosquitoes or other vectors will clearly increase the risk of house invasion by these species and will therefore require more careful measures to avoid them. Similarly, siting a house distant from any potential source of vectors implies that less rigorous measures may be adequate. In both cases, however, due consideration should be given to the likely mode of entry of the vectors. Entry can be by active crawling or flying, which may be inhibited by physical barriers such as window screens, or it may involve passive dispersal associated with man himself or other animals, or associated with materials brought into the house such as food and furniture. The second and third features relate to the physical elimination of potential resting and breeding sites within the domestic and peridomestic environment, and restriction of vector access to potential food sources (see Figure 2).

IV. LOW-COST DESIGN AND CONSTRUCTION

A. LOCATION OF THE HOUSE

The risk of vectors and vector-borne disease rarely influences the decision on where to site a house, although settlement plans have been disrupted by vector-borne diseases in some areas. For example, some new settlements in the Amazon region have been discouraged by the high malaria risk,[26,27] and some parts of West Africa have in the past become depopulated due to the high risk of onchocerciasis.[28] Clearly, however, the risk of invasion will influence the stringency of measures that should be applied to avoid vector invasion or colonization of houses.

Because water is needed for drinking and washing, houses must either be provided with a piped supply, a well, or rainwater collection system, or else dwellings must be sited near a suitable source such as a communal stand-pipe, well, pond, stream, lake, or river. Where these sites are also favored by man-biting mosquitoes, biting midges *(Culicoides, Leptoconops)*, or even blackflies *(Simulium)*, it is unwise to locate houses within the flight range of these insects unless the houses are adequately screened or the breeding sites are controlled by insecticidal or environmental methods.[29] Exposed water of any type is likely to serve as a breeding site for a range of dipterous insects, and some, even though not blood-suckers, may become an abundant nuisance, especially psychodid moth flies and chironomid midges whose cast pupal skins may provoke allergies.[30,31]

Different mosquito species tend to be associated with particular types of water. Some are efficient colonizers of even the most temporary rainwater pools, completing their development from egg to adult in as few as 7 d. Others prefer more permanent waters such as the margins of larger ponds and lakes. Generally, during the development of human settlements, the nature of exposed water bodies changes, often promoting a succession of mosquito species of different habits and different vectorial capacities. In the Amazon region, for example, new settlements have arisen at points where inadequate drainage of transverse streams and borrow pits along the highway have resulted in extensive shallow pools. Such pools have become good breeding sites for *Anopheles darlingi*, with a consequent rise in malaria transmission.[27,32] However, around older settlements where exposed waters have become successively more polluted through man's activities, fewer *An. darlingi* are now found in comparison to other mosquito species.

Despite the tendency of females of most anopheline malaria vectors to enter houses at night in search of a blood meal and often to rest indoors during the day, nearly all these *Anopheles* species breed in pools and swamps some distance from houses. Thus, it is usually people living in rural areas or on the outskirts of a town who are most vulnerable to the nightly immigration of these mosquitoes. An exception is *An. stephensi* in urban situations in the Middle East and India, which has adapted to breeding in wells and water-storage tanks both indoors and outdoors, giving rise to an increase in malaria incidence in cities such as Bombay and Karachi. Strict surveillance supported by legislation has proved effective in compelling householders in affected areas to cover water tanks and to treat wells with larvicides to eliminate mosquito breeding sites.[23] The versatile malaria vector *An. arabiensis* may also be adapting to urban conditions in some West African cities,[33] a process which could be retarded by appropriate house design.

B. RESTRICTING ACCESS OF VECTORS

Table 3 summarizes some of the factors affecting mosquitoes biting or resting indoors.[34] The tendency of most flying insects, especially mosquitoes, to fly upward on encountering an obstruction such as a house wall means that the eaves under a roof — as well as holes in the wall, open windows, and doors — are important points of entry into a house. Such apertures can be screened to prevent entry of flying insects, but efficient screens can severely restrict ventilation and for this reason are not generally used except in houses with alternative means of air-conditioning (as in parts of the southern U.S.) or in areas with constant year-round problems of mosquitoes and high incidence of mosquito-borne disease. For efficient screens, mesh apertures must be less than 1.2 mm to keep out mosquitoes, but the smaller biting midges *(Culicoides)* and some phlebotomine sand flies can penetrate much finer screens down to 0.2 mm aperture. Cotton netting, which is most efficient but prone to damage, can restrict airflow by up to 70%. Metal screens are less restrictive of airflow, but in humid areas can corrode rapidly (except for stainless steel mesh, which is prohibitively expensive). Plastic mesh is better for window screens, but still reduces airflow by up to 35%.[35,36]

TABLE 3
Factors Increasing (+) or Decreasing (−) the Likelihood of Mosquitoes Biting and Resting Inside Houses[34]

	Anophelines	Culicines
Elevation	+	−
Isolation of house	+	−
Water containers indoors	−[a]	+
Proximity to streams/lakes	+	+
Windows open/unscreened	+	+
Eaves open/unscreened	+	+
Walls open/damaged	+	+
Ceiling present	−	−
Bednets in use	−	−
Woodsmoke indoors	−	−
Pyrethrum burning	−	−
Use of other insecticides	−	−

[a] + for urban *Anopheles stephensi* in Middle East, India, and Pakistan.

Where the eaves of a house cannot be blocked or screened because of the need to ventilate the roof space, an alternative is to fit some form of ceiling. Even where the house structure or personal preferences of the householder would not permit construction of a solid ceiling of wood or of sticks and clay (Figure 3a), an effective alternative is to fit a lightweight ceiling of hessian-type cloth or woven matting (Figure 3b). This can bar mosquitoes from access to people sleeping in the room below, acting, in a sense, as a permanent communal bednet.[37] In houses with corrugated metal roofs, ceilings have the added advantage of reducing the heat radiated inside from the roof. In contrast, a ceiling can make a thatched house less comfortable by reducing ventilation. In a survey of 475 villagers in Tanzania, several other advantages and disadvantages of ceilings were given, with 86% of those who had fitted some form of ceiling citing exclusion of mosquitoes as one of the reasons for constructing it (Table 4). However, if the ceiling is of fairly solid construction, such as plywood or board, or sticks sealed with rendered clay, this can transform the roof space into a habitat for birds and small mammals, such as rats, mice, opossums, bush-babies, and bats. These are inconvenient and can cause structural damage to the house, as well as acting as reservoirs for pathogens and additional food sources for blood-sucking insects. In Latin America, the roof space above a ceiling can also provide an excellent habitat for triatomine bugs.

Many domestic arthropods can also be carried into the house on other animals, with baggage, or with items of food or furniture. Ectoparasites such as fleas, lice, and follicle mites *(Demodex)* gain access by their close association with domestic mammalian hosts. In Latin America, triatomine bugs can gain access to houses by active flight,[38,39] by passive carriage as small nymphs and eggs among the feathers of sinanthropic birds such as house sparrows,[40] and carried in the luggage and clothing of people visiting from other areas.[21,42] Cimicid bedbugs can also be transported in the sleeping mats of visitors and even in temperate climates can be passed from house to house in cracks in second-hand furniture. Cimicids and Triatominae can both survive for several months without feeding. As a result, they can disperse through the sale of second-hand furniture from infested premises and may even spread in furniture stores among different items offered for sale.[42]

C. RESTRICTION OF FOOD SOURCES

Those insects and mites that live in permanent contact with their vertebrate hosts, such

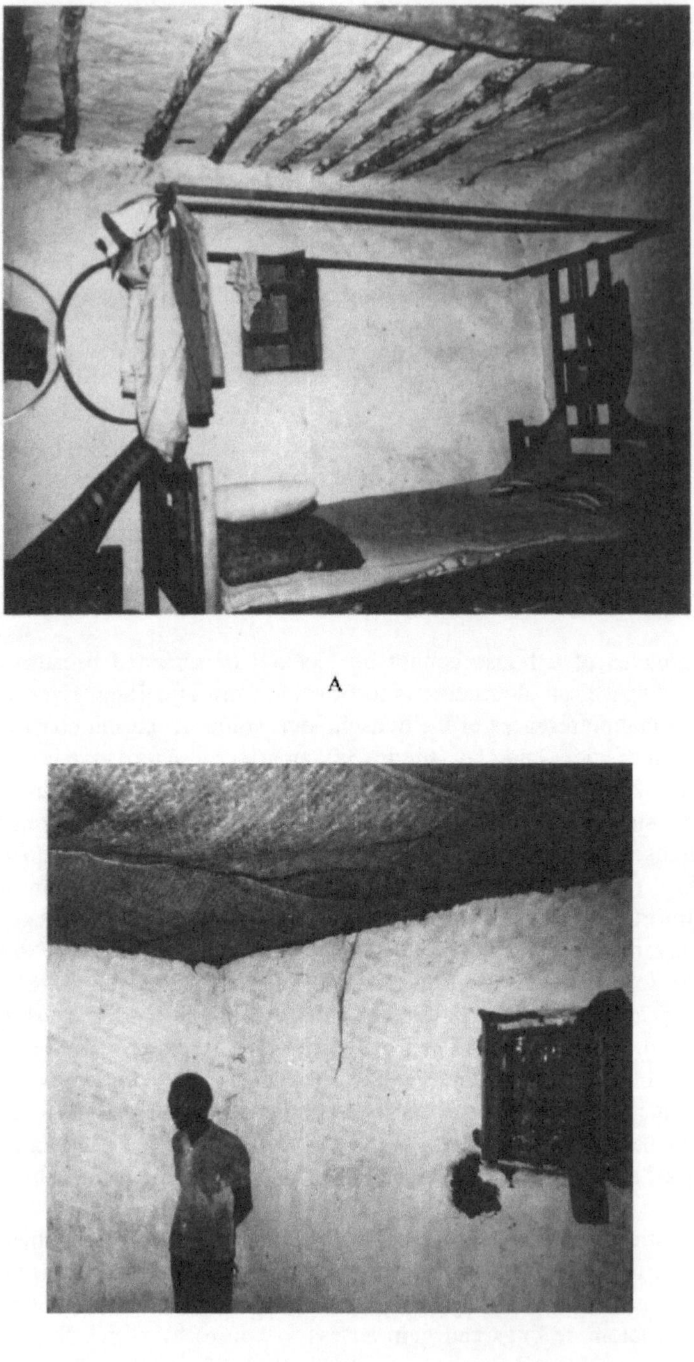

FIGURE 3. Examples of ceilings in houses near Tanga, Tanzania. (A) Traditional clay ceiling; (B) ceiling of mats woven from palm fibers. (Photographs by N. Kolstrup.)

as ectoparasitic mites and lice, are best treated as medical conditions to be dealt with by a clinician or similarly trained specialist. Prevention of further infestations, however, will be assisted by provision within the house of suitable washing facilities and by provision of adequate sleeping quarters such that, for example, children are not required to share the same bed or sleeping mat. Overcrowding, leading to close physical contact while ectopar-

TABLE 4
Opinions of Tanzanian Villagers on the Merits of Ceilings in Homes[37]

Factor	Reason for making ceiling (%)	Advantage of having ceiling: 51 Haves (%)	Advantage of having ceiling: 424 Have-nots (%)	Disadvantage of having ceiling: 51 Haves (%)	Disadvantage of having ceiling: 414[a] Have-nots (%)
Prevention of					
Mosquitoes	86	86	71	—	—
Dirt	34	42	50	—	—
Thieves	20	24	13	—	—
Fire	20	26	14	—	—
Animals	18	16	6	—	—
Smartness	2	6	0.2	—	—
Warmth	4	2	1	30	12
None	—	—	—	66	87
Possible collapse	—	—	—	4	1

[a] Also 10 "don't knows".

asites are at their most active, can facilitate transmission between family members. However, most ectoparasites are not actually transmitted on clothing or bed linen because they survive for only a very short time away from their vertebrate hosts.[43,44]

For other domestic and peridomestic arthropods, restricted access to food sources is an important way of minimizing their numbers. For scavenging insects, it is particularly important to disallow access to stored foodstuffs and to provide adequate disposal for organic wastes because scavenging insects, especially houseflies and cockroaches, can readily transmit pathogens from one to the other. Fecal pathogens carried by houseflies and cockroaches probably contribute significantly to health problems such as childhood diarrhea, which is a major cause of child mortality in many countries. In more affluent communities these hygienic measures can be achieved by the use of refrigerators and screened foodstores, with insecticidal fumigation where appropriate, and the provision of closed refuse containers (protected from physical damage by rats, mice, and other animals) and well-maintained toilets (see Chapter 10, Cairncross,[45] and Cairncross and Feachem[46]). Such measures combined with vigorous public health advertising, contributed to the decline in housefly populations in Europe during the 1950s. Elsewhere, traditional styles of housing and lower family income levels may limit the possibilities of pest control by these methods, but community health programs should inform and encourage householders to minimize insect pests by preventing access to both food and waste.

In agricultural communities, protection of stored products is primarily to prevent climatic damage and losses due to birds, rodents, and other small mammals. These small animals often act as reservoirs of zoonotic infections and are often associated with insects which can bring these infections into the domestic environment. Important examples include bubonic plague transmitted by rat fleas and sylvatic forms of *Trypanosoma cruzi* transmitted by triatomine bugs in Latin America. It is clearly desirable that stored products be adequately protected and should preferably be sited in outhouses away from living quarters.

Domestic animals should also be housed away from the house itself, thus removing one source of litter and waste from the house and also reducing the number of blood sources in a house available for blood-sucking insects. Care must be taken that the outhouses and animal enclosures do not also become resting and breeding sites for pest insects. For goat corrals in the Argentine chaco, for example, wire fencing should be preferred to traditional brushwood because the latter can provide a good habitat for blood-sucking triatomine bugs.[47,48] In Latin America, domestic populations of triatomine bugs are primarily regulated by the size of available host biomass, and removal of each domestic animal is predicted to cause a corresponding reduction in bug population size.[20,49] Domestic cats and dogs also serve as reservoir hosts for *T. cruzi,* so provision for these animals to sleep outside the house is likely to reduce both the size of the bug population and the proportion infected with *T. cruzi*.

The regulation of mosquito populations is less clearly understood, and it may be that most species are regulated primarily by access to aquatic breeding sites rather than to blood sources. Moreover, it is not clear if reduction in numbers of hosts within a house would reduce the numbers of mosquitoes attracted to feed in that house. However, it is sometimes possible to exploit the host preferences of some mosquito species for cattle and other livestock, and thus reduce the frequency with which they bite people. This tactic, known as "zooprophylaxis" is insufficient to interrupt transmission entirely, but can lead to reduced transmission of some mosquito-borne diseases. For example, many vectors of malaria are strongly attracted to cattle (e.g., *An. arabiensis* in Africa; *An. annulatus, An. culicifacies,* and *An. maculatus* in South Asia), so rates of transmission rise significantly when few livestock are present as alternative hosts to man. Some sand fly vectors of *Leishmania* are also less likely to enter the house to bite people if their preferred animal hosts are kept outside the house. Examples include the Indian *Phlebotomus argentipes* which is attracted to cattle, *P. papatasi* and the South American *Lutzomyia longipalpis* which are attracted to chickens and may breed in chicken houses.[50]

Screens and mosquito nets over beds at night can also be effective in preventing host access of nocturnal flying insects such as mosquitoes. Their effectiveness can be improved by impregnating the nets with insecticides such as permethrin (see Chapter 2). However, screens and nets are generally ineffective against insects that inhabit the bed or that gain access by crawling (such as argasid ticks or triatomine bugs). Such pests can be inhibited by hanging sleeping mats and bedclothes outside in the sunlight during the day, since many are unable to tolerate direct insolation.

D. RESTRICTION OF HIDING PLACES

All domestic arthropods require a shaded undisturbed resting site within the house for part of their lives. Even for exophilic mosquitoes, such as *An. darlingi,* at least a few minutes are spent resting on an interior surface before and after feeding.[31,51,52] Other domestic insects and ticks rest during their inactive phases in dark crevices in the floor, walls, roof, or under debris or similar type of habitat. In all cases, control campaigns may include fumigation or spraying the internal surfaces with residual insecticides, and the usefulness of this approach can be assisted by the use of housing materials that offer a good nonabsorbent surface for the action of such insecticides. Walls of unrendered earth, for example, tend to absorb the active ingredient very quickly, causing a rapid loss of insecticidal activity at the surface where the insect rests.[53,54] Similarly, thick roofs of thatch or brushwood can provide an open structure within which insects can rest but where insecticide sprays cannot reach. Some types of earth used in house walls, due to their extreme pH values, have the additional disadvantage of provoking chemical breakdown of some sprayed insecticides.

In areas of high risk for vector-borne diseases, where insecticide spraying campaigns may be required, housing materials and construction methods should be chosen to provide nonabsorbent surfaces suitable for the insecticide and to minimize areas where insecticide sprays would be unable to penetrate. The presence of domestic arthropods can also be discouraged directly by the use of materials that do not crack to provide resting sites, ensuring close fitting junctions between dissimilar materials, and allowing easy maintenance such as plastering over any cracks or holes that subsequently appear. Moreover, since most domestic arthropods are nocturnal and so require dark crevices in which to hide during the day, houses that are well ventilated and well lit (for example, by adequate window openings) will discourage indoor resting insects. These design considerations apply equally to all parts of the house.

A

FIGURE 4. The importance of foundations. (A) House wall undermined by water during the wet season in Tanzania. (Photograph by N. Kolstrup.) (B) Schematic plan for a stone foundation with or without cement.

E. THE FOUNDATIONS AND FLOOR

A primary defect in many traditional houses is the absence of secure foundations. This imposes limits to the height and weight of wall and roof that can be supported, resulting in low houses with inadequate ventilation and dark interiors favoring domestic pests. Overweight walls and roofs may shift, causing damage or partial collapse, although, paradoxically, this can sometimes discourage domestic insects because of the resulting increase in ventilation and illumination. However, rainfall damage concentrated at the base of a wall can open up crevices through which rodents and other pests may enter (see Figure 4A), and may undermine the walls sufficiently to cause total collapse.

Foundations serve two main functions. They transfer load from the walls and roof uniformly to firm ground, and they serve as a shield preventing water damage to the base of the wall and the floor. To some extent, they may also inhibit termites. Wall foundations would probably be the single most important improvement to traditional houses, making the buildings more permanent, structurally safer, and so encouraging householders to make other improvements. Foundation footings should be embedded deep enough to reach root-free soil which will not be washed away by water. Usually, 30- to 40-cm depth will suffice for a single storey building, but in areas prone to water erosion and soil-swelling a depth up to 1 m or more may be required. Materials for footings can be compacted stones, soil-cement mortar, anthill or termite mound mortar, fired clay bricks, or compressed earth blocks (see below). Figure 4B shows the arrangement of a low-cost cement-soil foundation now widely used in coastal regions of Kenya and Tanzania.

Good foundations will also allow a firm and level floor. In many styles of traditional houses the floor is of beaten earth. This is cheap, easy to construct, and has some functional advantages such as absorption of spilled liquids. Cracked and uneven earth floors can provide

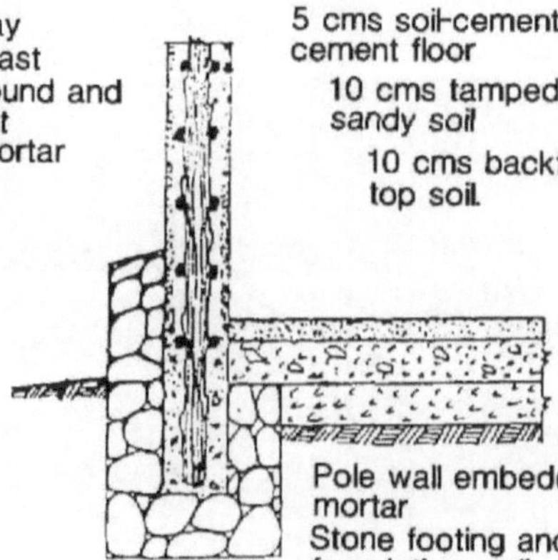

FIGURE 4B.

refuges for house dust mites and flea larvae among the debris that accumulates in the cracks, and for argasid ticks (especially under sleeping mats). In Central America, *Triatoma dimidiata*, an important vector of *Trypanosoma cruzi*, also occurs commonly in cracks in the floor, and replacement of mud floors with smooth cement can significantly reduce house infestation with this vector.[55] In areas where wood is readily available, floors may be made of wooden planks, often raised on stilts to avoid flooding, as in coastal parts of Southeast Asia. The space under the floor, which is often used for storage, can provide refuge for small mammals, snakes, scorpions, spiders, cockroaches, and other pests. In these cases, the underfloor space should be kept clear of stored products and debris, while the floor itself should be made of very close-fitting slats or slats with ventilated spaces more than 1 cm wide which are unsuitable for resting insects.

F. THE WALLS

Materials for house walls depend on the lifestyle of the householders, their income levels, and availability of local materials. Temporary shelters of nomads, such as the grass tents of some Amerindians or the leafy huts of West African Fulani tribesmen, are rarely colonized by arthropods because of their temporary nature, but they provide little protection against adventitious insects such as houseflies and mosquitoes. Dwellings of a more substantial and permanent nature often provide a range of niches for residential arthropods. In the high Andes of Chile and Bolivia, stones and earth are commonly used for the walls since timber is not readily available; the spaces between the stones are often occupied by cockroaches, spiders, and triatomine bugs. Similarly, houses of split palm logs in the humid chaco of Paraguay offer excellent habitats for triatomine bugs within the fibrous inside of the logs. In Ecuador, houses of split bamboo suffer the same problem, as do houses in parts of Southeast Asia which are sometimes colonized by *Triatoma rubrofasciata* (although Chagas disease is not transmitted outside the New World).[56] In contrast, houses with walls formed entirely from wooden poles, as with those of Mayans in Central America, are rarely colonized by arthropods. This is probably because (1) the poles are too widely spaced to offer suitably dark crevices for bugs and cockroaches and (2) the poles allow very good ventilation and good daytime illumination to the inside of the house. Darker houses of hewn planks or overlapping boards do provide good refuges for arthropods.

However, the most widely used material for the walls of traditional-style houses is earth, either as mud pressed to a framework of sticks and poles, or as adobe mud blocks held in place by a mud-based mortar. Such houses, common throughout savanna areas of Africa and Latin America, are very prone to cracking as the earth dries and shrinks, pulling away from wooden supports and the frames of doors and windows. This provides a multitude of dark, undisturbed cracks and crevices where insect pests and vectors thrive. Where the wall is otherwise sound, cracks can be filled and the wall rendered smooth with cement or lime-based plaster, often mixed with cow dung, sisal, straw, cotton waste, or other fiber to give added resistance to cracking and powdering off. Use of fresh cow dung in the mixture is traditional in the Indian subcontinent and minimizes the requirement for water in the mix. However, in experiments in Brazil it was found that this could give problems during certain seasons as dung beetles later emerged from the plaster leaving substantial holes.[57] Also, in experiments in Argentina, goat dung was found to be inferior to cow dung as an additive to low-cement plaster.[57] In addition to physically eliminating cracks in the walls, plaster or other rendering can improve the esthetic appearance of a house, and light-colored plaster will both reduce the rate of heat absorption on exterior surfaces as well as improve reflected illumination inside the house. Plaster also protects the wall surface from rain and wind, but for this reason is generally made of stronger material than the wall itself (e.g., with higher cement content). This means that the plaster and the wall will have different rates of movement in response to changing humidity, tending to cause the plaster to crack and flake off. Plaster rendering should therefore use material of a similar strength to the wall material, applied in two coats (with the first allowed to dry before applying the second). External plastering should also be protected from rain by an overhanging roof. As a rough guide, the roof overhang will protect down to three times its total width.

Where walls are unsound they should be replaced. Fired bricks and cement mortar can be used where available, and income levels permit their purchase. However, such materials are energy-expensive both in terms of fuel for the firing and in terms of distribution cost. There are several alternatives: using lightweight sticks and mud rendered with heavy-duty plaster — as in the improved *bahareque* in low-cost Venezuelan houses[58] — or using stabilized earth methods.[58-64]

Earth used in construction must contain clay for cohesion, sand as a filler, and silt to provide fine particles. However, the clay portion of a soil will swell when taking in moisture and shrink when drying out, causing shrinkage cracks to occur. For house construction, the earth can be stabilized against erosion, cracking, and linear shrinkage by the addition of bitumen (to make asfadobe), cement (to make soil-cement), or lime. Stabilized earth can be further improved by compaction, either *in situ* or to form blocks using a mechanical press.[58,62,63] *In situ* compaction is achieved by ramming successive layers of earth with wooden mallets between wooden formers, as in the French *pisé de terre,* widely used in parts of France, Spain, and North Africa. A similar technique, known as *pisa tapial* or *taipa pilão* is also used in parts of Latin America, often with milk or animal blood added to increase cohension of the rammed earth. In parts of France, especially around Grenoble, houses of two or three stories dating back a century or more testify to the usefulness of this technique, even in regions of high rainfall and winter frosts.

Block-making presses (Table 5) also produce building materials of a very high quality. All block-making presses follow a similar principle: sieved earth with up to 25% clay content stabilized with a small proportion of lime or cement is loaded into a steel mold and compressed by the action of a hand-operated lever. Some versions allow power-operated levers. One of the earliest and best known hand-operated block-making presses is the CINVA ram (Figure 5A), developed in Colombia, which applies a compacting pressure of about 2 MN/m^2 (about

TABLE 5
Compressed Earth Block-Making Machines[61]

Name	Origin	Introduction	Gross wt. (kg)	Pressure[a]	Hourly production[b]	No. men	Cost[c]
ABI	Côte d'Ivoire	Mid-1970s	120	3.0	40	?	900
Astram	India	Mid-1970s	110	5.0	?	3/4	540
AVM	W. Germany	Early 1980s	65	2.2	40	2	216
Block Press	Kenya	Mid-1970s	60	2.1	30	3/4	288
Bonner	Kenya	1982	85	3	40	4	630
BREPAK	U.K.	1980	140	10	40	3	2250
CENEEMA	Cameroon	Early 1960s	70	2.1	40	2	180
Ceraram	Belgium	Mid-1970s	350	2.1	300	2	2520
CETA	Paraguay	Early 1980s	90	1.5	180	4	540
CINVA Ram	Colombia	1952	60	2.0	40	3	414
Consolid AG	Switzerland	Late 1970s	1,600[d]	4.8	400	6	36,000
CTA	Colombia	Early 1970s	90	1.5	?	?	432
Dirt Press	Thailand	Late 1970s	55	1.9	40	3	
Ecobrick	Switz.	Late 1970s	650[d]	4.9	100	2	13,500
Ellson Blockmaster	S. Africa/India	Early 1950s	210	6.8	60	4	810
GEO 50	France	Late 1970s	100	1.6	25	2	1080
Groupe Unipress	France	Mid 1970s	2,400[d]	3.6	2,000	4	?
Giza	Italy	Early 1970s	5,000[d]	6.0	40	?	?
			12,000[d]		250/500		
Jesson Press	S. Africa	Mid-1970s	85	0.15	75	4	?
La Palafitte	France	Mid-1970s	95	2.8	40	3	216
Landcrete/Terstaram	S. Africa/Belgium	1960	350	4.0	120	4	1260
			2,100[d]		250		21,600
Latorex[e]	Denmark	Mid-1970s	?[d]	5.0	1,500	?	?
Lescha SBM	W. Germany	Late 1970s	2,300[d]	4.5	500/700	4	36,000
Lorev	Italy	Mid-1970s	150	3.0	?	?	?
Maquina	Colombia	Early 1970s	170	1.8	80	4	?
Maro	Switz.	Late 1970s	125	2.2	40	3	630
Meili 60	Switz.	Late 1970s	85	5.0	60	3	684
Meili Mechan	Switz.	Early 1980s	1,700[d]	5.2	1,000	4	32,400
Pact 500	France	Early 1970s	550[d]	9.8	300	4	12,600

PPB Sarot	France	Late 1950s	9,000[d]	8.6	?	?	?
Ranko	U.K.	1984	60	f	20	?	?
Riffon	Belgium	Early 1970s	150	f	80	3	?
Saturnia	Switz.	Early 1980s	200	3.2	200	4/5	1,530
Supertor	Brazil	Mid-1960s	1,100[d]	4.2	2,500	?	2,880
Tecmor	Brazil	1978	85	2.5	250	?	?
			2,000[d]		1,500		
Teck-Block	Ghana	Late 1960s	90	1.5	20	3	?
Terrablock	U.S.	1985	5,350[d]	12.0	600	4/6	90,000
Ultrablock	U.S.	Late 1970s	1,000[d]	?	300	?	?
Unata	Belgium	Late 1970s	80	1.8	20	2	540
Winget	U.K.	1948	1,100[d]	9.6	140	5	?
Zora	U.K.	1982	450	19	40	3	4,500
			850[d]		180		6,300

[a] Compacting pressure in MN/m^2.
[b] Hourly production rate is generally from manufacturer's literature.
[c] Ex-works cost in the U.S. $ equivalents.
[d] Power operation — all others are manually operated.
[e] Latorex blocks are steam cured — all others are atmospherically cured.
[f] Impact.

From Briceño-León, *Parasitol. Today,* 3, 384, 1987. With permission.

A

FIGURE 5. Hand-operated presses for producing stabilized soil building blocks. (A) The CINVA ram, first developed in Colombia in 1952. (Photograph by D. J. T. Webb.) (B) The BREPAK, developed in the U.K. in 1980. (Photograph by J. A. Noonan.) For other block-making presses, see Table 5. Sieved earth used in block-making is mixed with 4 to 10% Portland cement or hydrated lime (depending on its clay/sand content). After compression, blocks are "cured" by leaving them under shade for 7 to 28 d during which time the pozzolanic reaction within the block is completed and the block achieves its maximum strength.[63,64] Other stabilizers may be used in some circumstances, such as pulverized fuel ash, rice husk ash cement, or cementitious slag.[64] For a given soil, stabilized and cured appropriately, the strength of the resulting block reflects the compaction pressure applied. (C) Shows a typical strength/compaction curve for lime-stabilized blocks. (A) The CINVA ram typically applies about 2 MN/m^2 resulting in blocks with a wet compressive strength usually less than 1 MN/m^2. (B) In contrast, the BREPAK has a hydraulically assisted pump applying 10 MN/m^2, producing blocks that can exceed the British Standard minimum wet compressive strength (for fired clay bricks) of 2.8 MN/m^2.

8 tonnes per brick) and produces blocks of a standard size (290 × 140 × 100 mm) that can be used in construction in the same way as adobe blocks. More recently, the U.K. Building Research Establishment has developed an improved press, known as the BREPAK (Figure 5B), in which compression exerted by the hand-operated lever is hydraulically assisted to produce a compaction pressure of about 10 MN/m^2 (about 40 tonnes per brick). The resulting blocks, of the same size as adobe blocks but considerably heavier, actually exceed minimum British Standard Specifications (BS.1924:1975) for fired clay bricks in terms of load-bearing strength when wet[62] (see Figure 5C).

The superior performance of compressed soil blocks and their ease of manufacture in local communities could bring about considerable improvements in the quality of rural and urban housing throughout the world both in arid and humid climates, and extensive field trials are now being conducted in several countries. Moreover, experimental work suggests

FIGURE 5B.

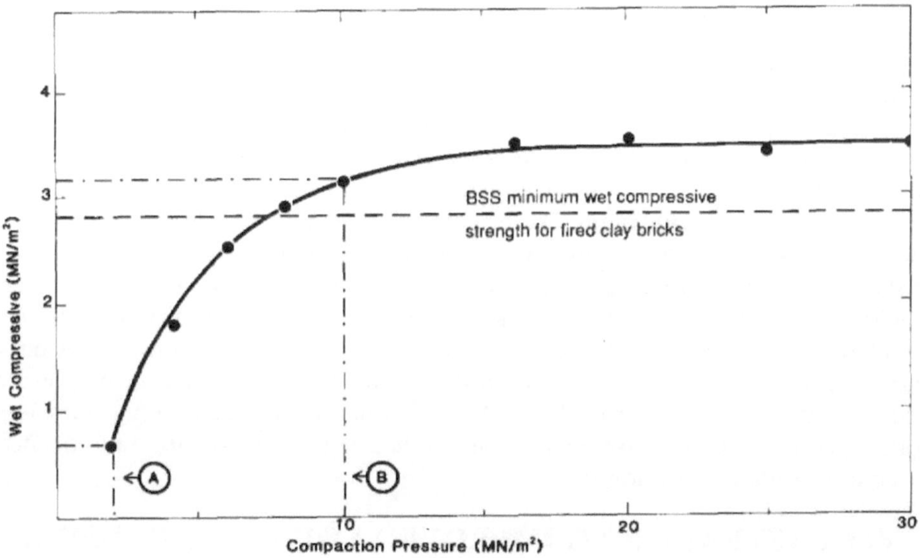

FIGURE 5C.

that by changing the type of mold in the press, blocks could be made of a size, shape, and hardness suitable for floor and roof tiles, and for the liners of pit latrines (Chapter 10) and irrigation canals.[63] Used in wall construction, the blocks do not crack, but bind well to soil-cement mortars to produce a structure without crevices where domestic pests could hide. Even where insecticides must be used against mosquitoes and other insects flying into a house to feed, the nonporous nature of the compressed soil blocks appears to offer a greatly improved surface for the action of residual insecticides, compared to adobe blocks made from the same material.[65] Inexpensive presses such as the CINVA ram and BREPAK are ideally suited to small-scale business enterprises or community cooperatives, where they can be moved from site to site to produce the blocks required for each building.[64] This is both simpler and cheaper in transport than distributing bulky construction materials from a central production site.

G. THE ROOF

Traditional styles of housing with an earth roof are still favored by some communities, especially in the arid Sudan savanna and sahelian belts of Africa. Like the earth walls, such roofs are prone to cracking, providing dark crevices that can be occupied by insect pests. More usually, however, a timber framework supports a roof of thatch, tiles, or corrugated metal or fibrocement sheets. Thin unshaded roofs, exposed to direct sunlight, can heat up considerably during the day. This discourages insects although it can also make the house intolerably hot for the householders.

Thicker roofs of thatch offer better insulating properties, but also provide good resting places for insects. A favored hiding place for many house-resting and man-biting mosquitoes is within the exposed strands of thatch under an unlined roof. These hiding places closely resemble those among straw and leaves outside the house, but offer greater climatic protection and proximity to blood sources. Omer and Cloudsley-Thompson[66] reported that in Sudan, females of *Anopheles arabiensis* could survive for up to 9 months during the extended dry season, sheltering in the roofing thatch during the day when outdoor temperatures were above their thermal death point. In more northerly latitudes, some malaria vectors could overwinter indoors by resting on rafters, walls, or under thatch, with intermittent blood-feeding that gave rise to the phenomenon of winter malaria in Europe before improved housing eliminated such problems.

In Venezuela, traditional roofs made of palm thatch pose a particular problem since the eggs of the Chagas disease vector *Rhodnius prolixus* are often stuck to the palm fronds used in thatching. This can be an important source of domestic infestation, and replacement of palm thatch with corrugated metal roofs has been an important factor in Chagas vector control in Venezuela.[67] In the Argentine chaco and parts of Bolivia, where nighttime temperatures can drop below 5°C, the traditional flat roofs are made from poles piled with brushwood and packed with earth. The resulting thick porous structure provides an ideal habitat for *Triatoma infestans,* into which insecticides cannot be sprayed even using motorized mist-blowers. Corrugated metal sheets are not favored because in addition to heating up intolerably during the day, they provide no protection from the cold at night. Insulating layers of earth over the metal offer no solution because rapid corrosion results. A possible compromise would be the use of locally made ceramic tiles of the type favored in Brazil and lowland parts of Bolivia and Peru. The increasing use of such roofing tiles was an important part of the progressive improvement of housing which contributed to the decline of malaria transmission in Europe.[68]

V. CONSTRAINTS TO IMPROVED LOW-COST HOUSING

The necessity to improve low-cost housing is increasingly recognized, both as a developmental goal in itself and also as an important contribution to improved health through improvements in hygiene and sanitation, reduction of domestic vectors of disease, and reductions in the stress and dissatisfaction associated with overcrowding and squalor. In parallel, there are welcome technical advances such as the use of compressed stabilized earth blocks for walls and floors, fiber-reinforced sheets for roofing, and recognition of the role that improved house design can play in providing comfort and freedom from vector-borne diseases. A few selected examples have been given here, but these factors do not complete the full spectrum of constraints to large-scale improvements to rural and urban housing. Cost is a major factor, and with this is the question of who will pay. Here, key questions concern the perceptions and priorities of individual householders and their relationship with state organizations which must motivate and assist in any large-scale housing program.

Human perceptions rarely differ in the generalities of what constitutes an ideal house, but often, low-income householders may feel that such standards are inappropriate or unobtainable for them. With this concept, they may accept their current living standards as appropriate to their particular class and so lack motivation for their own role in improving living standards for themselves and their families. This is often reflected in perceptions of disease, where, for example, disease agents may be recognized but accepted within the "natural order" as inevitable or not worth the investment or change in life-styles required to avoid them.[58,69]

Perceptions of specific features of a house vary in different societies. Some, for example, would not consider a house complete without a swimming pool, while at the other extreme, a simple space for social meetings is considered the most desirable feature. There seems little doubt, however, that improved housing is quite high on the list of priorities of many of the less privileged. A recent survey in Venezuela asked rural farmers what they would do if they won the lottery: 47% replied they would invest in house improvements or obtaining a new home.[70] A survey in Brazil gave similar results with 44% choosing "to get a better house or apartment", although in Mexico this response was given by only 19% compared to 28% who would prefer to open a business.[71] Clearly, however, improved housing falls below the need for employment, income, and food.

Studies in Venezuela on the role of house improvement as a means to limit domestic infestation by triatomine bugs show that the most important factor is to generate the belief in affected communities that they themselves can contribute to the improvement; i.e., that this is not merely a state intervention as shown by large-scale prefabricated constructions or by periodic visits by spraymen to apply insecticides to control the "bugs". The first task is to establish what are the main perceived requirements of the local communities — which might not include vector control or improved housing as key factors — and then to design participative programs which satisfy perceived needs while incorporating those activities useful for health improvement. During the course of such programs, health education can be carried out, leading to changes in perceived needs and greater emphasis on the health requirements. At the same time, those responsible for such programs will then be satisfying the beneficiaries, rather than imposing a preconceived model of what ought to be done.

The role of the state in large-scale house improvement programs for low-income communities is threefold: to establish secure tenure of land on which such programs can be based, to mobilize teachers and community workers trained to respond appropriately to this level of development, and to open accessible lines of credit to those involved. One of the greatest barriers to progress at this level, however, is the established approach of many bureaucracies which fail to accept this type of community program. Often, the concept of large-scale authoritarian state intervention is preferred, with emphasis on the most "modern" of styles and materials. Such emphasis is often reinforced by the interests of industrial cartels because of profits to be made from the manufacture or import of inappropriate materials such as steel and concrete. The result in many regions is that traditional skills of house design and construction may be lost, while innovative techniques using such materials are discouraged.[11,15,72] For similar reasons, the potential of low-cost house improvement for limiting vector-borne diseases and other health risks may also be dismissed by many poorly informed bureaucrats.

Some of the difficulties in sustaining effective vector control by insecticide spraying alone (Chapter 1) reflect the fact that no lasting change has been made to the environment exploited by the target insects. The finding of insects or other arthropods in a domestic or peridomestic environment indicates clearly that that environment is appropriate for them. A transient change, such as making the environment toxic with insecticides, may inhibit the insect populations, but if that change cannot be sustained, then those insects will have the opportunity to return.

For vector-borne diseases such as Chagas disease, where transmission is absolutely linked to the domestic environment, the potential of house improvement for vector control has long been recognized. Carlos Chagas himself raised the importance of this approach 80 years ago, soon after he first described the disease and its vectors. Similarly, house screening and personal protection with nets has long been recognized as a primary defense against mosquitoes (see Chapter 2 and Reference 73) and against malaria,[74] while the idea of "building out malaria" was strongly advocated by Wilson in 1957 for malaria control in tropical Africa.[75] Techniques are now available to improve, at very low cost, the quality of urban and rural houses. The impact of such improvements on vector-borne diseases and other health indices has been demonstrated in numerous studies, and sociologists are now tackling the political and social constraints to large-scale implementation of such developments. The emphasis must be on technical, state, and community participation, a process that is already exemplified by collaboration between vector-control specialists, sociologists, economists, epidemiologists, and engineers. House improvement, at least on a large scale, is difficult to justify on the basis of disease control alone, but there can be no doubt that house improvements can contribute substantially to the control of a very wide range of diseases.

REFERENCES

1. Uses of Epidemiology in Housing Programmes and in Planning Human Settlements, W.H.O. Tech. Rep. Ser., No. 544, World Health Organization, Geneva, 1974.
2. Global Review of Human Settlements, U.N. Conf. Human Settlements, Vancouver, A/CONF.70/A/3, United Nations, New York, 1976.
3. **Satterthwaite, D.,** *Three Years After Habitat,* International Institute for Environment and Development, London, 1979.
4. Housing — the Implications for Health, Rep. of a W.H.O. Consultation, WHO/EHE/RUD/87.2, World Health Organization, Geneva, 1987.
5. Housing and Health, An Agenda for Action, World Health Organization, Geneva, 1987.
6. **Harpham, T.,** Health of the urban poor in developing countries, *Parasitol. Today,* 2, 320, 1986.
7. **Lipton, M.,** *Why Poor People Stay Poor. A Study of Urban Bias in World Development,* Temple Smith, London, 1977.
8. **Tait, A. A., Gratz, W. L. M., and Eichengreen, B. J.,** International comparisons of taxation for selected developing countries, *Int. Monetary Fund Staff Pap.,* 26, 123, 1979.
9. **Harrison, P.,** *Inside the Third World. The Anatomy of Poverty* 2nd ed., Pelican/Penguin books, New York, 1982.
10. **Lindqvist, S.,** *Land and Power in South America,* Pelican/Penguin books, New York, 1979.
11. **Agarwal, A.,** *Mud, Mud, The Potential of Earth-Based Materials for Third World Housing,* Earthscan, London, 1981.
12. **Oliver, P., Ed.,** *Shelter and Society,* Barrie & Jenkins, London, 1969.
13. **Oliver, P., Ed.,** *Shelter in Africa,* Barrie & Jenkins, London, 1971.
14. **Kroeger, A.,** Housing and health in the process of cultural adaptation: a case study among jungle and highland natives of Ecuador, *J. Trop. Med. Hyg.,* 83, 53, 1980.
15. **Dethier, J.,** *Des Architectures de Terre,* Centre Georges Pompidou, Paris, 1981.
16. **Schofield, C. J. and White, G. B.,** House design and domestic vectors of disease, *Trans. R. Soc. Trop. Med. Hyg.,* 78, 285, 1984.
17. **Lea, R. G.,** *Reducing the Risk of Pest Infestations: Design Recommendations and Literature Review,* Building Research Establishment Digest, Princes Risborough, U. K., 1980.
18. **Mason, D.,** *Hospitals Can Damage your Health,* British Pest Control Association, London, 1985.

19. **Norman, A. R.,** Hygiene in hospital kitchens, *Lancet,* 8480 (8th March), 561, 1986.
20. **Schofield, C. J.,** Population dynamics and control of *Triatoma infestans, Ann. Soc. Belge Med. Trop.,* 65 (Suppl. 1), 149, 1985.
21. **Schofield, C. J.,** The behaviour of Triatominae (Hemiptera: Reduviidae): a review, *Bull. Entomol. Res.,* 69, 363, 1979.
22. **Schofield, C. J. and White, G. B.,** Density-dependent feeding success in triatomine bugs and other vectors, in *Ectoparasites of Veterinary and Medical Importance in Temperate Areas,* Burgess, N. H. R., Port, G. R., and Tichener, R. N., Eds., Royal Army Medical College, London, 1982, 1.
23. **Deobhanker, P. B.,** Malaria control in Bombay by legislative measures and other non-insecticidal methods of vector control, in *Community Participation for Disease Vector Control,* Sharma, V. P., Bang, Y. H., Rosenfield, P., and Pant, C. P., Eds., Indian Malaria Research Centre, 1986, 101.
24. **Chan Kai Lok,** *Singapore's Dengue Haemorrhagic Fever Control Programme: a Case Study on the Successful Control of* Aedes aegypti *and* Aedes albopictus *Using Mainly Environmental Measures as a Part of Integrated Vector Control,* Publ. No. 45, Southeast Asian Medical Information Centre, Tokyo, 1985.
25. **Reiter, P. and Sprenger, D.,** The used tire trade: a mechanism for the worldwide dispersal of container breeding mosquitoes, *J. Am. Mosq. Control Assoc.,* 3, 494, 1987.
26. Doenças e Migração Humana, S.U.C.A.M., Ministerio da Saude, Brasilia, 1982.
27. **Cruz Marques, A.,** Human migration and the spread of malaria in Brazil, *Parasitol. Today,* 3, 166, 1987.
28. Ten Years of Onchocerciasis Control in West Africa, W.H.O. doc. OCP/GVA/85.1B, World Health Organization, Geneva, 1985.
29. Manual on Environmental Management for Mosquito Control with Special Emphasis on Malaria Vectors, W.H.O. offset publ. No. 66, World Health Organization, Geneva, 1982.
30. **Cranston, P. S., Tee, R. D., Credland, P. F., and Kay, A. B.,** Chironomid haemoglobins: their detection and role in allergy to midges in the Sudan and elsewhere, *Mem. Am. Entomol. Soc.,* 34, 71, 1984.
31. **Cranston, P. S.,** Allergens of non-biting midges (Diptera: Chironomidae): a systematic survey of chironomid haemoglobins, *Med. Vet. Entomol.,* 2, 117, 1988.
32. **Schofield, C. J.,** Malaria and its Anopheline Vectors in the Amazon Region of Brazil, unpublished report to the World Bank, Washington, D. C., 1986.
33. **Coluzzi, M., Sabatini, A., Petrarca, V., and Di Deco, M. A.,** Chromosomal differentiation and adaptation to human environments in the *Anopheles gambiae* complex, *Trans. R. Soc. Trop. Med. Hyg.,* 73, 483, 1979.
34. **White, G. B.,** Factors Influencing Densities of Mosquitoes Resting Indoors, Ann. Rep. East African Institute of Malaria and Vector-Borne Diseases, 1969.
35. **Hooper, C.,** *Design for Climate, Guidelines for the Design of Low-Cost Houses for the Climates of Kenya,* Housing Research and Development Unit, University of Nairobi, Kenya, 1975.
36. **Koenigsberger, O. H., Ingersoll, T. G., Mayhew, A., and Szokalay, S. V.,** *Manual of Tropical Housing and Building, Part One: Climatic Design,* Longman, London, 1973.
37. **Kolstrup, N.,** Studies on the Transmission and Control of *Wuchereria bancrofti* in North East Tanzania, Ph.D. thesis, University of London, 1980.
38. **Lehane, M. J. and Schofield, C. J.,** Flight initiation in *Triatoma infestans* (Klug) (Hemiptera: Reduviidae), *Bull. Entomol. Res.,* 72, 497, 1982.
39. **Schofield, C. J. and Matthews, J. N. S.,** Theoretical approach to active dispersal and colonisation of houses by *Triatoma infestans, J. Trop. Med. Hyg.,* 88, 211, 1985.
40. **Forattini, O. P., da Rocha e Silva, E. O., Ferreira, O. A., Rabello, E. X., and Pattoli, D. G. B.,** Aspectos ecológicos da tripanossomose americana. III. Dispersão local de triatomíneos, com especial referência ao *Triatoma sordida, Rev. Saude Publica,* 5, 193, 1971.
41. **Soler, C. A., Schenone, H., and Reyes, H.,** Problemas derivadas de la reaparición de *Triatoma infestans* en viviendas desinsectadas y el concepto de reinfestación, *Bol. Chil. Parasitol.,* 24, 83, 1969.
42. **Schofield, C. J.,** unpublished data, 1985.
43. **Mellanby, K.,** *Scabies,* E. W. Classey, London, 1972.
44. **Busvine, J. R.,** *Insects and Hygiene,* Methuen, London, 1966.
45. **Cairncross, S.,** Low-cost sanitation technology for the control of intestinal helminths, *Parasitol. Today,* 3, 94, 1987.
46. **Cairncross, S. and Feachem, R. G.,** *Environmental Health Engineering in the Tropics: an Introductory Text,* John Wiley & Sons, Chichester, 1983.
47. **Bucher, E. H. and Schofield, C. J.,** Uso de la Tierra y Enfermedad de Chagas, Centro de Zoologia Aplicada, Cordoba, Publ. No. 8, 1984.
48. **Ronderos, R. A., Schnack, J. A., and Mauri, R. A.,** Resultados preliminares respecto de la ecologia de *Triatoma infestans* (Klug) y especies congenericas con referencia especial a poblaciones peridomiciliares, *Medicina (Buenos Aires),* 40, (Suppl. 1), 187, 1980.
49. **Schofield, C. J.,** Control of Chagas disease vectors, *Br. Med. Bull.,* 41, 187, 1985.
50. **Lane, R. P.,** Chicken house reservoirs of sandflies, *Parasitol. Today,* 2, 248, 1986.

51. **Charlwood, J. D.,** Observations on the bionomics of *Anopheles darlingi* Root (Diptera: Culicidae) from Brazil, *Bull. Entomol. Res.*, 70, 685, 1980.
52. **Hudson, J. E.,** *Anopheles darlingi* Root (Diptera: Culicidae) in the Surinam rain forest, *Bull. Entomol. Res.*, 74, 129, 1984.
53. **Barlow, F.,** Physical processes associated with formulations, *Br. Crop Protection Council Monogr.*, 17, 127, 1976.
54. **Barlow, F., Hadaway, A. B., Flower, L. S., Gross, J. E. H., and Turner, C. R.,** Some laboratory investigations relevant to the possible use of new pyrethroids in control of mosquitoes and tsetse flies, *Pestic. Sci.*, 8, 291, 1977.
55. **Zeledón, R. and Vargas, L. G.,** The role of dirt floors and of firewood in rural dwellings in the epidemiology of Chagas' disease in Costa Rica, *Am. J. Trop. Med. Hyg.*, 33, 232, 1984.
56. **Kalshoven, L. G. E.,** Observations on the blood-sucking reduviid *Triatoma rubrofasciata* (De Geer) in Java, *Entomol. Ber.*, 30, 41, 1970.
57. **Gorla, D. E. and Schofield, C. J.,** unpublished data, 1982.
58. **Briceño-León, R.,** Rural housing for control of Chagas disease in Venezuela, *Parasitol. Today*, 3, 384, 1987.
59. **Lunt, M. G.,** Stabilized soil blocks for building, Overseas Building Notes, No. 184, Building Research Establishment, Watford, U.K., 1980.
60. **Vildoso, A., Monzon, F. M., Hays, A., Matuk, S., and Vitoux, F.,** *Seguir Construyendo con Tierra*, CRATERRE, Eybens, France, 1984.
61. **Webb, D. J. T.,** Low-cost housing and parasite vectors, *Parasitol. Today*, 1, 65, 1985.
62. **Webb, D. J. T.,** Construction en terre stabilisée au Kenya, Proc. Int. Conf. Economical Housing in Developing Countries: Materials Construction Techniques Components, Paris, 1983, 137.
63. **Webb, D. J. T. and Lockwood, A. J.,** BREPAK Operators Manual, Building Research Establishment Client Rep. CR21/87, Watford, U.K., 1987.
64. Small-scale Manufacture of Stabilised Soil Blocks, I.L.O. Tech. Memo. No. 12, International Labour Organization, Geneva, 1987.
65. **White, G. B., Schofield, C. J., Williams, N. G., Webb, D. J. T. and Boase, C. J.,** The role of compressed soil blocks in low-cost housing for the control of vector-borne diseases, *Trans. R. Soc. Trop. Med. Hyg.*, 79, 282, 1985.
66. **Omer, S. M. and Cloudsley-Thomson, J. L.,** Survival of female *Anopheles gambiae* Giles through a 9-month dry season in Sudan, *Bull. W.H.O.*, 42, 319, 1970.
67. **Gamboa, C. J.,** El processo modificador de la vivienda en el medio rural del Estado Miranda (Venezuela). Su relación con la infestación por *Rhodnius prolixus*, *Arch. Venez. Med. Trop. Parasitol. Med.*, 5, 353, 1973.
68. **Bruce-Chwatt, L. J. and de Zulueta, J.,** *The Rise and Fall of Malaria in Europe. A Historico-Epidemiological Study*, Oxford University Press, 1980.
69. **Briceño-León, R.,** Aspectos psicosomaticos de la enfermedad de Chagas, *Bol. Asoc. Venez. Psicol. Soc.*, 9, 37, 1986.
70. **Briceño-León, R.,** *El Comportamiento Social del Venezolano ante el Trabajo y la Riqueza*, Universidad Central, Caracas, 1988.
71. **Kahl, J. A.,** *The Measurement of Modernism: a Study of Values in Brazil and Mexico*, University of Texas Press, Austin, TX, 1968.
72. **Hongvivatana, T.,** Human behaviour and malaria, in 29th SEAMED-TROPMED Seminar: Social and Economic Research in Tropical Diseases, Kuala Lumpur, 1986.
73. **Lindsay, S. and Gibson, M.,** Bednets revisited. Old idea, new angle, *Parasitol. Today*, 4, 237, 1988.
74. **Manson, P.,** Experimental proof of the mosquito-malaria theory, *Br. Med. J.*, ii, 949, 1900.
75. **Wilson, D. B.,** Construction, irrigation and malaria, *East Afr. Med. J.*, 34, 479, 1957.
76. **Briceño-León, R.,** Rural housing for control of Chagas disease in Venezuela, *Parasitol. Today*, 3, 384, 1987.

Chapter 12

VECTOR CONTROL THROUGH RADIO AND TELEVISION: APPROPRIATE TECHNOLOGY FOR THE FUTURE

J. D. Gillett

TABLE OF CONTENTS

I. INTRODUCTION

"Appropriate" is defined by Webster and the Oxford English Dictionary as *suitable*. In the present context it is usually taken to mean suitable to the state of development of the recipient population. Equally, however, it could be taken in the sense of being suitable for achieving the desired result effectively. In this chapter I am going to keep both senses in mind, that is, suitability to the people affected by vector borne disease and suitability toward achieving a lasting result over a wide area. The former emphasis is, I suspect, the one subsumed by most of those who write about appropriate technology. However, as I shall later show, unless the latter emphasis is also taken into account the former is likely to fail, except in isolated instances.

Today brave attempts are going on in this village and that, to teach the people to help themselves by simple means: the screening of houses and beds, impregnated eaves curtains, smoking coils, adult insecticides, larvicides, repellents, biological agents, tsetse traps, and various combinations of all these together with appropriate preventive behavior. We must also include suppressives, prophylactics, vaccination, and radical cures. Such devices and means are being used or encouraged in individual homes and communities scattered throughout the vast region of the tropical world.

Now, there is no question that the simple devices and practices instanced in this book are on the right lines. These valiant attempts certainly bring freedom or relief from the parasitic yoke to some of those that use them. However, the operative words in the above sentence are, "to *some* of those" and "that *use* them". So much depends on how well each contraption or chemical is applied and on how well the general principles are understood. Really, however, when we consider the problem tropico-worldwide, in the future or even today, then we must seriously question whether these brave, imaginative, and well-intentioned efforts will have anything more than a local or temporary impact. It is now more than 70 years since Mary Simpson (whose husband helped found the London School of Hygiene and Tropical Medicine) invented and patented her antimosquito headgear;[1] forgotten gadgets, alas, litter the literature.

II. EARLY 20TH CENTURY TECHNOLOGY

In the 1950s, when malaria eradication was all the rage, there were something like 2.6 billion people living in the world, 1.6 billion of whom resided in the tropics. To go out into the tropical world and preach the gospel of prevention at the personal level or even at the community level in the 1950s seemed, perhaps, to be not too unreasonable. Such an approach in the 1990s, however, when the number of people in the tropics has more than doubled, seems to me to be no longer reasonable; in fact, it is, surely, totally unrealistic. However, if this book is not merely to be of ephemeral interest and importance we must begin to plan strategies that will still be meaningful, say, in 30 years' time. By then, however, the human population living in the tropical regions will have increased by a further factor of 2.6 to reach something like 8.3 billion (plus 1.7 billion in the temperate regions). To contemplate a continuing piecemeal, somewhat simplistic approach in the year 2020 would seem to me to be more like wholly inappropriate technology, to put it mildly. This population forecast looks like being realized unless some, perhaps numbers-related, catastrophe occurs such as that now threatened by AIDS. More than 8 billion people, a high proportion of them illiterate, spread over vast areas (many of which are isolated and relatively inaccessible); others crammed into gigantic urban conglomerates. This surely means that the old-fashioned methods of contact are mostly doomed to failure from the start.

Attempts at propaganda and/or teaching, even by posters and the use of films and drama, and even when spread through pyramid education, will, alas, also affect only the fringes of the problem and then, perhaps, only for a limited period. We have seriously to ask ourselves if such antiquated methods of communication are any longer appropriate as we approach the 21st Century of our era. To be sure, these old methods may be justified and, indeed, can be highly effective in small, circumscribed communities. Three cases come to mind: Professor Chan Kai Lok's celebrated success of mosquito control in Singapore (see Chapter 6).[2] The huge 1981 outbreak of dengue fever and dengue hemorrhagic fever (DHF) in Cuba (see Chapter 6), which was brought under control by the mobilization of 10,000 volunteers organized in small groups of two or three people, each working in its own neighborhood, where it was known and trusted.[3] Finally, even less conventional and less well known, the control of a human outbreak of trypanosomiasis on Buvuma Island, Uganda, in 1943, following the daily buying of huge numbers of tsetse from the local people.[4] This novel strategy, that successfully controlled an epidemic at the ''cost'' of prosperity to the people concerned, is perhaps worthy of further consideration, although it would clearly be of dubious value in any system involving mosquitoes or other easily cultured group. In our present context, however, it is important to bear in mind that all three of these success stories concerned islands.

III. LATE 20TH CENTURY TECHNOLOGY

A. THE HARDWARE

If we want the people of the tropical world to embrace these simple methods to free themselves of this crippling yoke, then we, outside the affected regions but in full collaboration with them, must be prepared to shake off our early 20th Century attitudes and come to terms with the means of mass communication that are already at our disposal. I refer, of course, to radio and television (TV) through the agency of artificial communications satellites placed 35,784 km above the equator in geosynchronous orbit. The technology is already here, not just in theory but in practice, and has been since Telstar was launched with such success a quarter of a century ago.[5]

If we fight shy of this new technology it is mainly, I suspect, because it lies wholly outside our own training and experience. Whether we spread Paris green on anopheline breeding places or kerosene on the surface of ponds and puddles in our own backyard, or whether we apply temephos to stored drinking water or use fenithrothion, bendiocarb, or pirimiphos methyl elsewhere, we are really only using early 20th Century methods, whether or not they are given a ring of modernity by being more lasting (or less lasting) or more effective yet less toxic to the environment including ourselves. Of course, we shall need these newer developments, but the burden of my argument is that we must be certain that the benefits they bring are not confined to a fortunate few. We must accept that a revolution in communications technology has been going on over the past 25 years and incorporate it into our own schemes. Indeed, the new communications technology must become the *cornerstone* for an entirely new approach.[6,7] Whether we are going to administer the use of new vaccines against *Plasmodium falciparum* or teach the proper use of mosquito nets or of repellents or a combination of these, we shall need the new technology if we are not merely to make a temporary dent in the problem. We must decide now whether to go on fiddling with this trial or that in some isolated and remote village or whether to embrace this timely innovation that brings an entirely new dimension to the possibilities of prevention of tropical disease including, I must add, diarrheal diseases, measles, tetanus, and whooping cough,[8] and, I must also add, perhaps especially to the growing problem of AIDS in tropical Africa.

We surely cannot for much longer maintain the present attitude of *we* the rich do-gooders and *they* the poor afflicted. Mankind shares this lovely and, within our part of the universe at least, uniquely rich but dangerous planet that we call Earth. All of us are affected and share the responsibility. We must forget politics and pet political theories and religious barriers with the hatreds that these engender, and glory in the very richness of our differences. *Vive la différence* must apply to race, religion, customs, and views as well as to sex. We are one species struggling to survive on this planet, and together we must wipe out this blot on our tenancy or stewardship: the absurd obscenity presented by the picture of tellurian man in the late 20th Century still riddled with parasites brought to him by insects and arachnids.

We must face up to the fact that the simple time-honored methods of prevention, prophylaxis, and cure have simply not worked. We have waited in vain for some wonder insecticide, some wonder prophylactic or cure, and the search must continue for who knows what new hope awaits discovery (the use of inhibitors of polyamine biosynthesis may offer one possible new approach, the use of verapamil with chloroquine, another).[9,10] However, even when such are found, what use will they be if the people concerned simply have not heard of them, or, if they have, do not really know how they should be used? The same goes for vaccination against malaria, when it comes.

Fortunately, we already have a new ''wonder'' technology, and we must now begin to harness it if we are to come to grips with these problems. Of course, the insecticides, the prophylactics (including any vaccines) and the cures for overt attacks must continue to be used and new ones sought. These, however, will make little or no difference to the overall picture unless they are brought to the attention of all concerned, even in the most isolated and remote areas of illiteracy. The chemical industry need not fear a drying up of its funds, for on this rich earth of ours the problem of the prevention of vector-transmitted parasitic disease is likely, for the most part, to remain a continuing one. In the final analysis it comes down to cost. In another article on this subject I asked if we can afford to harness this new technology in order to get rid of what some still regard as other people's problems. I then asked if, in the long run, we can afford not to?[11]

3. THE SOFTWARE

Many will agree that useful ideas can be instilled by radio and TV. People not only need to be informed of the dangers that beset them; they need to be told how they can avoid these dangers by the proper use of protective screening, impregnated nets, vaccination, new drugs, and so on. They must also be informed how to treat properly those that are unlucky enough to become infected. This is not going to be done at school or by visiting health teams or by mobile cinemas and the like. As I have stressed earlier, we are not only considering present problems, but also those that will face us in the 21st Century.

Until 1980, science programs on the Italian TV service, RAI, where shown late at night and viewed mostly by the cognoscenti. Since 1981 RAI has introduced a weekly popular science program known as *Quark* and broadcast it at peak viewing hours. Each program lasts for 40 min and is eagerly watched by nearly 6 million viewers, interested laymen and scientists alike. What a triumph of good programming!

This triumph was achieved where the population had a high rate of literacy. What we have to consider is how to get the message across to people that are, for practical purposes, illiterate. To be sure, a few will be capable of signing their names, and most will be adept in matters of currency recognition. Many will be able to quote long passages from the Koran in Arabic (without understanding what it means in their own tongue) or from other religious tracts (Buddhist, Christian, and so on). Some will even be able to recite a list of the seven wonders of the world of antiquity, taught them by narrow-minded pedants.[12] However, virtually none of these will be literate in the sense of being able to read with *understanding*. To illiterates, teaching by means of stories has long been recognized as the most effective means of getting the message across,[13] but today it is now possible to contact whole populations simultaneously. This is the exciting new prospect.

The people must be informed indirectly through the use of continuing, compelling, realistic stories; in other words (whatever we may now think) by regular injections of some sort of "soap opera". I am no addict myself, but, as the eminent historian, Sir Kenneth Clark, has pointed out, "the power of these stories should not be underestimated".[14] They must show successful and charismatic heroines and heroes, who, because of their way of life, remain free of disease; and, equally important, uncharismatic villains and ridiculous buffoons, who, weakened by parasitic disease, fail to make their pile or to win the love of the one they desire.

Even if vaccination comes to play a much more important part than at present, and new drugs and insecticides become available, we shall still need to keep the scattered populations informed. Good results, it might be supposed, will tend to spread geometrically across the affected regions. Or will they? Good habits and the benefits they bring are not likely to spread unless there is an obvious return in cash or kind; sickness and poor health will count for very little in communities where these are accepted as the norm. These are the unpalatable facts that will maintain the *status quo,* unless money and sex (and perhaps parental pride and love) come into the equation. Promotion of knowledge is one thing; what is seen as wisdom is another, wisdom being defined by Maxwell as "the capacity to solve problems of living so as to achieve what is of value, for ourselves and others (wisdom, thus, including but going beyond knowledge)".[15]

The lightening of the yoke of disease so often follows simple changes in behavior, rather than from scientific breakthrough, important though this may be to our understanding of the problem. Tuberculosis in Britain dropped from 16.4% of all deaths in 1850 to only 4.9% of all deaths by 1940; this was some years before antibiotics came on the scene. Similarly, bubonic plague disappeared from Europe long before anyone understood that it was a vector-borne disease. Malaria transmission virtually disappeared from the U.S. and from western Europe before the advent of DDT and chloroquine. Most of these improvements resulted from changes in the way of life of the people.[7]

IV. THE COST

The means to remove this blot are already with us. If we still feel that the technologies of radio and TV are far too complex to be considered along with insecticides, drugs, vaccination, and the rest, let us try to put these into some sort of perspective. Of course, it is cheap and easy and, therefore, *appropriate* to spray curtains, bed-nets, or tsetse traps with insecticides. However, the technology that lies behind the production of a suitable or appropriate insecticide is very far from being cheap and easy. The cost to develop and screen a new insecticide is variously estimated to be U.S. $20—40 million. Similarly it is cheap and easy to take a prophylactic pill each night or each week (or both), but the cost of producing that appropriate pill is variously estimated to be U.S. $50—100 million. Thus, the application of these tools may be simple in the extreme, but the infrastructure supporting their production is both complex and costly.

In the same way, one has merely to switch the radio or TV on or off. What could be simpler? Of course, the instruments have to be purchased by the people (or by the club to which the people flock each day), but simple radio receivers can be made available at a very low cost, indeed. Even the cost of TV could be brought down to a realistic level. In fact, it is not the cost of the hardware that is the problem. It is the software that is going to be expensive: the cost of producing suitable programs for transmission, programs that are meaningful and, just as important, not offensive to all the many cultures concerned. The cost of turning out *continuing* programs in a thousand different languages. It is these aspects that are going to be very expensive, but so, too, is the continuing search for new remedies for prevention and cure by the universities and chemical industry. These must, of course, continue, but in step and in harmony with the plans that I am suggesting. Indeed, the chemical industry should finance both these approaches, for it would surely be in its own interest to do so.[16] These two approaches are not mutually exclusive; rather, one is reciprocal to the other. However, whereas the successful outcome in the search for new remedies could lead to a reduction in future costs, the need to spread the news by appropriate means is likely to remain an ongoing expense. My answer to this is, that without it, malaria, trypanosomiasis, leishmaniasis, the filariases, and many other parasitic diseases and the death, misery, and poverty that go in their wake, will remain ongoing problems.

It has been pointed out by Pletsch[17] that it would be sad and somewhat ironic if health education through TV were to lead to an increase in transmission of mosquito-borne disease via the mass target presented by the watching populace. Obviously, there is a real problem, and either the TV halls (or rooms) will have to be properly screened or the timing of the programs must fall outside the natural circadian activity periods of likely vectors. This obvious risk must not be allowed to militate against the use of TV in the way proposed. After all, a United Nations report[18] estimates that pesticides now kill 40,000 people each year in the Third World and injure a further 2 million; this does not mean that their use is wrong, only that they are wrongly used.

V. POVERTY AND IGNORANCE

It could be argued that change follows economic improvement and this is undoubtedly true, up to a point. I have spent nearly half my working life in tropical regions, in Africa and elsewhere. Why, then, did I and my expatriate colleagues not have trypanosomiasis, bubonic plague, yellow fever, o'nyong-nyong, onchocerciasis, and the rest when others around us were suffering and some dying from these? It is true that I had falciparum malaria a score of times or so, but I survived. My freedom from most of these vector-borne diseases can be attributed wholly to one simple factor: my own behavior. I, along with others privileged to be already briefed in these dangers, took the necessary avoiding action; forewarned, we protected ourselves against the vectors of these diseases; those of us unlucky enough to become infected took the best available treatment and maintained it for the recommended period. Contrast this with the more usual pattern in the local population. Born and bred in regions abounding in tsetses, rats and their fleas, mosquitoes, or blackflies and combinations of these and others, and, perhaps more important, accustomed or immune to the discomfort that these bring to the newcomer, avoiding action is, unfortunately, not taken seriously. Where mosquito nets are used they are more often than not in a sad state of disrepair, functioning more as mosquito traps than as protective barriers. The point of this diversion is that the local people, too, could be forewarned of the dangers, and they, too, could take correct avoiding action. It is true that a proportion of the local people will have been taught at school about insect-transmitted disease, but, all too familiar with these pests since early childhood, few will have taken the lesson to heart. Moreover, those that have done so will find that avoidance of these pests and freedom from the afflictions they bring, costs money. Thus, to ignorance we must add poverty, poverty leading to ignorance, ignorance leading to poverty. This wretched cycle not only leads to ill health, but is constantly being fueled by ill health. Money will not break this downward helix unless it is used to escape from ignorance.

VI. THE IMMEDIATE FUTURE

A trial of this new approach might be most fruitful in India, where there is not only an effective administrative infrastructure serving the whole subcontinent, but also a fine and flourishing film industry. To begin with, the programs might have to be restricted to Hindi.

Later, however, they could perhaps also include Bengali, Kannada, Malayalam, Punjabi, Tamil, and Telugu. Other languages in other parts of Southeast Asia might also be appropriate, for example, Tok Pisin in Papua New Guinea. In Africa the problems are greater and the obstacles against successfully dealing with them more complex: Africa consists of 59 separate, independent states, most of them with a multitude of tribes with their own distinct languages, loyalties, and customs. The sheer size of that continent, too, is a daunting prospect: Sudan and Zambia together are the same size as India (usually referred to as a subcontinent); while Angola, Chad, Ethiopia, Mali, Sudan, and Zaire together are larger than the U.S. or Canada. Africa, however, must not be pushed to one side as previously in the malaria eradication program; it should be regarded as an urgent challenge that must be faced, however difficult. To start with, we would probably have to confine ourselves to a few widely understood languages: perhaps Arabic, English, French, Kiswahili, and possibly, West African Pidgin.

As I have already said, it is no longer appropriate to consider the problem of vector-borne diseases as one concerning "us" and "them". Today, for the first time, we are able to view the whole of our planet as a single entity; the concentration of parasitic disease in the warmer parts of this globe is not only a tragedy to those affected: it is a disgrace to us all. If the World Health Organization means what it claims, that is, health for all by the year 2000, then it really must prepare itself to embrace this new and wholly appropriate technology, if this problem is not to drag on into the mid-21st Century and beyond. We now have the means to tackle it worldwide. At last we are in a position to erase this dirty blot that continues to disfigure our own dear planet. The effort will have to be maintained, probably for a very long time — perhaps for ever. And money will have to be spent, but what, in heaven's name, is this man-made commodity for? If this question exposes a never-never land naïveté on the part of the writer in matters of industriofinancial reality, how, one may ask (and posterity will surely do so), is one to judge the apparent hard-bitten attitude of industry that seems all too often to regard the misfortune of our neighbors as fair grist to the mill — so much fortune-fodder? Perhaps, as our earth grows warmer during the next century and vector-borne disease extends its paralyzing grip northward,[19] industrial and scientific reality will come to meet on more equal terms.

If these ideas are now put to peer review, they are likely to be turned down, if only because, subconsciously or consciously, they will be seen to threaten the holy cows of conventional methods of control and the gigantic investment in teaching and industry that underwrites the present effort. The fact that conventional methods have been tried for almost a century and largely found to be wanting does not seem to matter; always there is the hope of science and technology coming to the rescue. However, the plans that I suggest are not really meant for those with minds largely closed by earlier teaching and practice; they are aimed at those who see that the generally accepted equation is incomplete and whose minds are flexible enough to allow the mixing of what some would regard as art with science and technology.

REFERENCES

1. **Simpson, W. J. R.,** The sanitary aspects of warfare in southeastern Europe, *Trans. Soc. Trop. Med. Hyg.*, 12, 1, 1918.
2. **Chan, K. L.,** *Singapore's Dengue Haemorrhagic Fever Control Programme: a Case Study on the Successful Control of* Aedes aegypti *and* Aedes albopictus *Using Mainly Environmental Measures as a Part of Integrated Vector Control*, Southeast Asian Medical Information Center, Tokyo, 1985.
3. **Ruiz, W.,** Cuba's campaign against deadly dengue, *IDRC Rep.*, 16, 16, 1987.

4. **Gillett, J. D.,** Medical entomology, past, present and future: a personal view, *Antenna*, 9, 66, 1985.
5. **Gillett, J. D.,** The harnessing of artificial satellites for the prevention of disease in the tropics: flights of fancy or of fact, *Trans. R. Soc. Trop. Med. Hyg.*, 81, 350, 1987.
6. **Gillett, J. D.,** *Mosquitos* (World Naturalist Series), Weidenfeld and Nicolson, London, 1971, 206.
7. **Gillett, J. D.,** The behaviour of *Homo sapiens*, the forgotton factor in the transmission of tropical disease, *Trans. R. Soc. Trop. Med. Hyg.*, 79, 12, 1985.
8. *The State of the World's Children*, UNICEF Report, Oxford University Press, New York, 1988.
9. **Walters, D. R.,** Polyamines: the cinderellas of cell biology, *Biologist*, 34, 73, 1987.
10. **Martin, S. K., Oduola, A. M. J., and Milhous, W. K.,** Reversal of chloroquine resistance in *Plasmodium falciparum* by verapamil, *Science*, 235, 899, 1987.
11. **Gillett, J. D.,** The universities and health in the tropics: the problem and an answer, in Pressures and Priorities, Craig, T., Ed., Rep. Proc. 12th Congr. Universities of the Commonwealth, Vancouver, British Columbia, 1979, 173.
12. **Sutton, F. X.,** Africa's educational needs and opportunities, in *Man and Africa*, Wolstenholme, G. and O'Connor, M., Eds., CIBA Foundation, J. & A. Churchill, London, 1965, 194.
13. **Gillett, J. D.,** Education for health by *opéras saponiques*, *Trans. R. Soc. Trop. Med. Hyg.*, 81, 351, 1987.
14. **Clark, Sir K.,** Television, in *Art, Science and Human Progress*, McConnell, R. B., Ed., John Murray, London, 1983, 26.
15. **Maxwell, N.,** Wanted: a new way of thinking, *New Sci.*, 114(1560), 63, 1987.
16. **Gillett, J. D.,** The maintenance and spread of insect-borne disease by the agency of man, in *Demography and Vector-Borne Diseases*, Vol. 1, Service, M. W., Ed., CRC Press, Boca Raton, FL, 1989, chap. 4.
17. **Pletsch, D. J.,** Mosquitoes and TV, *IDRC Rep.*, 15, 2, 1986.
18. Development/Environment Trends in Asia and the Pacific: A Regional Overview, Committee on Industry, Technology, and Human Settlements, United Nations Economic and Social Commission for Asia and the Pacific, Bangkok, 1983, 2.
19. **Gillett, J. D.,** Increased atmospheric carbon dioxide and the spread of parasitic disease, in *Parasitological Topics*, Canning, E., Ed., Spec. Publ. No. 1, Society of Protozoologists, Lawrence, Kansas, 1981, 106.

Chapter 13

APPROPRIATE VECTOR CONTROL IN PRIMARY HEALTH CARE

Carol P. MacCormack

TABLE OF CONTENTS

I. CONTROL THROUGH PRIMARY HEALTH CARE

One of the principal reasons for seeking appropriate technologies is to sustain control at relatively low cost over a long period of time through community action. Much of that community action may ultimately be guided, supervised, supplied, and sustained through the national primary health-care structure.

The literature on successful community involvement must be assessed carefully. Much of it is about small pilot programs with special resources. Often a manufacturer will provide chemicals or other materials free in the hope that successful trials will lead to large purchases by government or international agencies. Also there may be enthusiastic leadership by someone with a humanitarian vision and good practical leadership qualities. The community responds; the dream of a better life is kindled. If the pilot project is evaluated and publicized at this stage it seems an alluring model to follow, but in such projects the "hot" stage usually settles into a steady state, often cooling down to a lower level of enthusiasm and efficiency. People may opt out, coverage may become patchy, supplies irregular, or volunteered time diverted to a new scheme which is being promoted by another charismatic person or group. The original donors inevitably shy away from the prospect of supporting long-term recurrent expenditure. At this point in the life cycle of the project a well-funded and well-managed national primary health-care program may be its salvation.

We do not know much about how to "scale up" from a pilot project to a national program supported by government. Bureaucracy replaces the charismatic leader and the nature of interaction with the community changes. However, the advantage of embedding vector control within primary health care is that virtually all countries now have a system that is functioning (with varying degrees of success) at the community level. There are vehicles, training schools, supervisory staff, and other essentials funded on a long-term basis from public funds.

Interaction between communities and primary health-care vector-control workers requires planning. Technical workers' training must include the skills of community work, just as community members need to be taught as much as possible about vectors, their risks, and control techniques. First, there must be a study to learn what people in the community already know about vectors and what control methods they might be encouraged to continue using. What other skills can be added to that base of indigenous knowledge? Effective action will more likely be sustained if it grows out of local knowledge, and if there is at least one person in the community whose indigenous knowledge is augmented with primary health-care training, including vector control.

II. VOLUNTARY COMMUNITY HEALTH WORKERS

Can a vector control program, however appropriate its technologies, be sustained on a permanent basis with voluntary community labor? Much of the literature on enthusiastic pilot projects implies that it can, but more systematic analysis of community health-worker programs in a number of developing countries suggest it cannot.[1] In China the service of "barefoot doctors" arose out of the political ideology, the economics, and the rural social structure of the postrevolutionary period. In Burma, a profoundly Buddhist country, the ideologies of religion and socialism reinforce each other with relative harmony, and volunteers grow in the spiritual and social esteem of their community. However, the idea of voluntary community health workers did not grow out of the culture or traditional organization of rural Bangladesh, India, or Pakistan, for example. It was rather dropped from the heights of international and national policy into a vacuum and has not flourished (see, for example, Chen).[2]

When communities are markedly stratified by class, caste, ethnicity, and other social categories there is usually tacit agreement by all, that landowners, moneylenders, political leaders, and other social superiors will benefit more; that the volunteers will serve their needs first. If vectors are efficiently controlled only around the better houses, public health of the community will not noticeably improve, and community health workers lose their credibility. If there is a cost-recovery aspect of their work, as social inferiors the volunteer cannot easily collect money from their superiors (see, for example, Cham et al.).[3] If costs are not recovered, who will sustain the supply of vector control materials?

Another subtle problem that is always in the background is professional rivalry. Technical experts who train and give backup to community volunteers are often reluctant to give up much useful information. If they did, uneducated villagers might gain as much social status as they. The professional may teach some preventive techniques but not give curative skills which are usually more valued in the community (see, for example, Anita).[4] The voluntary worker may have a vague job description which does not give clear authority to mobilize community resources or may have such a broad set of tasks as to be quite defeating.

In summary, volunteer programs often suffer from high attrition rates as community financial support dwindles, little in-service training to augment useful skills is provided, and villages which generally are not well serviced by the national administrative system languish. Insecticides, curative medicines, and other supplies may arrive irregularly or in inadequate quantities. Professional rivalry may cause busy professionals to give half-hearted training and backup to volunteers. Volunteers become caught up in local social and political struggles, flounder with unclear job descriptions, and have difficulty keeping the community interested in control activities, especially if their tasks are purely preventive and educational when people want curative services. Therefore, preventive and curative work should be closely linked together within primary health care, and village level workers should not be expected to work without pay except in a few exceptional postrevolutionary countries.

III. PARTICIPATION

In the most narrow sense, participation occurs when communities carry out aspects of a control strategy which disease specialists, planners, and administrators have defined. In the more comprehensive sense, participation is a process in which local people plan, implement, monitor, and evaluate activities which affect their lives. In the process social injustice may decrease, and innate creative energies of the community may be released.[5-8]

Short of an ideal process of social transformation, much practical attention must be given to see that politically weak groups are not expected to give a disproportionate amount of labor or wealth, and that the costs to them do not outweigh the benefits they might receive. If the costs are too high, they will not sustain the program over time. For example, Anita[4] described a preventive and curative health project in 30 villages near Bombay. Professionals in the government health service, and to a lesser extent allopathic and non-allopathic private practioners, publicly praised the project and privately undermined it. Some village leaders also feared the process of empowering poor laborers to organize and carry out activities, with alternative power structures arising in the villages. Health professionals colluded with local leaders and, after threats and some open violence to project staff, evicted the project from their area, retaining its assets. The poor, the major beneficiaries, did not actively oppose the threats because their extreme poverty made them dependent on the local land-owning power structure for their survival. Health was a lower priority for them than avoiding starvation. The participatory process in health has led to violence and even murder in other parts of the world as well.[9,10]

However, lest we present too negative a picture, we should recall the initiatives of the Vector Control Research Centre (V.C.R.C.) (see Chapter 7) and Malaria Research Centre in India.[11] Workers from the centers not only trained broad-based community health workers in a range of vector control techniques, but helped provide other substantial benefits for the rural poor as well. In poor villages that had no cattle, they helped people to obtain water buffaloes and milk cows as economic assets which were also alternative mosquito hosts. Weed-choked ponds were cleared, and the profits from fish farming paid laborers to keep ponds clear, and even expand the fish farming. Prawn culture and paper-making from algae provided the economic stimulus to convert other forms of mosquito-breeding habitats to income-generating ends. In urban Pondicherry, filariasis control was the V.C.R.C.'s lever for insisting that laws which had been in the statute books for many years be actually enforced. Center staff were also intersectoral "brokers" working with the Department of Education to include practical health and vector-control information in school books and also in adult education curricula. Information was linked to community vector-control activities and carried out by children's science clubs, youth clubs, and women's clubs (see Young and Durston[12] for community health education and Hongvivatana et al.[13] for a malaria case study).

Chapter 7 does not include a detailed description of the political "broker" role of V.C.R.C. staff played in reconciling their transformative activities with the traditional privileges and prerogatives of local elites. How, for example, did local workers resist demands by landowners to divert vector-control insecticides to agricultural uses? Descriptions usually leave out this crucial political dimension, often for very good reasons. Or, descriptions may only be from an early stage of activity, before political opposition hardens, and evaluations done 10 to 15 years on would give a more instructive case study.

Technologies are not on offer in a value-free political vacuum, but in recognizing the social dynamics of vector control we must not ignore scientific considerations. What technologies or combinations of technologies are reasonable to advocate for community control? Table 1 (based on the World Health Organization[14] with modifications) summarizes chemical, biological, environmental, and health service control strategies for six diseases. Those marked with the number 1 are possible strategies which might be carried out with a maximum of community activity. However, to avoid over-simplified prescription, some are qualified by the need to know levels of transmission and the local vector species. If transmission levels are very high, local technologies may not make sufficient impact to warrant mobilizing community enthusiasm.

IV. COMMUNITY

In the most narrow sense, the community is a population which is to receive a targeted intervention. In the more comprehensive sense, a community is a group of people who are engaged in relatively stable social relationships, usually in a local area. An operational definition of community will have to be made for each control program since communities usually do not have distinct social boundaries. Even in stable agricultural villages people go away for schooling, work, marriage, or other reasons and may come back. If they do not come back, they may continue to define themselves as being at one with the spirit of the community, making contributions to it, and drawing resources from it. For analytical purposes we might define an urban neighborhood, an agricultural village, a collection of rural homesteads, or a nomadic group by its membership, social structure, and its functions at the time of survey.[15]

Communities function in different ways. Some are relatively egalitarian, cohesive, moral communities, but others are stratified by class, caste, ethnicity, and other factors (see, for example, Cham et al.).[3] Some are democratic, with matters discussed at large until a consensus is reached; some are governed in an authoritarian manner, or may even be under military rule. Democratic communities are often slow to take a decision, and the process of stimulating community participation is very labor intensive for vector-control workers. How-

TABLE 1
Some Methods of Control with Community Participation

Disease Principal vectors	Malaria *(Anopheles)*	Filariasis (Mosquitoes)	Leishmaniasis *(Phlebotomus)*	Oncho. *(Simulium)*	Sleeping sickness (Tsetse)	Chagas' disease (Triatomines)
Chemical						
Larviciding	1[a]	1	5	3	5	5
Residual spraying	1[a]	1[a]	3[a]	5	3[a]	1
Coils, aerosols, etc.	1[a]	1[a]	1[a]	5	5	1[a]
Space spraying	3[a]	4	4	4	3[a]	4
Biological						
Fish	1[a]	1[a]	5	5	5	5
Others	2	2	5	2	2	2
Environmental						
Drainage & filling	1[a]	1[a]	5	5	5	5
Sanitation	5	1[a]	5	5	5	5
House improvement	1[a]	1[a]	1[a]	5	5	1
Aquatic weeding	1[a]	1[a]	5	5	5	5
Bed nets	1	1	1[a]	5	5	1[a]
Trapping	5	5	5	5	1	4
Health services						
Education	1	1	1	1	1	1
Detection	1	1	1	1	1	1
Treatment	1	1	3	3	3	5

Note: 1 = possible with community participation; 2 = more research needed; 3 = method used, but requires professional involvement; 4 = not recommended; 5 = not applicable.

[a] Depending on vector species and epidemiological conditions.

ever, the potential for genuine commitment and long-term maintenance of control strategies may be high once a consensus is reached. In more authoritarian communities activities to meet imposed targets may take place swiftly, but sustained behavioral changes may not occur.

Since all communities have internal social divisions within them, and particular categories of people rather than the whole community usually carry out vector-control strategies, a further level of community analysis is needed. Social categories include class, in which the poor are often expected by local leaders to give labor. Or, a significant category may be an age group especially at risk, or men compared with women who have sex-segregated economic tasks. Women traditionally have primary responsibility for child health and maintenance of domestic hygiene.[16]

In urban areas, squatter neighborhoods or slums may have a strong sense of community. They are often located in areas of poor drainage and poor health because flooded land may not be valuable enough for owners to bother with evicting squatters. Residents often give very high priority to drainage, and once flooding is under control they tend to improve spontaneously other aspects of their environment and housing.[17] Giving people clear title to land will also enhance their propensity to improve their environment — a political remedy for a vector-control problem.

Poor urban communities in developing tropical countries often have coherent organization and social solidarity at the neighborhood level. It may arise negatively in response to such things as eviction threats, or it may arise positively through the work of religious or political groups. The best methods for organizing vector control in such urban communities is to identify religious, political, and other groups and understand their structure and function. This learning stage is essential for realistic planning and sustained work with these community groups to improve drainage and carry out other vector-control strategies.[17]

V. SUSTAINING COMMUNITY PARTICIPATION

The literature suggests two basic and complementary ways to sustain participation in vector control (see, for example, World Health Organization[18]). First is to link it to well-organized government programs and fund them adequately. This would include some salary and career structure for community health workers. However, case studies in primary health care suggest this is more easily said than done. Urban elites who want expensive curative services still have a strong political voice in allocating government health expenditure. Restructuring being imposed on debtor nations by lending nations and international financial agencies usually requires decreased government spending with a shift to the private sector, which is usually little interested in prevention at low cost.

The complementary approach is to link vector control to income-generating activities at the local level. The program in Pondicherry is a good example (Chapter 7). Planting water-avid trees in marshes as a cash crop has supported the local cost of malaria vector control activities in the Kheda area of Gujarat.[11] Costs, effectiveness, and benefits of a community approach must be calculated from two points of view: (1) the way community members perceive their costs and benefits from participating and (2) the way those who fund the program assess effectiveness of control against costs. Except through coercion (which is expensive to sustain), vector-control activities will not be sustained if people perceive that the costs to them outweigh the benefits they receive. Seldom will people work against their economic interests. For example, in dry areas convenient cattle-watering borrow pits will never be filled in by villagers who find them convenient or essential. House spraying cannot be tolerated if it endangers the financial survival or households keeping silkworms or honeybees.[11]

Funds for any initiative must be secure before the community-participation process begins. If, for example, in a control program people are persuaded to dig pits for improved latrines, but reinforcing cement does not arrive before the rains and the pits collapse, people will be very wary of the next scheme that comes along. Similarly, reliable technical skills, including epidemiological skills, must be available from the beginning. If people expend effort, time, and wealth on a vector-control strategy that does not reduce the basic case reproduction rate for the disease, or even reduce the nuisance of the vector, cynicism and noncooperation in the community will deepen.[19]

Larvivorous fish need not be large enough to produce fillets in order to be a local cash crop (Chapter 8). Local markets in tropical countries often offer for sale small fish, either fresh, dried, salted, or pickled. Fish predation on mosquito larvae can be enhanced if both larvivorous fish and herbivorous fish which feed on vegetation that protects larvae are encouraged. The strategy may even be enhanced by substituting mosquito-indigestible phytoplankton for the type already there.[20] To introduce fish that are effective in vector control, well adapted to local conditions, and have local market value a preliminary study of what local people already know and want is essential. Fish may not only generate funds but decrease vector-control costs (Chapter 8). In Afghanistan larvivorous fish appear to have reduced the mosquito population by 75% at a cost equivalent to $0.02 per person protected compared to $0.50 per person for residual spraying with DDT.[21]

Long-burning mosquito coils made as a cottage industry from locally grown pyrethrum flowers mixed with local scrap wood dust and bound with vegetable starch generate income and provide effective protection in China and other parts of Asia (see Chapter 4 and Chadwick).[22] These protective and income-generating skills might be spread throughout the tropics. The skill of obtaining and expanding polystyrene beads for vector control in latrines and soakage pits (see Chapter 10) might be extended to a small local industry in bead-filled cushions, or even energy efficient "hay box" cooking aids in fuel scarce countries. If insecticide-impregnated bednets become a primary health care strategy in malaria control

(Chapter 2), local income will be generated if tailors are encouraged to make them, perhaps at a subsidized price.[23] Local industries for repellent-dipped cloth or body lotion for personal use might also be encouraged (Chapter 4). There may be a small market for tsetse traps made from local materials which people may wish to buy and place where the flies are a particular nuisance (see Chapter 3 and Okoth).[24]

To conclude, community participation in vector control is most likely to be sustained if there is the right ratio of costs to their perceived benefits. Costs must include subtle opportunity costs, that is, the value of what people would have done if they had not given time and effort to vector control[25] Long-term community committment is also enhanced if activities meet, or are integrally linked with, people's priorities, if skills training is in conjunction with practical community activities, and preventive work is linked to curative and self-care activities within a primary health-care or income-generating framework.[26]

REFERENCES

1. **Walt, G.,** Community health workers: are national programmes in crisis?, *Health Policy Plann.*, 3, 1, 1988.
2. **Chen, L.,** Coping with economic crisis: policy development in China and India, *Health Policy Plann.*, 2, 138, 1987.
3. **Cham, K., MacCormack, C., Touray, A., and Baldeh, S.,** Social organization and political factionalism: primary health care in The Gambia, *Health Policy Plann.*, 2, 214, 1987.
4. **Anita, N. H.,** The Mandwa project: an experiment in community participation, *Int. J. Health Serv.*, 18, 153, 1988.
5. **Freire, P.,** *Educacão como Práctica da Liberdade*, Paz e Terra, Rio de Janeiro, 1967.
6. **Werner, D.,** The village health worker: lackey or liberator?, *W. Health Forum*, 2, 46, 1981.
7. **Werner, D.,** *Helping Health Workers Learn*, Hesperian Foundation, Palo Alto, 1983.
8. **MacCormack, C.,** Community participation in primary health care, *Trop. Doctor*, 13, 51, 1983.
9. **Chowdhury, Z.,** The good health worker will inevitably become a political figure, *W. Health Forum*, 2, 55, 1981.
10. **Heggenhougen, K.,** Will primary health care efforts be allowed to succeed?, *Soc. Sci. Med.*, 19, 217, 1984.
11. **Sharma, V. P.,** Community based malaria control in India, *Parasitol. Today*, 3, 222, 1987.
12. **Young, B. and Durston, S.,** *Primary Health Education*, Longman, London, 1987.
13. **Hongvivatana, T., Leerapan, P., and Chaiterranuwatsiri, M.,** Knowledge, Perception and Behaviour in Malaria, Centre for Health Policy Studies Monogr. Mahidol University, Bangkok, 1985.
14. Community Participation in Tropical Disease Control: Social and Economic Research, WHO TDR/SER-SWG (4)/CP/83.3, World Health Organization, Geneva, 1983.
15. **MacCormack, C.,** Anthropological research in African primary health care, in *Rural Transformation in Tropical Africa*, Rimmer, D., Ed., Francis Pinter, London, 1988, 36.
16. **Elmendorf, M.,** *Women, Water and Waste: Beyond Access*, W.A.S.H., Arlington, Virginia, 1981.
17. **Cairncross, S.,** Urban drainage in developing countries, *Parasitol. Today*, 2, 200, 1986.
18. Vector Control in Primary Health Care, *W.H.O. Tech. Rep. Ser.*, No. 755, World Health Organization, Geneva, 1987.
19. **MacCormack, C.,** Human ecology and behaviour in malaria control in the Afro-Tropical region, *Bull. W.H.O.*, 62 (Suppl.), 81, 1984.
20. **Marten, G.,** Indigestible phytoplankton for mosquito control, *Parasitol. Today*, 2, 150, 1986.
21. **Wickramasinghe, N. and Costa, H.,** Mosquito control with larvivorous fish, *Parasitol. Today*, 2, 228, 1986.
22. **Chadwick, P. R.,** Mosquito coils protect against bites, *Parasitol. Today*, 1, 90, 1985.
23. **MacCormack, C., Snow, R., and Greenwood, B.,** Use of impregnated bed nets in Gambian primary health care: economic aspects, *Bull. W.H.O.*, in press.
24. **Okoth, J.,** Community participation in tsetse control, *Parasitol. Today*, 2, 88, 1986.
25. **Bos, R. and Mills, A.,** Financial and economic aspects of environmental management for vector control, *Parasitol. Today*, 3, 160, 1987.
26. **Tanner, M., Lwihula, G., Burnier, E., de Savigny, D., and Degremont, A.,** Community participation within a primary health care programme, *Trop. Med. Parasitol.*, 37, 164, 1986.

INDEX

N

O

P

R